iOS 4 Programming Cookbook

Vandad Nahavandipoor

Beijing · Cambridge · Farnham · Köln · Sebastopol · Tokyo

iOS 4 Programming Cookbook
by Vandad Nahavandipoor

Copyright © 2011 Vandad Nahavandipoor. All rights reserved.
Printed in the United States of America.

Published by O'Reilly Media, Inc., 1005 Gravenstein Highway North, Sebastopol, CA 95472.

O'Reilly books may be purchased for educational, business, or sales promotional use. Online editions are also available for most titles (*http://my.safaribooksonline.com*). For more information, contact our corporate/institutional sales department: (800) 998-9938 or *corporate@oreilly.com*.

Editors: Andy Oram and Brian Jepson
Production Editor: Kristen Borg
Copyeditor: Audrey Doyle
Proofreader: Andrea Fox
Production Services: Molly Sharp

Indexer: Fred Brown
Cover Designer: Karen Montgomery
Interior Designer: David Futato
Illustrator: Robert Romano

Printing History:

January 2011: First Edition.

RepKover™ This book uses RepKover™, a durable and flexible lay-flat binding.

ISBN: 978-1-449-38822-5

[M]

1294844599

To Agnieszka Marta Dybowska.

Table of Contents

Preface

I started developing iPhone applications in late 2007. Between then and now, I have worked on various iPhone OS applications for different companies across the globe. As you might have already guessed, iOS is my favorite platform and Objective-C is my favorite programming language. I find that Objective-C helps programmers write clean code and iOS helps developers write user-friendly and useful applications.

I have coded in other programming languages such as Assembly (using NASM and TASM) and Delphi/Pascal for many years, and I still find myself going through disassembled Objective-C code to find out which method of doing a certain thing or accomplishing a certain result in Objective-C is better optimized on a certain device and/or operating system.

After becoming comfortable with the iOS SDK, I gradually built up a thirst to write a book about the iOS SDK, and with the help of wonderful people at O'Reilly, you are now reading the result of the several hundred hours that have been put into writing new material for iOS 3 and iOS 4, editing, reviewing, revising, and publishing.

So, please go ahead and start exploring the recipes. I hope you'll find that they are easy to cook and digest!

Audience

I assume you are comfortable with the iOS development environment and know how to create an app for the iPhone or iPad. This book does not get novice programmers started, but presents useful ways to get things done for iOS programmers ranging from novices to experts.

Organization of This Book

In this book, we will discuss frameworks and classes that are available in iOS 3 and iOS 4. In some recipes, you will find code that runs only on iOS 4 and later; in those recipes, I note that you will need the iOS 4 SDK or later to compile the example code.

Here is a concise breakdown of the material each chapter covers:

Chapter 1, *Working with Objects*
Explains how Objective-C classes are structured and how objects can be instantiated. The chapter talks about properties and delegates as well as memory management in Objective-C. Even if you are competent in Objective-C, I strongly suggest that you go through this chapter, even if you are skimming through it, to understand the basic material that is used in the rest of the chapters.

Chapter 2, *Implementing Controllers and Views*
Describes various approaches to constructing your iOS application's user interface by taking advantage of different tools the SDK provides. This chapter also introduces you to features that are only available on the iPad, such as the popover and split view controllers.

Chapter 3, *Constructing and Using Table Views*
Shows how you can work with table views to create professional-looking iOS applications. Table views are very dynamic in nature, and as a result, programmers sometimes have difficulty understanding how they should work with them. By reading this chapter and having a look at and trying out the example code, you will gain the knowledge that is required to comfortably work with table views.

Chapter 4, *Core Location and Maps*
Describes how you should use Map Kit and Core Location APIs to develop location-aware iOS applications. First you will learn about maps, and then you will learn how to detect a device's location and tailor your maps with custom annotations. You will also learn about geocoding and reverse geocoding, as well as some of the methods of the Core Location framework, which are only available in the iOS 4 SDK and later.

Chapter 5, *Implementing Gesture Recognizers*
Demonstrates how to use gesture recognizers, which enable your users to easily and intuitively manipulate the graphical interface of your iOS applications. In this chapter, you will learn how to use all available gesture recognizers in the iOS SDK, with working examples tested on iOS 3 and iOS 4 on different devices such as the iPhone 3GS, iPhone 4, and iPad.

Chapter 6, *Networking and XML*
Demonstrates how to download data from a URL and parse XML files. You will learn about synchronous and asynchronous connections and their pros and cons. You will also learn about caching files in memory and on disk to avoid consuming the possibly limited bandwidth of an iOS device on which your application could be running.

Chapter 7, *Operations, Threads, and Timers*
Provides details regarding operations, threads, and timers. Using the material in this chapter, you can develop modern multithreaded iOS applications. In addition,

you will learn about operations and operation queues, and how to avoid implementing your own threads and instead let iOS do it for you.

Chapter 8, *Audio and Video*

Discusses the AV Foundation and Media Player frameworks that are available on the iOS SDK. You will learn how to play audio and video files and how to handle interruptions, such as a phone call, while the audio or video is being played on both iOS 3 and iOS 4. This chapter also explains how to record audio using an iOS device's built-in microphone(s). At the end of the chapter, you will learn how to access the iPod Library and play its media content, all from inside your application.

Chapter 9, *Address Book*

Explains the Address Book framework and how to retrieve contacts, groups, and their information from the Address Book database on an iOS device. The Address Book framework is composed entirely of C APIs. Because of this, many Objective-C developers find it difficult to use this framework compared to frameworks that provide an Objective-C interface. After reading this chapter and trying the examples for yourself, you will feel much more confident using the Address Book framework.

Chapter 10, *Camera and the Photo Library*

Demonstrates how you can determine the availability of front- and back-facing cameras on an iOS device. Some of the recipes in this chapter are specific to iOS 4, with the rest working on both iOS 3 and iOS 4. You will also learn how to access the Photo Library using the Assets Library framework which is available in iOS 4 and later. At the end of the chapter, you will learn about editing videos right on an iOS device using a built-in view controller.

Chapter 11, *Multitasking*

Explains, with examples, how to create multitasking-aware applications that run beautifully on iOS 4. You will learn about background processing, from playing audio and retrieving users' locations in the background, to downloading content from a URL while your application is running in the background.

Chapter 12, *Core Data*

Describes how to maintain persistent storage for your iOS applications using Core Data. You will learn how to add to, delete from, and edit Core Data objects and how to boost access to data in a table view. In addition, you will learn how to manage relationships between Core Data objects.

Chapter 13, *Event Kit*

Demonstrates the use of the Event Kit and Event Kit UI frameworks, which are available on iOS 4 and later, in order to manage calendars and events on an iOS device. You will see how to create, modify, save, and delete events. You will also learn, through examples, how to add alarms to calendar events and how to set up CalDAV calendars so that you can share a single calendar among multiple devices.

Chapter 14, *Graphics*

Introduces the Core Graphics framework. You will learn how to draw images and text on a graphics context, grab the contents of a graphics context and save it as an image, and much more.

Chapter 15, *Core Motion*

Explains the Core Motion framework, which is new in iOS 4. Using Core Motion, you will access the accelerometer and the gyroscope on an iOS device. You will also learn how to detect shakes on a device. Of course, not all iOS devices are equipped with an accelerometer and a gyroscope, so you will also learn how to detect the availability of the required hardware.

Additional Resources

From time to time, I refer to official Apple documentation. Some of Apple's descriptions are right on the mark, and there is no point in trying to restate them. Throughout this book, I have listed the most important documents and guides in the official Apple documentation that every professional iOS developer should read.

For starters, I suggest that you have a look at the "iPhone Human Interface Guidelines" and the "iPad Human Interface Guidelines." These two documents will tell you everything you should know about developing engaging and intuitive user interfaces for the iPhone/iPod and the iPad. Every iOS programmer must read these documents. In fact, I believe these documents must be read by the product design and development teams of any company that develops iOS applications.

iPhone Human Interface Guidelines
http://developer.apple.com/library/ios/#documentation/userexperience/conceptual/mobilehig/Introduction/Introduction.html

iPad Human Interface Guidelines
https://developer.apple.com/library/ios/#documentation/General/Conceptual/iPadHIG/Introduction/Introduction.html

I also suggest that you skim through the "iOS Application Programming Guide" in the iOS Reference Library for some tips and advice on how to make great iOS applications:

https://developer.apple.com/library/ios/#documentation/iPhone/Conceptual/iPhoneOSProgrammingGuide/Introduction/Introduction.html

One of the things you will notice when reading Chapter 11 is the use of block objects. This book concisely explains block objects, but if you require further details on the subject, I suggest you read "A Short Practical Guide to Blocks," available at this URL:

https://developer.apple.com/library/ios/#featuredarticles/Short_Practical_Guide_Blocks/index.html#//apple_ref/doc/uid/TP40009758

In Chapter 7, you will learn about operations. To be able to implement custom operations, as you will see later, you must have a basic knowledge of key-value coding (KVC). If you require more information about KVC, I recommend that you read the "Key-Value Coding Programming Guide," available at the following URL:

https://developer.apple.com/library/ios/#documentation/Cocoa/Conceptual/KeyValueCoding/KeyValueCoding.html

Throughout this book, you will see references to "bundles" and loading images and data from bundles. You will read a concise overview about bundles in this book, but if you require further information, head over to the "Bundle Programming Guide," available at this URL:

https://developer.apple.com/library/ios/#documentation/CoreFoundation/Conceptual/CFBundles/Introduction/Introduction.html

Conventions Used in This Book

The following typographical conventions are used in this book:

Italic
: Indicates new terms, URLs, filenames, file extensions, and directories

`Constant width`
: Indicates variables and other code elements, the contents of files, and the output from commands

`Constant width bold`
: Highlights text in examples that is new or particularly significant in a recipe

`Constant width italic`
: Shows text that should be replaced with user-supplied values

 This icon signifies a tip, suggestion, or general note.

Using Code Examples

This book is here to help you get your job done. In general, you may use the code in this book in your programs and documentation. You do not need to contact us for permission unless you're reproducing a significant portion of the code. For example, writing a program that uses several chunks of code from this book does not require permission. Selling or distributing a CD-ROM of examples from O'Reilly books *does* require permission. Answering a question by citing this book and quoting example code does not require permission. Incorporating a significant amount of example code from this book into your product's documentation *does* require permission.

We appreciate, but do not require, attribution. An attribution usually includes the title, author, publisher, and ISBN. For example: "*iOS 4 Programming Cookbook*, by Vandad Nahavandipoor (O'Reilly). Copyright 2011 Vandad Nahavandipoor, 978-1-449-38822-5."

If you feel your use of code examples falls outside fair use or the permission given here, feel free to contact us at *permissions@oreilly.com*.

We'd Like to Hear from You

Every example and code snippet in this book has been tested on the iPhone 4, iPad, iPhone 3GS, and iPhone/iPad Simulator, but occasionally you may encounter problems—for example, if you have a different version of the SDK than the version on which the example code was compiled and tested. The information in this book has also been verified at each step of the production process. However, mistakes and over-sights can occur, and we will gratefully receive details of any you find, as well as any suggestions you would like to make for future editions. You can contact the author and editors at:

> O'Reilly Media, Inc.
> 1005 Gravenstein Highway North
> Sebastopol, CA 95472
> (800) 998-9938 (in the United States or Canada)
> (707) 829-0515 (international or local)
> (707) 829-0104 (fax)

We have a web page for this book, where we list errata, examples, and any additional information. You can access this page at:

> *http://oreilly.com/catalog/9781449388225*

There is also a companion website to this book where you can see all the examples with color-highlighted syntax:

> *http://www.ios4cookbook.com*

To comment or ask technical questions about this book, send email to the following address, mentioning the book's ISBN number (9781449388225):

> *bookquestions@oreilly.com*

For more information about our books, conferences, Resource Centers, and the O'Reilly Network, see our website at:

> *http://www.oreilly.com*

Safari® Books Online

Safari Books Online is an on-demand digital library that lets you easily search more than 7,500 technology and creative reference books and videos to find the answers you need quickly.

With a subscription, you can read any page and watch any video from our library online. Read books on your cell phone and mobile devices. Access new titles before they are available for print, and get exclusive access to manuscripts in development and post feedback for the authors. Copy and paste code samples, organize your favorites, download chapters, bookmark key sections, create notes, print out pages, and benefit from tons of other time-saving features.

O'Reilly Media has uploaded this book to the Safari Books Online service. To have full digital access to this book and others on similar topics from O'Reilly and other publishers, sign up for free at *http://my.safaribooksonline.com*.

Acknowledgments

I have always loved writing and running my programs on computers. I look at programming as a way to speak to whatever computer the program runs on. To me, programming is a way to actually connect with the computer and give it instructions and listen to what it says in return.

I have been exceptionally lucky to have almost always found the right people to help me find the right path in whatever journey I've started in my life. First and foremost, I would like to thank my beloved fiancée, Agnieszka Marta Dybowska, for her unconditional love and support throughout the years and for the many hours she had to spend without me while I was working on this book. Your strong enthusiasm for writing and journalism has greatly moved me, and I sincerely hope that one day you will gather enough material to be able to write your book.

I also want to thank Brian Jepson (whose many hidden talents are yet to be discovered!) for giving me a chance to work on this book. This book would have been impossible if it wasn't for Brian's consistent urge to improve the outline and the table of contents that I originally sent him. This reminds me to thank Andy Oram, whom I would like to call the virtual second writer of this book. Andy's perfectionism and his undeniable desire to finely form every single sentence you read in this book are absolutely impressive. I must also thank Sarah Kim and Rachel James for helping me update my profile page on O'Reilly's website. I also appreciate Meghan Blanchette's help in doing the initial paperwork for this book.

I want to say a big thank you to my technical reviewers, Eric Blair and Alasdair Allan, for all their helpful insight. Kirk Pattinson, Gary McCarville, and Sushil Shirke are among the people who have greatly influenced me to become who I am today. Thank you to Sushil for being a great mentor and colleague and for providing continuous support. Thanks to Kirk for believing that I was up to the challenge of working on some high-profile iOS applications. Thank you to Gary for his support while I worked on this project, and for being a wonderful mentor.

Last but not least, I would like to sincerely thank Apple and its employees for making such a wonderful operating system and SDK. It's truly a great pleasure to work with the iOS SDK, and I hope you, the reader, will enjoy working with it as much as I do.

CHAPTER 1

Working with Objects

1.0 Introduction

Objective-C, the language in which you program the iOS SDK, is an object-oriented programming language. We create, use, work with, and extend objects; we construct them, customize them to our specific needs, and destroy them when they are no longer needed. Object-oriented programming languages also work with primary data types such as integers, strings, and enumerations. However, the distinctive feature of such languages is their ability to create and manage the lifetime of objects during execution.

Objects are defined in *classes*, and therefore these two terms are commonly used interchangeably. But actually, a class is just a specification for defining objects; each object is said to be an *instance* of its class. Each class—and therefore the objects that are created from that class—is a set of properties, tasks and methods, enumerations, and much more. In an object-oriented programming language, objects can inherit from each other much like a person can inherit certain traits and characteristics from his parents.

 Objective-C does not allow multiple inheritance. Therefore, every object is the direct descendant of, at most, one other object.

The root class of most Objective-C objects is the NSObject class. This class manages the runtime capabilities offered by iOS; as a result, any class that directly or indirectly inherits from NSObject will inherit these capabilities as well. As we will see later in this chapter, objects that inherit from NSObject can take advantage of Objective-C's distinguished memory management model.

1.1 Implementing and Using Custom Objects

Problem

You want to create a unique Objective-C class because none of the classes in the frameworks shipped with the iOS SDK offers the methods and properties you need.

Solution

Follow these steps:

1. In Xcode, while your project is open, choose the group (on the lefthand side of the screen) where you would like to create a new Objective-C class. This is normally the Classes group.

2. Select File→New File in Xcode. You will be presented with a dialog similar to that shown in Figure 1-1.

3. In the New File dialog, make sure Cocoa Touch Class is chosen on the lefthand side of the screen. Choose the Objective-C Class item on the righthand side of the New File dialog, and make sure NSObject is selected in the "Subclass of" drop-down menu. Click Next, as shown in Figure 1-1.

4. Choose the name of the new class you are about to create and where it has to be stored. Make sure you enter **MyObject.m** in the File Name box and that the "Also create 'MyObject.h'" checkbox is checked, as shown in Figure 1-2. In this dialog, you can also manage which project target your new class file will be added to.

5. Click Finish in the New File dialog.

Discussion

If you take a close look at the New File dialog, you will notice that it is divided into two major sections. The section on the left is where you can browse various templates based on their categories, and the section on the right is where you can find the different types of files you can create in the selected category. For instance, by choosing the Cocoa Touch Class category, you will be able to create a new Objective-C class. The bottom-right pane of this dialog is where you can choose other options, if available, such as the superclass of the object you are creating. Not all templates have extra options.

When creating a new Objective-C class, Xcode will ask for the name you would like to assign to both the class and the object, as shown in Figure 1-2. You will also need to specify which target you would like to add your class to. By default, your current target is selected, so you will not need to change any further settings.

Figure 1-1. The New File dialog

After you click Finish, Xcode will create the new class in the currently selected location in the project hierarchy:

```
#import <Foundation/Foundation.h>


@interface MyObject : NSObject {

}

@end
```

As you can see, the `MyObject` object inside the `MyObject` class file inherits from `NSObject`. Now you can import this class into your other class files and instantiate an object of type `MyObject`:

```
#import "ObjectsAppDelegate.h"
#import "MyObject.h"

@implementation ObjectsAppDelegate

@synthesize window;
```

```
- (BOOL)               application:(UIApplication *)application
  didFinishLaunchingWithOptions:(NSDictionary *)launchOptions {

  MyObject *someObject = [[MyObject alloc] init];
  /* Do something with the object, call some methods, etc. */
  [someObject release];

  [window makeKeyAndVisible];

  return YES;
}

- (void)dealloc {
  [window release];
  [super dealloc];
}

@end
```

Figure 1-2. Choosing a name for the class that is to be created

I am a big supporter of writing reusable code. In other words, I try to avoid writing the same line of code twice, if possible. Sometimes you have to write code here and there that feels like repeated code. But if you believe a block of code can be reused, you can put it in an object and keep it in a separate project. After you have built up a library of reusable code, you will want to, at some point, reuse that code in your new projects. In such cases, it's best to keep your reusable code in one place and refer to it in your Xcode project instead of duplicating that code by copying it into your new project.

To reuse that code, you can create a new group in Xcode by right-clicking on your project in Xcode and choosing Add→New Group, as shown in Figure 1-3. You can then give this new virtual folder or group a name, such as "Shared Libraries" or "Shared Code." Once you are done, locate your reusable code in the Finder and drag and drop it into the Shared Libraries group in Xcode. This way, Xcode will create the required paths so that you can instantiate new objects of your reusable classes.

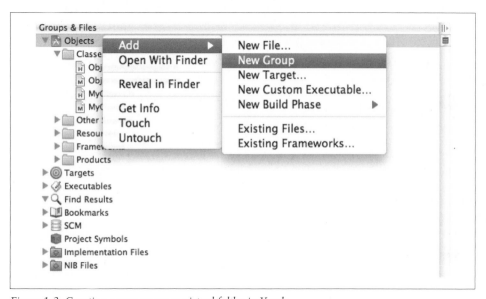

Figure 1-3. Creating a new group or virtual folder in Xcode

 Creating a new group does not create the corresponding folder in the filesystem.

See Also

Recipe 1.6; Recipe 1.7

1.2 Allocating and Initializing Objects

Problem

You want to create an instance of a new object, but you don't understand the difference between *allocation* and *initialization* and why you should have to both allocate and initialize an object before you can use it.

Solution

You must both allocate and initialize an object before using it. An object must be allocated using the `alloc` instance method. This class method will allocate memory to hold the object and its instance variables and methods. Each object must have one designated initializer, which is normally the initialization method with the most parameters. For instance, the `initWithFrame:` method is the designated initializer of objects of type `UIView`. Always allocate and initialize your objects, in that order, before using them.

When implementing a new object, do not override the `alloc` method. This method is declared in `NSObject`. Instead, override the `init` method and create custom initialization methods that handle required parameters for the specific object you are working on.

Discussion

An object that inherits from `NSObject` must be prepared for use in two steps:

1. Allocation
2. Initialization

The allocation is done by invoking the `alloc` method, which is implemented in the `NSObject` class. This method creates the internal structure of a new object and sets all instance variables' values to zero. After this step is done, the `init` method takes care of setting up the default values of variables and performing other tasks, such as instantiating other internal objects.

Let's look at an example. We are creating a class named `MyObject`. Here is the *.h* file:

```
#import <Foundation/Foundation.h>

@interface MyObject : NSObject {

}

- (void) doSomething;

@end
```

The implementation of this class is as follows (the *.m* file):

```
#import "MyObject.h"

@implementation MyObject

- (void) doSomething{

  /* Perform a task here */
  NSLog(@"%s", __FUNCTION__);

}

@end
```

The doSomething instance method of the MyObject object will attempt to print the name of the current function to the console window. Now let's go ahead and invoke this method by instantiating an object of type MyObject:

```
MyObject *someObject = [[MyObject alloc] init];
/* Do something with the object, call some methods, etc. */
[someObject doSomething];
[someObject release];
```

This code will work absolutely fine. Now try to skip initializing your object:

```
MyObject *someObject = [MyObject alloc];
/* Do something with the object, call some methods, etc. */

[someObject doSomething];
[someObject release];
```

If you run this code now, you will realize that it works absolutely fine, too. So, what has happened here? We thought we had to initialize the object before we could use it. Perhaps Apple can explain this behavior better:

> An object isn't ready to be used until it has been initialized. The init method defined in the NSObject class does no initialization; it simply returns self.

Simply put, this means the init method is a placeholder for tasks that some classes need to perform before they are used, such as setting up extra data structures or opening files. NSObject itself—along with many of the classes you will use—does not have to initialize anything in particular. However, it is a good programming practice to always run the init method of an object after allocating it in case the parent of your class has overridden this method to provide a custom initialization. Please bear in mind that the return value for initializer methods of an object is of type id, so the initializer method might even return an object that is not the same object that the alloc method returned to you. This technique is called *Two-Stage Creation* and is extremely handy. However, discussing this technique is outside the scope of this book. For more information about Two-Stage Creation, please refer to *Cocoa Design Patterns* by Erik M. Buck and Donald A. Yacktman (Addison-Wesley Professional).

1.3 Defining Two or More Methods with the Same Name in an Object

Problem

You would like to implement two or more methods with the same name in one object. In object-oriented programming, this is called *method overloading*. However, in Objective-C, method overloading does not exist in the same way as it does in other programming languages such as C++.

 The techniques used and explained in this recipe's Solution and Discussion are merely ways to create methods that have different numbers of parameters or have parameters with different names, just to give you an idea of how you can have methods whose first name segments are the same.

Solution

Use the same name for your method, but keep the *number* and/or the *names* of your parameters different in every method:

```
- (void) drawRectangle{

  [self drawRectangleInRect:CGRectMake(0.0f, 0.0f, 4.0f, 4.0f)];

}

- (void) drawRectangleInRect:(CGRect)paramInRect{

  [self drawRectangleInRect:paramInRect
                  withColor:[UIColor blueColor]];

}

- (void) drawRectangleInRect:(CGRect)paramInRect
          withColor:(UIColor*)paramColor{

  [self drawRectangleInRect:paramInRect
                  withColor:paramColor
                  andFilled:YES];

}

- (void) drawRectangleInRect:(CGRect)paramInRect
               withColor:(UIColor*)paramColor
               andFilled:(BOOL)paramFilled{
```

```
    /* Draw the rectangle here */

}
```

Discussion

Method overloading is a programming language feature supported by Objective-C, C++, Java, and a few other languages. Using this feature, programmers can create different methods with the same name, in the same object. However, method overloading in Objective-C differs from that which can be used in C++. For instance, in C++, to overload a method the programmer needs to assign a different number of parameters to the same method and/or change a parameter's data type.

In Objective-C, however, you simply change the name of at least one parameter. Changing the type of parameters will not work:

```
- (void) method1:(NSInteger)param1{

    /* We have one parameter only */

}

- (void) method1:(NSString *)param1{

    /* This will not compile as we already have a
       method called [method1] with one parameter */

}
```

Changing the return value of these methods will not work either:

```
- (int) method1:(NSInteger)param1{

    /* We have one parameter only */
    return(param1);

}

- (NSString *) method1:(NSString *)param1{

    /* This will not compile as we already have a
       method called [method1] with one parameter */
    return(param1);

}
```

As a result, you need to change the *number of parameters* or the *name* of (at least) one parameter that each method accepts. Here is an example where we have changed the number of parameters:

```
- (int) method1:(NSInteger)param1{

    return(param1);

}
```

```
- (NSString*) method1:(NSString *)param1
          andParam2:(NSString *)param2{

  NSString *result = param1;

  if (param1 != nil &&
      param2 != nil){
    result = [result stringByAppendingString:param2];
  }

  return(result);

}
```

Here is an example of changing the name of a parameter:

```
- (void) drawCircleWithCenter:(CGPoint)paramCenter
                       radius:(CGFloat)paramRadius{

  /* Draw the circle here */

}

- (void) drawCircleWithCenter:(CGPoint)paramCenter
                       Radius:(CGFloat)paramRadius{

  /* Draw the circle here */

}
```

Can you spot the difference between the declarations of these two methods? The first method's second parameter is called **radius** (with a lowercase *r*) whereas the second method's second parameter is called **Radius** (with an uppercase *R*). This will set these two methods apart and allows your program to get compiled. However, Apple has guidelines for choosing method names as well as what to do and what not to do when constructing methods. For more information, please refer to the "Coding Guidelines for Cocoa" Apple documentation available at this URL:

http://developer.apple.com/iphone/library/documentation/Cocoa/Conceptual/Co dingGuidelines/CodingGuidelines.html

Here is a concise extract of the things to look out for when constructing and working with methods:

- Have your method names describe what the method does clearly, without using too much jargon and abbreviations. A list of acceptable abbreviations is in the Coding Guidelines.

- Have each parameter name describe the parameter and its purpose. On a method with exactly three parameters, you can use the word *and* to start the name of the last parameter if the method is programmed to perform two separate actions. In any other case, refrain from using *and* to start a parameter name.

- Start method names with a lowercase letter.
- For delegate methods, start the method name with the name of the class that invokes that delegate method. For more information about delegates, please refer to Recipe 1.7.

See Also

Recipe 1.6

1.4 Defining and Accessing Properties

Problem

You would like to create an object that has properties of different data types.

Solution

Use the `@property` directive in the *.h* file of your object:

```
#import <Foundation/Foundation.h>

@interface MyObject : NSObject {
@public
    NSString    *stringValue;
@private
    NSUInteger  integerValue;
}

@property (nonatomic, copy)   NSString   *stringValue;
@property (nonatomic, assign) NSUInteger integerValue;

@end
```

Only properties that are objects can be of type **copy** or **retain**. Usually only properties that are scalars (such as `NSUInteger`, `NSInteger`, `CGRect`, and `CGFloat`) can have the `assign` setter attributes. For more information about this, please refer to the "Declared Properties" section of Apple's "The Objective-C Programming Language" guide, available at the following URL:

http://developer.apple.com/iphone/library/documentation/cocoa/conceptual/objecti vec/Introduction/introObjectiveC.html

Now use the `@synthesize` directive in the *.m* file of your object, like so:

```
#import "MyObject.h"

@implementation MyObject

@synthesize stringValue;
@synthesize integerValue;
```

```
- (id) init {
  self = [super init];

  if (self != nil){
    stringValue = [@"Some Value" copy];
    integerValue = 123;
  }

  return(self);
}

- (void) dealloc{
  [stringValue release];
  [super dealloc];
}

@end
```

Discussion

In Objective-C, you can create and access variable data within an object in two ways:

- Declaring properties
- Declaring variables

It is important to note that properties are not simply variables. They are in fact, depending on their declaration (whether they are synthesized or not), methods. Simple variables can also be defined in Objective-C for an object, but they need to be managed and accessed in a different way. A property can be created, managed, and released in much more flexible ways than a mere instance variable. For example, a synthesized and retained property of type NSObject will retain its new value and release the previous value whenever we assign a new value to it with dot notation, whereas an instance variable that is being managed by the programmer (not synthesized and not defined as an @property) will not be accessible with dot notation and will not have a getter or a setter method created for it. So, let's focus on properties for now. You should declare a property in three steps:

1. Declare the variable.
2. Declare the property.
3. Declare the implementation method of that property.

After step 2, our variable becomes a property.

By synthesizing our properties, Objective-C will create the setter and getter methods (depending on whether the property being synthesized is read/write or read-only). Whenever the program writes to the property using dot notation (object.property), the setter method will be invoked, whereas when the program reads the property, the getter method will be invoked.

1.5 Managing Properties Manually

Problem

You want to be able to control the values that are set for and returned by your properties. This can be particularly useful, for instance, when you decide to save the values of your properties to disk as soon as they are modified.

Solution

Declare your property in the *.h* file of your object and implement the setter and getter methods in the *.m* file manually:

```
#import <Foundation/Foundation.h>

@interface MyObject : NSObject {
@public
  NSString  *addressLine;
}

@property (nonatomic, copy) NSString  *addressLine;

@end
```

In your *.m* file, implement your getter and setter methods:

```
#import "MyObject.h"

@implementation MyObject

@synthesize addressLine;

- (void) setAddressLine:(NSString *)paramValue{

  if (paramValue != addressLine){
    [addressLine release];
    addressLine = [paramValue copy];
  }

}

- (void) dealloc {
  [addressLine release];
  [super dealloc];
}

@end
```

Discussion

The `@synthesize` directive creates setter and getter methods for a property. A property, by default, is readable and writable. If a synthesized property is declared with the readonly directive, only its getter method will be generated. Assigning a value to a property without a setter method will throw an error.

This means that, by referring to a property in Objective-C using dot notation, you are accessing the setter/getter methods of that property. For instance, if you attempt to assign a string to a property called myString of type NSString in an object using dot notation, you are eventually calling the setMyString: method in that object. Also, whenever you read the value of the myString property, you are implicitly calling the myString method in that object. This method must have a return value of type NSString; in other words, the return value of the getter method of this property (called myString) must return a value of the same data type as the property that it represents. This method must not have any parameters.

Instead of using the @synthesize directive to generate getter and setter methods for your properties automatically, you can create the getter and setter methods manually as explained before. This directive will generate the getter and setter methods of a property if they don't already exist. So, you can synthesize a property but still specify its getter, its setter, or both manually.

You might be asking: why should I create my own getter and setter methods? The answer is that you may want to carry out custom operations during the read or write. A good way to explain this is through an example.

Imagine you have an object with a property called addressLine of type NSString. You want this address to accept only strings that are 20 characters or less in length. You can define the setter method in this way:

```
#import <Foundation/Foundation.h>

@interface MyObject : NSObject {
@public
  NSString  *addressLine;
}

@property (nonatomic, copy) NSString  *addressLine;

@end
```

Here is the implementation of the validation rule on our property's setter method:

```
#import "MyObject.h"

@implementation MyObject

@synthesize addressLine;

- (void) setAddressLine:(NSString *)paramValue{

  if ([paramValue length] > 20){
    return;
  }
```

```
    if (paramValue != addressLine){
      [addressLine release];
      addressLine = [paramValue copy];
    }

}

- (void) dealloc {
  [addressLine release];
  [super dealloc];
}

@end
```

Now we will attempt to set the value of this property to one string that is 20 characters and another string that is 21 characters in length:

```
MyObject *someObject = [[MyObject alloc] init];

someObject.addressLine = @"12345678901234567890";

NSLog(@"%@", someObject.addressLine);

someObject.addressLine = @"123456789012345678901";

NSLog(@"%@", someObject.addressLine);

[someObject release];
```

What we can see in the console window proves that our validation rule is working on our custom setter method (see Figure 1-4).

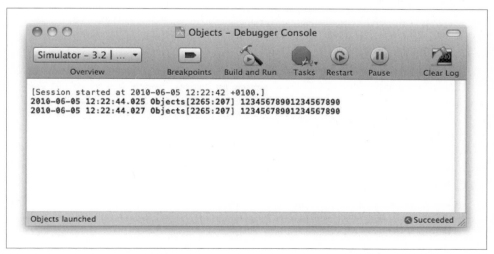

Figure 1-4. The 21-character string was not set

As you can see, the value of our property stayed intact after we tried to change that property's value to a string of 21 characters.

See Also

Recipe 1.4

1.6 Reusing a Block of Code

Problem

You want to be able to reuse a block of code you've written or call another reusable block of code somewhere else in your application.

Solution

Create instance or class methods for your classes in order to create reusable blocks of code, or simply call a method in your program.

Discussion

In programming languages such as C, we create procedures and functions. A procedure is a block of code with a name and an optional set of parameters. A procedure does not have a return value. A function is a procedure with a return value. Here is a simple procedure (with an empty body) written in C:

```
void sendEmailTo(const char *paramTo,
                 const char *paramSubject,
                 const char *paramEmailMessage){

    /* send the email here ... */
}
```

This procedure is named sendEmailTo and has three parameters, namely paramTo, paramSubject, and paramEmailMessage. We can then call this procedure in this way:

```
sendEmailTo("somebody@somewhere.com",
            "My Subject",
            "Please read my email");
```

Turning this procedure into a function that returns a Boolean value, we will have code similar to this:

```
BOOL sendEmailTo(const char *paramTo,
                 const char *paramSubject,
                 const char *paramEmailMessage){

    /* send the email here ... */
```

```
if (paramTo == nil ||
    paramSubject == nil ||
    paramEmailMessage == nil){
    /* One or some of the parameters are nil */
    NSLog(@"Nil parameter(s) is/are provided.");
    return(NO);
}

return(YES);
}
```

Calling this function is similar to calling the `sendEmailTo` procedure except that with a function, we can retrieve the return value, like so:

```
BOOL isSuccessful = sendEmailTo("somebody@somewhere.com",
                                "My Subject",
                                "Please read my email");

if (isSuccessful == YES){
    /* Successfully sent the email */
} else {
    /* Failed to send the email. Perhaps we should display
      an error message to the user */
}
```

In Objective-C, we create methods for classes. Creating Objective-C methods is quite different from writing procedures and functions in a programming language such as C. As mentioned before, we can have either instance or class methods. Instance methods are methods that can be called on an instance of the class, and class methods are methods that get called on the class itself and do not require an instance of the class to be created by the programmer. To create a method in Objective-C, follow these steps in the *.m* file of your target class:

1. Type - if you want an instance method or + if you want a class method.

2. Choose the return type of your method and enclose it within parentheses—for instance, (`void`) for no return value, (`BOOL`) for a Boolean value, (`NSObject *`) to return an instance of `NSObject`, and so on.

3. Choose a name for your method. Start the name with a lowercase letter. It is common in Objective-C to start method names with a lowercase letter—for instance, `sendEmailTo` instead of `SendEmailTo`.

4. If you do not want any parameters for your method, jump to step 9.

5. Choose two names for your parameter. One name becomes a part of the method name and will be used from outside the method (this is an optional name). The other name will be used as a parameter name inside the method. There is an exception to this in which the first name of the first parameter of a method is part of the name of the method that you chose in step 3. For this first parameter, you must only choose a second name, which becomes the parameter name used inside the method itself.

6. Once you are done choosing the name for your parameter, choose the data type of the method and enclose this within parentheses.

7. Put a colon after your parameter's first chosen name (if any), and put the parentheses that carry the data type of your method followed by the second name for your parameter.

8. Repeat steps 5 through 7 for any other parameters that you might have.

9. Insert an open curly brace ({) after the method name and parameter names (if you have parameters) and a closing curly brace (}) at the end.

Going back to the `sendEmailTo` procedure example that we saw earlier, let's attempt to create the same procedure as a method in Objective-C:

```
- (BOOL) sendEmailTo:(NSString *)paramTo
        withSubject:(NSString *)paramSubject
    andEmailMessage:(NSString *)paramEmailMessage{

    /* Send the email and return an appropriate value */

    if (paramTo == nil ||
        paramSubject == nil ||
        paramEmailMessage == nil){
        /* One or some of the parameters are nil */
        NSLog(@"Nil parameter(s) is/are provided.");
        return(NO);
    }

    return(YES);

}
```

This is an instance method (-) that returns a Boolean value (`BOOL`). The name of this method is `sendEmailTo:withSubject:andEmailMessage:` and it has three parameters. We can then call this method in this way:

```
[self sendEmailTo:@"someone@somewhere.com"
      withSubject:@"My Subject"
  andEmailMessage:@"Please read my email."];
```

As mentioned previously, the first name of every parameter (except the first parameter) is optional. In other words, we can construct the `sendEmailTo:withSubject:andEmail Message:` method in another way with a different name:

```
- (BOOL) sendEmailTo:(NSString *)paramTo
                    :(NSString *)paramSubject
                    :(NSString *)paramEmailMessage{

    /* Send the email and return an appropriate value */

    if (paramTo == nil ||
        paramSubject == nil ||
        paramEmailMessage == nil){
```

```
    /* One or some of the parameters are nil */
    NSLog(@"Nil parameter(s) is/are provided.");
    return(NO);
}

return(YES);

}
```

We can call this method like so:

```
[self sendEmailTo:@"someone@somewhere.com"
                 :@"My Subject"
                 :@"Please read my email."];
```

As you can see, the first implementation is easier to understand when you look at the invocation calls since you can see the name of each parameter in the call itself.

Declaring and implementing a class method is similar to declaring and implementing an instance method. Here are a couple of things you have to keep in mind when declaring and implementing a class method:

- The method type identifier of a class method is + instead of the - type identifier for instance methods.
- You can access self in a class method. However, the class methods of self can be accessed only inside a class method's implementation.
- Class methods are useful when you want to provide new methods of instantiation for your classes. For example, a class method named allocAndInit could both allocate and initialize an object and return the object to its caller.

1.7 Communicating with Objects

Problem

You want to pass values from one object to another without introducing tightly coupled objects.

Solution

Define a protocol using the @protocol directive, like so:

```
@protocol TrayProtocol <NSObject>
@required
  - (void) trayHasRunoutofPaper:(Tray *)paramSender;
@end
```

Conform to this protocol in another object, which will then be responsible for handling the protocol's messages.

Discussion

The more experienced programmers become, the better they understand the value of decoupling in their code. *Decoupling* simply means making two or more objects "understand" each other without needing each another in particular; in other words, if Object A has to talk to Object B, in a decoupled design, Object A won't know if it is Object B that it is talking to. All it knows is that whatever object is listening understands the language in which it is speaking. Enter protocols!

The best way to describe the value of protocols is to go through a real-life example. Let's imagine we have a printer and we intend to print a 200-page essay. Our printer has a paper tray that, when empty, tells the printer to ask us to insert more paper. The application, in effect, passes a message from the tray to the printer.

Let's go ahead and create a printer and a tray object. The printer owns the tray, so the printer object, upon initialization, must create an instance of the tray object. We will define our printer class and object in this way:

```
#import <Foundation/Foundation.h>
#import "Tray.h"

@interface Printer : NSObject {
@public
  Tray  *paperTray;
}

@property (nonatomic, retain) Tray  *paperTray;

@end
```

We will also implement the printer object in this way:

```
#import "Printer.h"

@implementation Printer

@synthesize paperTray;

- (id) init {
  self = [super init];

  if (self != nil){
    Tray *newTray = [[Tray alloc] initWithPrinter:self];
    paperTray = [newTray retain];
    [newTray release];
  }

  return(self);
}
```

```
- (void) dealloc {
  [paperTray release];
  [super dealloc];

}

@end
```

We have not yet written any code for the paper tray object. The printer and paper tray objects inherit from NSObject. The printer object can speak to the tray object by directly invoking its various selectors. However, when the tray object is created, it has no way of finding its corresponding printer and has no way of speaking to this printer because no communication protocol is defined for these two objects.

This is when protocols play a major role. A protocol is a set of *rules* that an object that *conforms* to that protocol must agree with. For instance, humans conform to various protocols such as eating, sleeping, and breathing. In the eating protocol, there is a rule that governs the input of food. In sleeping, there is a rule that governs going to bed or finding a comfortable place to sleep.

Now, going back to our example, there are two things that we want our paper tray object to do:

1. Contain 10 pieces of paper by default.
2. Let the printer know when it runs out of paper.

To meet the second requirement, we must give the paper tray access to the printer.

Our first step will be to give the paper tray a reference to the printer. Then we will make sure the paper tray talks to the printer, in a well-defined manner, using a protocol. For this, our paper tray must define the protocol since it is the object that wants to talk to the printer in its own way. The printer, in turn, must conform to that protocol.

This is how we will define the header for our paper tray object:

```
#import <Foundation/Foundation.h>

@class    Tray;

@protocol TrayProtocol <NSObject>
@required
  - (void) trayHasRunoutofPaper:(Tray *)paramSender;
@end

@interface Tray : NSObject {
@public
  id<TrayProtocol>            ownerDevice;
@private
  NSUInteger                  paperCount;
}

@property (nonatomic, retain) id<TrayProtocol>  ownerDevice;
@property (nonatomic, assign) NSUInteger        paperCount;
```

```
- (BOOL)  givePapertoPrinter;
/* Designated Initializer */
- (id)     initWithOwnerDevice:(id<TrayProtocol>)paramOwnerDevice;

@end
```

 The trayHasRunoutofPaper: method defined in the TrayProtocol protocol is a required method and the printer object must implement it.

Now we will implement the paper tray:

```
#import "Tray.h"

@implementation Tray

@synthesize paperCount;
@synthesize ownerDevice;

- (id) init {
  /* Call the designated initializer */
  return([self initWithOwnerDevice:nil]);

}

- (id) initWithOwnerDevice:(id<TrayProtocol>)paramOwnerDevice {

  self = [super init];

  if (self != nil){
    ownerDevice = [paramOwnerDevice retain];
    /* Initially we have only 10 sheets of paper */
    paperCount = 10;
  }

  return(self);
}

- (BOOL) givePapertoPrinter{

  BOOL result = NO;

  if (self.paperCount > 0){
    /* We have some paper left */
    result = YES;
    self.paperCount--;
  } else {
    /* We have run out of paper */
    [self.ownerDevice trayHasRunoutofPaper:self];
  }
```

```
    return(result);

}

- (void) dealloc {
  [ownerDevice release];
  [super dealloc];
}

@end
```

The givePapertoPrinter instance method is the only method in the paper tray's implementation that needs a description. Whenever the printer wants to print on a piece of paper, it calls this method in the paper tray to ask for a sheet of paper. If the paper tray doesn't have enough paper, it will call the trayHasRunoutofPaper: method in the printer. This is the important bit indeed. The trayHasRunoutofPaper: method is defined in the TrayProtocol protocol and the printer object is expected to conform to this protocol. Simply said, the tray object is telling the printer what methods the printer object needs to implement in order to let the paper tray speak to the printer.

The paper tray has 10 sheets of paper initially. Whenever the printer asks for a sheet of paper, the tray will reduce this number by one. Eventually, when there is no paper left, the tray lets the printer know.

Now it is time to define the printer object (the *.h* file):

```
#import <Foundation/Foundation.h>
#import "Tray.h"

@interface Printer : NSObject <TrayProtocol> {
@public
  Tray  *paperTray;
}

@property (nonatomic, retain) Tray  *paperTray;

- (BOOL) printPaperWithText:(NSString *)paramText
            numberofCopies:(NSUInteger)paramNumberOfCopies;

@end
```

Finally, we will implement the printer object (the *.m* file):

```
#import "Printer.h"

@implementation Printer

@synthesize paperTray;

- (void) trayHasRunoutofPaper:(Tray *)Sender{
  NSLog(@"No paper in the paper tray. Please load more paper.");
}

- (void) print {
```

```
    /* Do the actual printing here after we have a sheet of paper */

}

- (BOOL) printPaperWithText:(NSString *)paramText
            numberofCopies:(NSUInteger)paramNumberOfCopies{

  BOOL result = NO;

  if (paramNumberOfCopies > 0){

    NSUInteger copyCounter = 0;
    for (copyCounter = 0;
         copyCounter < paramNumberOfCopies;
         copyCounter++){
      /* First get a sheet of paper from the tray */
      if ([self.paperTray givePapertoPrinter] == YES){
        NSLog(@"Print Job #%lu", (unsigned long)copyCounter+1);
        [self print];
      } else {
        /* No more paper in the tray */
        return(NO);
      }
    }
    result = YES;
  } /* if (paramNumberOfCopies > 0){ */

  return(result);

}

- (id) init {
  self = [super init];

  if (self != nil){
    Tray *newTray = [[Tray alloc] initWithOwnerDevice:self];
    paperTray = [newTray retain];
    [newTray release];
  }

  return(self);
}

- (void) dealloc {
  [paperTray release];
  [super dealloc];

}

@end
```

The `printPaperWithText:numberofCopies:` method gets called with a number of pieces of paper to print. Every time a page needs to be printed, the printer object will perform the following tasks:

1. Ask the tray to provide one sheet of paper to the printer.
2. If the tray has run out of paper, it will return from this method.
3. If the paper tray can give the printer a sheet of paper, the printer will call its `print` method.

Now let's go ahead and instantiate an object of type `Printer` and attempt to print a couple of sheets of paper:

```
Printer *myPrinter = [[Printer alloc] init];

[myPrinter printPaperWithText:@"My Text"
            numberofCopies:100];

[myPrinter release];
```

Figure 1-5 shows the output we will get in the console window.

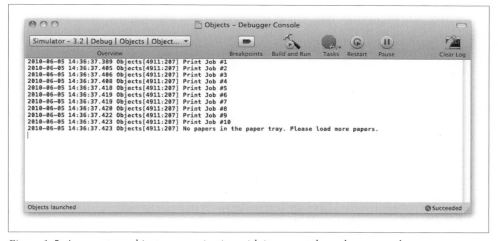

Figure 1-5. A paper tray object communicating with its parent through a protocol

The 11th page cannot be printed because the paper tray has reported to the printer that it has run out of paper. Great, isn't it?

1.8 Invoking the Selectors of an Object Dynamically

Problem

You want to be able to dynamically call any method in any object given the name of the method and its parameters.

Solution

Use the NSInvocation class, like so:

```
- (NSString *) myMethod:(NSString *)param1
            withParam2:(NSNumber *)param2{

  NSString *result = @"Objective-C";

  NSLog(@"Param 1 = %@", param1);
  NSLog(@"Param 2 = %@", param2);

  return(result);
}

- (void) invokeMyMethodDynamically {

  SEL selector = @selector(myMethod:withParam2:);

  NSMethodSignature *methodSignature =
  [[self class] instanceMethodSignatureForSelector:selector];

  NSInvocation *invocation =
  [NSInvocation invocationWithMethodSignature:methodSignature];
  [invocation setTarget:self];
  [invocation setSelector:selector];

  NSString *returnValue = nil;
  NSString *argument1 = @"First Parameter";
  NSNumber *argument2 = [NSNumber numberWithInt:102];

  [invocation setArgument:&argument1
                  atIndex:2];

  [invocation setArgument:&argument2
                  atIndex:3];

  [invocation retainArguments];
  [invocation invoke];
  [invocation getReturnValue:&returnValue];

  NSLog(@"Return Value = %@", returnValue);

}
```

Discussion

Objective-C is so dynamic that you can invoke a method in another object just by knowing the name of that object, the name of the method, and its parameters. Then you can form a selector out of the method name and parameters, form a method signature, and assign parameters to that invocation. Eventually you can, depending on how your method is defined, attempt to fetch the return value of that method as well. NSInvocation, used in this recipe, is not the optimal way of invoking a method in another object. This technique is usually used when you want to forward a method invocation to another object. `NSInvocation` is also useful if you are constructing and calling selectors dynamically and on the fly for dynamic methods. However, for an average iOS application, you probably will not need to use such a dynamic mechanism to call methods on an object.

To do this, you need to follow these steps:

1. Form a `SEL` value using the name of the method and its parameter names (as explained in Recipe 1.7).
2. Form a method signature of type `NSMethodSignature` out of your `SEL` value.
3. Form an invocation of type `NSInvocation` out of your method signature.
4. Tell the invocation what object you are targeting.
5. Tell the invocation what selector in that object you want to invoke.
6. Assign any arguments, one by one, to the invocation.
7. Invoke the method using the invocation object and demand a return value (if any).

After calling the `invokeMyMethodDynamically` method, as seen in this recipe's Solution, we will observe the following values printed in the console window:

```
Param 1 = First Parameter
Param 2 = 102
Return Value = Objective-C
```

See Also

Recipe 1.6; Recipe 1.7

1.9 Managing Memory with the iOS SDK

Problem

You want to allocate, deallocate, and use *autorelease* objects in Objective-C without causing memory leaks.

Solution

Each object allocated in the Objective-C runtime has a retain count. The retain count can be thought of as the number of references that have been made to the current object. Once the retain count is zero, the object will be deallocated.

```
/* After this line:
 foo's retain count is 1 */
NSObject *foo = [[NSObject alloc] init];
NSLog(@"%lu", (unsigned long)[foo retainCount]);
```

```
/* After this line:
 foo's retain count = 2
 bar is equal to foo */
NSObject *bar = [foo retain];
NSLog(@"%lu", (unsigned long)[foo retainCount]);

/* After this line:
 foo's retain count = 2
 bar is equal to foo
 baz is equal to bar and foo */
NSObject *baz = bar;
NSLog(@"%lu", (unsigned long)[foo retainCount]);

/* After this line:
 baz's retain count = 1
 bar's retain count = 1
 foo's retain count = 1 */
[baz release];

/* After this line:
 foo, bar and baz are deallocated here and
 should not be referenced */
[bar release];
```

Autorelease objects get deallocated at the end of the current run loop's cycle (when the autorelease pool gets released). If you create autorelease objects in a thread, you must provide your own autorelease pool object in order to deallocate the autorelease objects at the end of the thread's life. Autorelease objects are similar to the **foo** object in the previous example, with one difference: as long as there is an autorelease pool around, the autorelease object should not be released directly in code.

Discussion

When an object is allocated using the `alloc` method, its retain count is set to 1. For every time an object is retained, it must be released using the `release` method. An object will never get deallocated if the retain count of that object is greater than or equal to 1. Therefore, to prevent memory leaks, you must release the objects that you allocate for as many times as they are retained.

You usually don't need to call the `retainCount` method of objects directly. You might do it as a part of your logging or debugging process. Just keep in mind the rule in Objective-C that you must issue a release on an object for every time that it is retained. In the case of autorelease objects, they are sent a `release` message by the autorelease pool that contains them.

 Autorelease objects require autorelease pools. Incorrect use of these objects can cause dramatic performance hits on your application. Because of this, I strongly suggest that you refrain from using autorelease objects in iOS, and release the objects that you have retained or copied manually as demonstrated in various places in this book.

1.10 Managing Untyped Objects

Problem

You have an untyped object and you want to find its class name.

Solution

Use the `class` instance method of the object to find that object's class name:

```
/* The MyBundle variable is of type NSBundle */
NSBundle *myBundle = [NSBundle mainBundle];

/* The UntypedObject is of type id, an untyped object indeed */
id untypedObject = myBundle;

/* We now get the class of our untyped object */
Class untypedObjectClass = [untypedObject class];

/* Do a comparison here */
if ([untypedObjectClass isEqual:[NSBundle class]] == YES){
  /* This is an object of type NSBundle */
} else {
  /* Process other classes... */
}
```

Discussion

There are various ways to determine whether a particular object inherits from a certain other object. For instance, you can invoke the `NSStringFromClass` function in order to retrieve the class name of an object as a string. However, there are certain cases where neither `NSStringFromClass` nor direct comparison of class names is effective:

```
NSNumber *integerNumber = [NSNumber numberWithInt:10];
NSNumber *boolNumber = [NSNumber numberWithBool:10];
NSNumber *longNumber = [NSNumber numberWithLong:10];

NSLog(@"%@", NSStringFromClass([integerNumber class]));
NSLog(@"%@", NSStringFromClass([boolNumber class]));
NSLog(@"%@", NSStringFromClass([longNumber class]));
```

The output of this code is:

```
NSCFNumber
NSCFBoolean
NSCFNumber
```

This example demonstrates that the class instance method does not always return the actual name of the class from which a certain object is instantiated. For this reason, the `NSObject` protocol defines an instance method called `isKindOfClass:` that tells you whether a certain object is of a certain type or class:

```
NSNumber *integerNumber = [NSNumber numberWithInt:10];
NSNumber *boolNumber = [NSNumber numberWithBool:10];
NSNumber *longNumber = [NSNumber numberWithLong:10];

if ([integerNumber isKindOfClass:[NSNumber class]] &&
    [boolNumber isKindOfClass:[NSNumber class]] &&
    [longNumber isKindOfClass:[NSNumber class]]){
  NSLog(@"They are all of type NSNumber.");
} else {
  NSLog(@"They are not all of type NSNumber.");
}
```

The output of this code is:

```
They are all of type NSNumber.
```

CHAPTER 2

Implementing Controllers and Views

2.0 Introduction

We write iOS applications using the MVC architecture. MVC is an abbreviation for Model-View-Controller. These are the three main components of an iOS application from an architectural perspective. The model is the brain of the application. It does the calculations; it creates a virtual world for itself that can live without the views and controllers. In other words, think of a model as a virtual copy of your application, without a face!

Controllers in Xcode usually refer to view controllers. Think of view controllers as a bridge between the model and your views. A view is the *window* through which your users interact with your application. It displays what's inside the model most of the time, but in addition to that, it accepts users' interactions. Any interaction between the user and your application is sent to a view, which then can be captured by a view controller and sent to the model.

In this chapter, you will learn how the structure of iOS applications is created and how to use views and view controllers to create intuitive applications.

2.1 Getting and Using the Application Delegate

Problem

An object needs to access your application's delegate.

Solution

The UIApplication class has a method called delegate. You can retrieve the value of this property in order to get a reference to your application's delegate:

```
Project1AppDelegate *delegate = (Project1AppDelegate *)[UIApplication
sharedApplication].delegate;
```

Not all applications use *Project1AppDelegate* as the delegate object's class. Use the proper class for your application.

After obtaining the reference to your application's delegate, you can use it in different ways, such as to access the window that currently represents your application's contents:

```
- (UIWindow *) delegateWindow{

  UIWindow *result = nil;

  ViewsAndVCAppDelegate *delegate =
    (ViewsAndVCAppDelegate *)[[UIApplication sharedApplication] delegate];

  if (delegate != nil){
    result = [delegate window];
  }

  return(result);

}

- (void) logMainWindowRect{

  UIWindow *mainWindow = [self delegateWindow];

  if (mainWindow != nil){

    NSLog(@"Window Rect = %@", NSStringFromCGRect(mainWindow.frame));

  }

}
```

Discussion

iOS uses an application's delegate to send and receive important information to and from the application. For instance, if the application is getting sent to the background, the delegate of the application receives a message from iOS, and based on this message, the application can make decisions—for instance, to stop threads.

The application delegate is a simple object of type NSObject that conforms to the UIApplicationDelegate protocol. Another example is the applicationWillTerminate: selector that gets called in an application delegate when the user chooses to terminate the application. Usually, this method is used to perform operations such as storing critical application data to disk.

2.2 Managing the Views in Your Application

Problem

You would like to take advantage of multiple view controllers inside your application.

Solution

Create an object of type UIViewController and a corresponding XIB file. This can be done as follows.

In Xcode, go to File→New File and choose UIViewController Subclass, as shown in Figure 2-1.

Figure 2-1. Xcode's New File dialog

Make sure the "With XIB for user interface" option is selected in the New File dialog and click Next (see Figure 2-2). Make sure the "Also create 'FirstViewController.h'" option is selected, as shown in Figure 2-1. Of course, this name will be different if you choose another name for your filename. If you intend to create an XIB file for an iPad application, check the "Size view for iPad" checkbox as well. This will ensure that the view created for your XIB file is sized appropriately for an iPad screen. Click Finish.

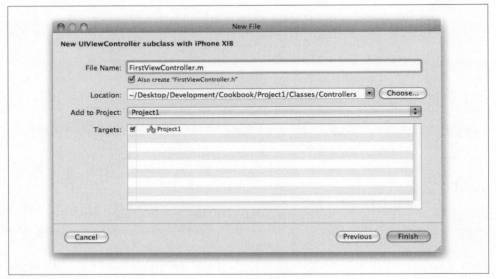

Figure 2-2. Selecting a name for your new view controller object

Discussion

A view controller is an object of type `UIViewController`. Its purpose, as its name suggests, is to choose which view is displayed at particular times, and to control aspects of its display.

iOS has great capabilities for working with view controllers. The number of view controllers that can be created in an application is limited only by the amount of memory left in the system for that particular application.

After you create a view controller as shown in this recipe's Solution, you can instantiate and use it:

```
FirstViewController *firstController =
[[FirstViewController alloc]
 initWithNibName:@"FirstViewController"
 bundle:nil];
SecondViewController *secondController =
[[SecondViewController alloc]
 initWithNibName:@"SecondViewController"
 bundle:nil];
```

```
/* Do something with the view controllers here */

[firstController release];
[secondController release];
```

 The initWithNibName:bundle: method accepts the name of the XIB file (without the file extension) and the instance of the bundle containing that XIB file. You can create a view controller with a name that's different from its XIB file, but we follow a common convention by giving the XIB file the name of the view controller. In any case, Xcode uses the name of the view controller for the view controller's .h and .m files.

2.3 Creating Your Application's GUI

Problem

You want to create an XIB file and create your GUI from that particular XIB file.

Solution

In Xcode, choose File→New File. In the template section (on the lefthand side of the screen), choose User Interface, then choose View XIB on the righthand side. The screen will look like Figure 2-3.

Choose the View XIB item on the righthand side, and in the Product drop-down menu choose the type of XIB file you want to create, such as a View XIB file for the iPhone or iPad, then click Next. On the next screen, choose the name and the location for your XIB file, as shown in Figure 2-4. Click Finish to create the View XIB file.

When you created the view controller, Xcode by default also created a view for that XIB file. Because that view is defined inside the XIB file, there are no separate .m and .h files to define it.

Discussion

After creating the View XIB file, you can double-click on it to open it and Xcode will execute Interface Builder, showing the contents of your XIB file on the screen. You can then use Interface Builder to drag and drop components onto views. Initially, the contents will look similar to Figure 2-5.

As you can see, a View XIB file by default consists of three components, one of which is View. File's Owner usually refers to the view controller that will load this XIB file from a specific bundle (usually the application's main bundle).

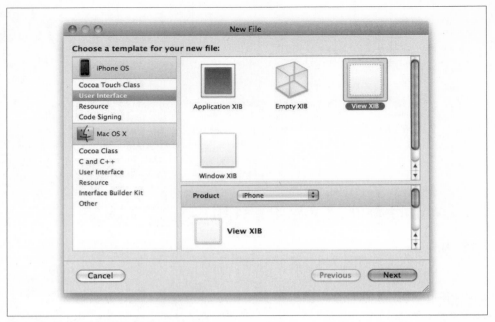

Figure 2-3. The New File dialog in Xcode

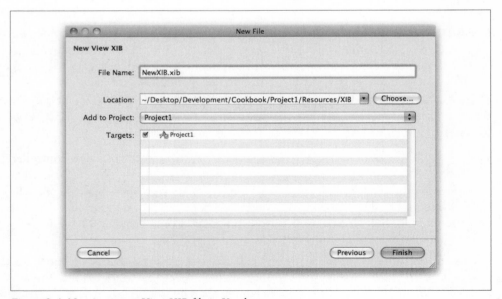

Figure 2-4. Naming a new View XIB file in Xcode

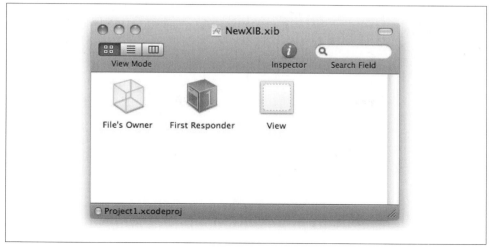

Figure 2-5. A View XIB file opened in Interface Builder

By opening View and selecting Tools→Library in Interface Builder, you can see various components that can simply be dragged and dropped on any view. Go ahead and give it a try: drag a text field, a web view, and any other component that you like onto your view.

See Also

Recipe 2.4; Recipe 2.5

2.4 Making Your View Controller Available at Runtime

Problem

You have an XIB file and a view controller object. Now you want your view controller to be able to load the XIB file's contents on the screen.

Solution

Follow these steps:

1. Open your XIB file by double-clicking it.
2. In Interface Builder, you will see the three objects created by default inside your View XIB file, as shown in Figure 2-5.
3. Make sure File's Owner is selected.
4. In Interface Builder, select Tools→Identity Inspector.
5. In the Class Identity section of the Identity Inspector pane, find the Class box.

6. Enter the class name of your view controller in the Class box. If Interface Builder recognizes that class name, it will help you write it by guessing your class name based on the recognized class names inside your current application. This is shown in Figure 2-6.

7. Save your XIB file and close Interface Builder.

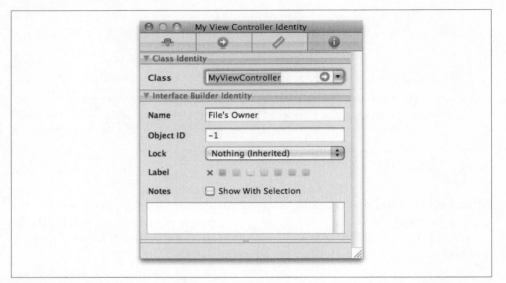

Figure 2-6. Assigning a class name to File's Owner of a View XIB file in Interface Builder

Discussion

The UIViewController is a subclass of NSObject. When Xcode creates a new View XIB file for you, as shown in Figure 2-5, the new XIB file will not initially have a definition (*.h*) and implementation (*.m*) file associated with it. What we need now is to be able to instantiate an object of type UIViewController that is associated with the XIB file. To do this, we need to open our View XIB file in Interface Builder, and as explained in this recipe's Solution, associate the File's Owner class name (Figure 2-6) with the class name of a custom view controller object. For information on how you can create a new view controller object, refer to Figure 2-1 and Figure 2-2 in Recipe 2.2.

Once we have the File's Owner of the XIB file associated with the class name of the custom view controller class we created, we can instantiate the custom view controller and as the XIB file, provide the name of the XIB file we associated with our view controller.

2.5 Using a View in Your Application's GUI

Problem

You have an XIB file connected to a view controller and you want to associate the XIB file's view to the view controller.

Solution

Follow these steps:

1. Open your XIB file with Interface Builder by double-clicking the XIB file in Xcode.

2. In Interface Builder, depending on the type of XIB file you have created, you may or may not see a view. Figure 2-5 in Recipe 2.3 shows a View XIB file that already has a View object. If you can see the view in Interface Builder, skip to step 5.

3. In Interface Builder, choose Tools→Library.

4. In the Library pane, select View as shown in Figure 2-7. Drag and drop a view from the Library into Interface Builder.

5. Double-click the view in Interface Builder, as shown in Figure 2-8.

6. In Interface Builder, select Tools→Connections Inspector and make sure the title of the Connections Inspector screen reads View Connections. If it doesn't, make sure your view is the currently selected item in Interface Builder.

7. Under the Referencing Outlets section of the Connections Inspector pane, find the item that reads New Referencing Outlet and drag and drop the empty circle next to it onto your File's Owner (view controller).

8. Interface Builder will find all accessible properties of File's Owner of type UIView* and display them in a pop-up window. Once you are sure which outlet you want to plug your UIView into, select that outlet.

Congratulations, you have created and associated a view with your view controller.

Discussion

When a new View XIB file is created with Xcode (as shown in Figure 2-3 in Recipe 2.3), the declaration and implementation files (.h and .m, respectively) are not created with it; because of this, the File's Owner item (the view controller object), as shown in Figure 2-8, will be of type NSObject. An instance of NSObject does not have a property called view, whereas an instance of type UIViewController does. If you create a new View XIB file using Xcode and then open this file using Interface Builder, you will notice that the View object (shown in Figure 2-8) is not connected to the File's Owner object merely because the File's Owner object itself has no identity and is not connected to a view controller. To solve this, first we associate the File's Owner object with a view controller class (as explained in Recipe 2.4) and then connect the view to the File's Owner object. The File's Owner object in a View XIB file is usually the view controller.

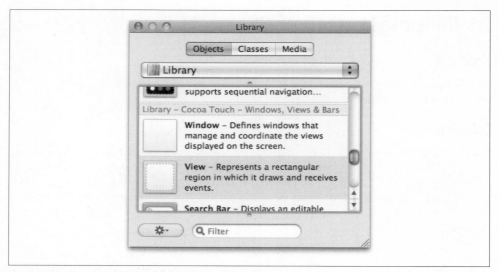

Figure 2-7. The View object in Interface Builder's Library pane

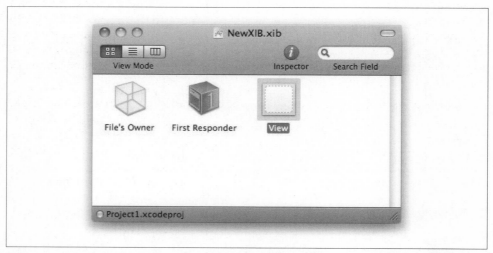

Figure 2-8. The View object after being dragged and dropped from the Library pane in Interface Builder into the XIB file

2.6 Managing Master-Detail Views

Problem

You would like to present two side-by-side views in your application, one view presenting master content and the other presenting details.

Solution

Use a split view controller. This is available only for applications running on the iPad, not the iPhone. In the following examples, I will create two conventional view controllers for the left and right views, and I will place them in an array that I pass to the split view controller.

Discussion

A split view controller is a view controller that can present two view controllers on the screen simultaneously. The split view controller has a property called `viewControllers` of type `NSArray`. You must assign two objects of type `UIViewController` (this array can also include objects of type `UINavigationController` since this class derives from `UIViewController`) to this array. These view controllers will be the master and detail view controllers, with the master view controller visible only in landscape mode and appearing on the lefthand side of the screen, and the detail view controller visible in all orientations. Figure 2-9 shows a split view controller running on the iPhone Simulator (for iPad) in landscape mode.

The split view controller does not facilitate communication between the left and right view controllers. It is your responsibility to find the optimal way for them to communicate, depending on your needs. Delegates are usually the best way to implement this communication tunnel between the two view controllers. For this, let's jump right into our application delegate and create our split view controller:

```
#import <UIKit/UIKit.h>

@interface ViewsAndVCAppDelegate : NSObject
            <UIApplicationDelegate> {
  UIWindow                   *window;
  UISplitViewController      *splitViewController;
}

@property (nonatomic, retain)
IBOutlet UIWindow *window;

@property (nonatomic, retain)
UISplitViewController *splitViewController;

@end
```

Figure 2-9. Split view controller in landscape orientation

We will implement our application delegate in this way:

```
#import "ViewsAndVCAppDelegate.h"
#import "LeftViewController.h"
#import "RightViewController.h"

@implementation ViewsAndVCAppDelegate

@synthesize window;
@synthesize splitViewController;

- (BOOL) isiPad{

  BOOL result = NO;

  NSString *classAsString =
    NSStringFromClass([UISplitViewController class]);

  if (classAsString == nil ||
      [classAsString length] == 0){
    return(NO);
  }

  UIDevice *device = [UIDevice currentDevice];

  if ([device respondsToSelector:
      @selector(userInterfaceIdiom)] == NO){
```

```
      return(NO);
   }

   if ([device userInterfaceIdiom] != UIUserInterfaceIdiomPad){
      return(NO);
   }

   /* Do extra checking based on screen size for
    instance if you want */
   result = YES;

   return(result);

}

- (BOOL)            application:(UIApplication *)application
   didFinishLaunchingWithOptions:(NSDictionary *)launchOptions {

   /* First make sure this is an iPad running this application */

   if ([self isiPad] == YES){  ❶
      /* Create the View Controller that is shown on
       the left side */
      LeftViewController *leftVC =
      [[LeftViewController alloc]
       initWithNibName:@"LeftViewController"
       bundle:nil];

      /* Create a Navigation Controller for the
       View Controller on the left */
      UINavigationController *leftNC =
      [[UINavigationController alloc]
       initWithRootViewController:leftVC];

      /* Create the right-side View Controller now */
      RightViewController *rightVC =
      [[RightViewController alloc]
       initWithNibName:@"RightViewController"
       bundle:nil];

      leftVC.delegate = rightVC;  ❷
      [leftVC release];

      /* And a Navigation Controller for the View
       Controller on the right */
      UINavigationController *rightNC =
      [[UINavigationController alloc]
       initWithRootViewController:rightVC];

      [rightVC release];

      /* Put all the Navigation Controllers in one
       array to be passed to the Split
       View Controller */
      NSArray *navigationControllers =
```

```
                  [NSArray arrayWithObjects:leftNC, rightNC, nil];

          [leftNC release];
          [rightNC release];

          /* Create the Split View Controller now. */
          UISplitViewController *splitController =
          [[UISplitViewController alloc] init];

          self.splitViewController = splitController;
          [splitController release];

          /* Place the Navigation Controllers (which are linked
           to our View Controllers), into the Split View Controller */
          [self.splitViewController
           setViewControllers:navigationControllers];

          /* Show the View of the Split View Controller
           which is now the mixture of the left
           and right View Controllers */
          [window addSubview:self.splitViewController.view];
      } else {
          /* Choose another interface path if the device is not an iPad */
      }

      [window makeKeyAndVisible];

      return YES;
  }

  - (void)dealloc {
      [splitViewController release];
      [window release];
      [super dealloc];
  }

  @end
```

❶ Our application delegate makes a decision based on whether the device is an iPad or not, and it will create the user interface based on this decision. Bear in mind that split view controllers are only available on the iPad, so we have to make sure we instantiate the UISplitViewController class just on an iPad.

❷ The right view controller becomes the delegate of the left view controller. It's the job of the application delegate (which in this case creates both view controllers) to set the right view controller as the delegate of the left view controller. This way, the left view controller can contact the right view controller whenever the user picks an item on the list.

Now it's time to declare the left view controller:

```
  #import <UIKit/UIKit.h>

  @class LeftViewController;
```

```
@protocol LeftViewControllerDelegate <NSObject>
@required
- (void)            leftViewController:(LeftViewController*)paramSender
    itemIsSelectedWithItemIndexPath:(NSIndexPath*)paramItemIndexPath
                            itemText:(NSString*)paramItemText;
@end

@interface LeftViewController : UITableViewController {
@public
    id<LeftViewControllerDelegate> delegate;
}

@property (nonatomic, assign)
id<LeftViewControllerDelegate> delegate;

@end
```

This code defines a protocol named `LeftViewControllerDelegate`. Our plan is to have the right view controller conform to this protocol and have the left view controller send messages to the right view controller through this protocol. Notice that the left view controller defines a `delegate` property.

The left view controller is just a table view controller with a few items prepopulated in it. We will implement the left view controller in this way:

```
#import "LeftViewController.h"

@implementation LeftViewController

@synthesize delegate;

- (void)        tableView:(UITableView *)tableView
didSelectRowAtIndexPath:(NSIndexPath *)indexPath{

    Protocol *neededProtocol = @protocol(LeftViewControllerDelegate);

    SEL neededSelector =
    @selector(leftViewController:itemIsSelectedWithItemIndexPath:itemText:);

    if (self.delegate == nil ||
        [self.delegate conformsToProtocol:neededProtocol] == NO ||
        [self.delegate respondsToSelector:neededSelector] == NO){
      return;
    }

    UITableViewCell *selectedCell =
    [tableView cellForRowAtIndexPath:indexPath];

    [self.delegate leftViewController:self
      itemIsSelectedWithItemIndexPath:indexPath
                            itemText:selectedCell.textLabel.text];

}
```

```objc
- (NSInteger) numberOfSectionsInTableView:(UITableView *)tableView{
  return(1);
}

- (NSInteger) tableView:(UITableView *)table
  numberOfRowsInSection:(NSInteger)section{
  return(10);
}

- (UITableViewCell*) tableView:(UITableView *)tableView
          cellForRowAtIndexPath:(NSIndexPath *)indexPath{

  static NSString *CellIdentifier = @"MyCellStyle";

  UITableViewCell *result =
  [tableView dequeueReusableCellWithIdentifier:CellIdentifier];

  if (result == nil){
    result = [[[UITableViewCell alloc]
              initWithStyle:UITableViewCellStyleDefault
              reuseIdentifier:CellIdentifier] autorelease];
  }

  result.textLabel.text = [NSString stringWithFormat:@"Item %lu",
                          (unsigned long)indexPath.row];

  return(result);

}

- (void)viewDidLoad {
  [super viewDidLoad];
  self.title = NSLocalizedString(@"Left View Controller", nil);
}

- (void) viewDidUnload{
  [super viewDidUnload];
}

- (BOOL)shouldAutorotateToInterfaceOrientation:
  (UIInterfaceOrientation)interfaceOrientation {
  return(YES);
}

- (void) dealloc{
  [super dealloc];
}

@end
```

The left view controller simply passes the cell-selected event to its delegate object without knowing what the delegate object really is. The left view controller only asks for the delegate object to conform to the LeftViewControllerDelegate protocol through which the left view controller can speak to the delegate object.

The corresponding code to respond to the `LeftViewControllerDelegate` protocol is in the right view controller. We will write the header file of this view controller in this way:

```objc
#import <UIKit/UIKit.h>
#import "LeftViewController.h"

@interface RightViewController : UIViewController
        <LeftViewControllerDelegate> {

}

@end
```

After specifying that our view controller conforms to the `LeftViewControllerDelegate` protocol, we must implement the methods that are required by this protocol. We will implement our right view controller in this way:

```objc
#import "RightViewController.h"

@implementation RightViewController

- (id) initWithNibName:(NSString *)nibNameOrNil
               bundle:(NSBundle *)nibBundleOrNil{

  self = [super initWithNibName:nibNameOrNil
                       bundle:nibBundleOrNil];

  if (self != nil){
    self.title = NSLocalizedString(@"Right View Controller", nil);
  }

  return(self);

}

- (void)         leftViewController:(LeftViewController*)paramSender
    itemIsSelectedWithItemIndexPath:(NSIndexPath*)paramItemIndexPath
                         itemText:(NSString*)paramItemText{

  self.title = [NSString stringWithFormat:
               @"(%@) is selected on the left side", paramItemText];

}

- (void)viewDidLoad {
  [super viewDidLoad];
}

- (void) viewDidUnload{
  [super viewDidUnload];
}

- (BOOL)shouldAutorotateToInterfaceOrientation:
  (UIInterfaceOrientation)interfaceOrientation {
  return YES;
```

```
}

@end
```

As you can see, it's all very simple. Now whenever the user selects an item in the table of the left view controller, a delegate message is called on the delegate object of that view controller (in this case, the delegate is the right view controller) and the delegate object will take an appropriate action to respond to this event.

The results are shown in Figure 2-10.

Figure 2-10. Split view with delegate tunnel for transferring information from the master to the detail view controller

See Also

Recipe 2.17

2.7 Managing Multiple Views

Problem

You have a view controller with an XIB file, but you want to create a view for the view controller dynamically.

Solution

In your view controller's object implementation (inside the *.m* file of your view controller), create a new view and assign or add it to the **view** property of your view controller, like so:

```
- (void)loadView {

  [super loadView];

  UIView *myNewView = [[UIView alloc] init];

  CGRect currentViewFrame = self.view.bounds;
  CGRect newViewFrame = CGRectMake(0.0f,
                                   0.0f,
                                   currentViewFrame.size.width / 2.0f,
                                   currentViewFrame.size.height / 2.0f);

  myNewView.frame = newViewFrame;
  myNewView.backgroundColor = [UIColor blueColor];

  [self.view addSubview:myNewView];

  [myNewView release];

}
```

 If you do *not* want to use the default view of your view controller, a call to **super** is unnecessary as that creates the default view. If you want to use the default view but, for instance, add another view to it, call the **loadView** method of **super** *before* you do anything else in the **loadView** method of your view controller. Then you can be sure that **super** has created your default view and that you can now safely modify it the way you want.

Discussion

A view controller has a property called **view** that is the main view being maintained by the view controller. This view gets loaded and initialized when the **loadView** method is called on an instance of the **UIViewController** class. If you override this method, you can take control over how the view has to be loaded. A call to the **super** object on the same method will create your default view for you, as mentioned before.

2.8 Incorporating and Using Models in the GUI

Problem

You are using the MVC architecture to create your program and you are wondering how you can access your model from inside your view controllers, views, or other objects that are not incorporated in and are not part of the model.

Solution

There are two convenient ways to do this:

- Implement the top class in the model hierarchy as a singleton (shared instance) so that it can be accessed from anywhere.

- Create an object containing a property that points to the model's highest-level object. The object that points to the model must now be the parent of every other object that needs to access the model. Thus, if you want views to access the model, create a high-level view with a pointer to the model and subclass the other views from it. Alternatively, you can access the model from inside a view controller and pass the values required to each view object controlled by that view controller.

Discussion

In Objective-C, you can define a shared instance using a static variable, just as in C++. For instance, assume we want to instantiate an object of type **NSObject**, named **MyWorld**, and let it have a shared instance (shared instances in Objective-C can be accessed through a class method). The interface for this object can be created in this way:

```
#import <Foundation/Foundation.h>

@interface MyWorld : NSObject {

}

+ (MyWorld *)    sharedInstance;

@end
```

You can implement the **MyWorld** object as follows.

 The name **MyWorld** was arbitrarily chosen and has nothing to do with this object being a singleton or shared instance. In real-life code, you'd probably choose a name denoting what the object does for you.

```
#import "MyWorld.h"

@implementation MyWorld
```

```
static MyWorld      *sharedInstance = nil;

- (id) init {

  self = [super init];

  if (self != nil){

    /* Do NOT allocate/initialize other objects here
       that might use the MyWorld's sharedInstance as
       that will create an infinite loop */

  }

  return(self);

}

- (void) initializeSharedInstance{

  /* Allocate/initialize your values here as we are
     sure this method gets called only AFTER the
     instance of MyWorld has been created through
     the [sharedInstance] class method */

}

+ (MyWorld *)      sharedInstance{
  @synchronized(self){
    if (sharedInstance == nil){
      sharedInstance = [[self alloc] init];
      /* Now initialize the shared instance */
      [sharedInstance initializeSharedInstance];
    }
    return(sharedInstance);
  }

}

- (NSUInteger) retainCount{
  return(NSUIntegerMax);
}

- (void) release{
  /* Don't call super here. The shared instance should
     not be deallocated */
}

- (id) autorelease{
  return(self);
}

- (id) retain{
  return(self);
}
```

```
- (void) dealloc {
  [super dealloc];
}

@end
```

In your views and view controllers or other nonmodel objects, you can retrieve an instance to your model using the `sharedInstance` class method of `MyWorld`:

```
- (void)viewDidLoad {
  [super viewDidLoad];

  MyWorld *bigWorld = [MyWorld sharedInstance];
  MyWorld *theSameWorld = [MyWorld sharedInstance];

  if ([bigWorld isEqual:theSameWorld] == YES){
    NSLog(@"The Big World and the Same World are the same!");
  } else {
    NSLog(@"These two worlds are different.");
  }

}
```

Both `bigWorld` and `theSameWorld` now point to the same instance in memory. This is one of the reasons you might consider creating a shared instance—or what is traditionally called a *singleton*—whenever you are designing a model or storage of some type that has to be accessed from multiple objects across a project and you do not want to keep a reference to the same instance of the object anywhere specific. Using a class method providing you with a shared instance of an object, you will easily be able to call this method anywhere using only the name of your class, as shown previously, in order to obtain a reference to the one and only copy of this object in memory.

 A shared instance of an object has a retain count of 1 and should not, in any circumstances, be retained since that will impose complications on how the shared instance is maintained.

If you had a good look at the implementation of `MyWorld`, you must have noticed the comments in the `init` and `initializeSharedInstance` methods. The reason we should *not* allocate and initialize important resources in the `init` method of `MyWorld` is that if any of the classes that might be allocated and initialized in the `init` method of `MyWorld` access the `sharedInstance` method of `MyWorld`, the `sharedInstance` class method will be called prematurely without having already returned from its previous `init` method. Therefore, we will be in an infinite loop of calling the `sharedInstance` class method since none of the `sharedInstance` class methods can complete without depending on another `sharedInstance` call. This can be demonstrated better with an example:

1. A view controller calls the `sharedInstance` class method of `MyWorld` for the first time in our application. We have not called this class method anywhere else yet.

2. The `sharedInstance` class method determines that the `sharedInstance` static variable is `nil`, and therefore attempts to allocate and initialize this static variable by calling the `alloc` and `init` methods of the `MyWorld` object.

3. Imagine we instantiate an object of some type in the `init` instance method of `MyWorld`, by calling the `alloc` and `init` methods of that object. Now imagine that that object's `init` method attempts to initialize itself using a variable that is inside `MyWorld`. Therefore, the object will call the `sharedInstance` class method of `MyWorld` to get a reference to `MyWorld`. However, the problem here is that we are still not out of the `init` instance method of `MyWorld` explained in step 2. We are still trying to finish the `init` method, but the initialization of our new object was in the `init` method and needs to be finished, and since this new initialization depends on `MyWorld` itself, another `sharedInstance` class method will be called and `MyWorld` will be allocated and initialized again. Another initialization means an infinite loop to steps 2 and 3, until the end of time!

To demonstrate this fragile fact, change the `init` instance method of `MyWorld` to the following implementation:

```
- (id) init {

  self = [super init];

  if (self != nil){

    /* Do NOT allocate/initialize other objects here
       that might use the MyWorld's sharedInstance as
       that will create an infinite loop */
    [MyWorld sharedInstance];

  }

  return(self);

}
```

Now if you try to run your application, it will crash. To fix this issue, add the code shown in bold to the `initializeSharedInstance` instance method:

```
- (void) initializeSharedInstance{

  /* Allocate/initialize your values here as we are
     sure this method gets called only AFTER the
     instance of MyWorld has been created through
     the [sharedInstance] class method */

  [MyWorld sharedInstance];

}
```

```
+ (MyWorld *)       sharedInstance{
  @synchronized(self){
    if (sharedInstance == nil){
      sharedInstance = [[self alloc] init];
      /* Now initialize the shared instance */
      [sharedInstance initializeSharedInstance];
    }
    return(sharedInstance);
  }

}
```

2.9 Implementing Navigation Bars

Problem

You want to be able to present options to your users and maintain a stack-like array of screens that users can traverse easily.

Solution

Use a UINavigationController class to organize the hierarchy of view controllers in your application. A navigation controller will create a navigation bar for you automatically:

```
#import <UIKit/UIKit.h>

@class ViewsAndVCViewController;

@interface ViewsAndVCAppDelegate : NSObject
          <UIApplicationDelegate> {
@public
  UIWindow *window;
  UINavigationController  *navigationController;
}

@property (nonatomic, retain)
IBOutlet UIWindow *window;

@property (nonatomic, retain)
UINavigationController  *navigationController;

@end
```

As you can see, we have declared our navigation controller in our application delegate. You do not have to do this in the application delegate. In fact, you can instantiate an object of type UINavigationController anywhere in your application. We will now implement the application delegate:

```
#import "ViewsAndVCAppDelegate.h"
#import "FirstViewController.h"

@implementation ViewsAndVCAppDelegate
```

```
@synthesize window;
@synthesize navigationController;

- (BOOL) application:(UIApplication *)application
didFinishLaunchingWithOptions:(NSDictionary *)launchOptions {

    // Override point for customization after application launch.
    FirstViewController *controller =
    [[FirstViewController alloc]
     initWithNibName:@"FirstViewController"
     bundle:nil];

    UINavigationController *theNavigationController =
    [[UINavigationController alloc]
     initWithRootViewController:controller];

    self.navigationController = theNavigationController;

    [theNavigationController release];

    [self.navigationController setNavigationBarHidden:NO
                                             animated:YES];

    [controller release];

    // Add the view controller's view to the window and display.
    [window addSubview:self.navigationController.view];
    [window makeKeyAndVisible];

    return YES;
}

- (void)dealloc {
    [navigationController release];
    [window release];
    [super dealloc];
}

@end
```

Discussion

There are two good ways to create navigation controllers in an application for the iPhone or iPad:

- Creating a navigation-based application
- Creating a `UINavigationController` object manually in the application delegate

I choose the second method over the first, for various reasons. One is that I feel I get more control over how the structure of my application is created from the beginning; another is that the second method makes it easier for me to control the lifetime of the navigation controller.

The first method also gives you control over navigation, but makes you feel less like you can do anything you want.

Both methods will be described in this recipe.

To create a navigation-based application, follow these steps:

1. In Xcode, choose File→New Project.
2. In the project templates, choose Navigation-based Application on the righthand side of the screen, as shown in Figure 2-11.
3. Click the Choose button.
4. Give your project a name and choose where you want to save it, as shown in Figure 2-12.
5. Click Save and Xcode will create your project for you.

Figure 2-11. Creating a navigation-based application in Xcode

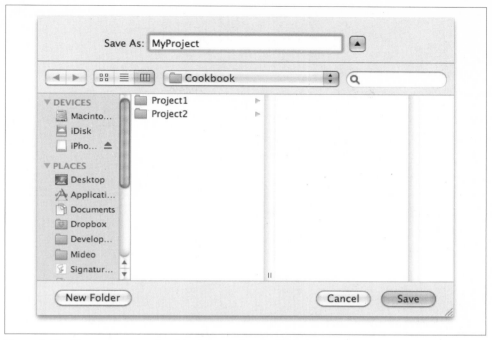

Figure 2-12. Choosing a name for the project and saving it to disk

Now that you have created a navigation-based application, you might wonder how that project template differs, for instance, from the utility application. Well, open the *.h* file of your application's delegate; you will see code similar to this:

```
#import <UIKit/UIKit.h>

@interface MyProjectAppDelegate : NSObject <UIApplicationDelegate> {

    UIWindow *window;
    UINavigationController *navigationController;
}

@property (nonatomic, retain) IBOutlet UIWindow *window;
@property (nonatomic, retain) IBOutlet UINavigationController
*navigationController;

@end
```

Xcode created a navigation controller for your delegate (which, as I mentioned before, is the preferred place to create your navigation controllers). Now you can go ahead and add view controllers to this navigation controller (I will explain this process shortly).

The other way to create a navigation-based application is to use any application template you want and create your own navigation controller. For instance:

1. In Xcode, choose File→New Project.
2. From the project templates, choose the View-based Application template, as shown in Figure 2-13.
3. Click the Choose button.
4. Choose a name and location for your project, as shown in Figure 2-14.
5. Click the Save button.

Figure 2-13. Creating a view-based application in Xcode

Now you need to find your application's delegate. If you named your project "X," the application's delegate by default will be called *X*AppDelegate. The project created in Figure 2-14 is called Project3; therefore, the application delegate is called Project3App Delegate. Once you find the delegate, open the *.h* file first and add a navigation controller to it:

```
#import <UIKit/UIKit.h>
```

```
@class Project3ViewController;

@interface Project3AppDelegate : NSObject
            <UIApplicationDelegate> {
@public
  UIWindow                *window;
  Project3ViewController  *viewController;
  UINavigationController  *navigationController;
}

@property (nonatomic, retain)
IBOutlet UIWindow *window;

@property (nonatomic, retain)
IBOutlet Project3ViewController *viewController;

@property (nonatomic, retain)
UINavigationController  *navigationController;

@end
```

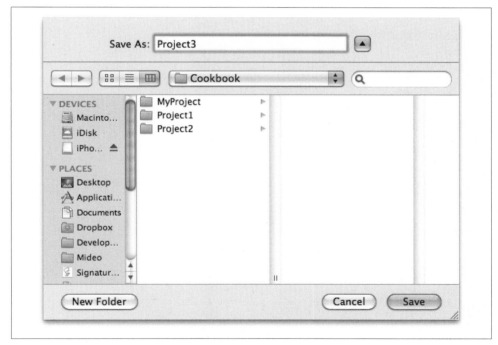

Figure 2-14. Choosing a name for the view-based application

Open the application delegate's *.m* file and create your navigation controller:

```objc
#import "Project3AppDelegate.h"
#import "Project3ViewController.h"

@implementation Project3AppDelegate

@synthesize window;
@synthesize viewController;
@synthesize navigationController;

- (BOOL)              application:(UIApplication *)application
   didFinishLaunchingWithOptions:(NSDictionary *)launchOptions {


    UINavigationController *theNavigationController =
    [[UINavigationController alloc]
     initWithRootViewController:self.viewController];

    self.navigationController = theNavigationController;

    [theNavigationController release];

    [self.navigationController setNavigationBarHidden:NO
                                             animated:YES];

    [window addSubview:self.navigationController.view];
    [window makeKeyAndVisible];

    return YES;
}

- (void)dealloc {
    [navigationController release];
    [viewController release];
    [window release];
    [super dealloc];
}

@end
```

 You must release the navigation controller whenever you are done with it. Some programmers prefer to release the navigation controller in the `applicationWillTerminate` delegate message received by the application delegate. But you might find it more convenient to release the navigation controller elsewhere.

See Also

Recipe 2.10; Recipe 2.11

2.10 Switching from One View to Another

Problem

You have two or more view controllers and a navigation controller. You want to switch from one view controller to the other.

Solution

If you already have a navigation controller, you are halfway there. If not, you need to create a navigation controller first (see Recipe 2.9). Once you have a navigation controller, you can use the pushViewController: method to push another view controller onto the hierarchy:

```
- (void) goToSecondViewController{

    SecondViewController *controller =
    [[SecondViewController alloc]
     initWithNibName:@"SecondViewController"
     bundle:nil];

    [self.navigationController pushViewController:controller
                                        animated:YES];

    [controller release];

}

- (void) viewDidLoad{
    [super viewDidLoad];

    /* Show the second View Controller 3 seconds
     after this View Controller's
     view is loaded */

    [self performSelector:@selector(goToSecondViewController)
            withObject:nil
            afterDelay:3.0f];

}

- (void) viewDidUnload{
    [super viewDidUnload];

    [NSObject
     cancelPreviousPerformRequestsWithTarget:self
       selector:@selector(goToSecondViewController)
         object:nil];
}
```

This code runs on a view controller and attempts to push a view controller of class `SecondViewController` into the stack maintained by the navigation controller, three seconds after the first view controller's view is loaded.

Discussion

The `pushViewController:animated:` method of the navigation controller takes two parameters. One is the view controller to be pushed onto the stack. The other specifies whether to animate the transition between the current top view controller and the new view controller getting pushed onto the stack (a Boolean value).

Once the new view controller is pushed onto the stack, the view associated with that view controller will get shown as the top view, and that view controller will become the topmost view controller on the stack.

The `title` property of an instance of the `UIViewController` sets the title of the navigation bar of a navigation controller displaying that view controller. A tab bar also uses this property, but we will discuss tab bars in Recipe 2.16.

See Also

Recipe 2.9; Recipe 7.4

2.11 Setting the Title on a Navigation Bar

Problem

You want your navigation bar to have a title.

Solution

Each view controller is represented on a navigation controller by a navigation item. Think of a navigation item as a placeholder for the current view controller (the owner of the navigation item) on the navigation controller. Therefore, every instance of the `UIViewController` class comes with a property named `navigationItem`. You can access the `title` property of the `navigationItem` in order to set the title of the current view controller on the navigation controller:

```
self.navigationItem.title = @"My View Controller";
```

Discussion

The navigation item controls the buttons and the title that appear on the navigation controller, and these items are specific to the top view controller of the navigation controller. Once a new view controller is pushed onto the stack, the navigation bar will automatically update itself to reflect the changes requested by the new top view controller, such as the title of the navigation bar and right and left navigation bar buttons.

Aside from setting the `title` property of the `navigationItem` of every view controller, you can directly set the `title` property of the view controller itself, which will propagate the title throughout the related GUI elements such as the tab bar (if present). Bear in mind that in this example, `self` refers to an instance of `UIViewController` that is being shown in a navigation controller.

```
self.navigationItem.title = @"My View Controller";
self.title = @"My View Controller";
```

 The `navigationItem` property of every view controller by default is set to `nil` and will be dynamically allocated the first time this property is used. You must not use this property if you do not have a navigation controller present.

See Also

Recipe 2.9

2.12 Displaying an Image for the Title of the Navigation Bar

Problem

You want to display an image instead of text as the title of the current view controller on the navigation controller.

Solution

Use the `titleView` property of the view controller's navigation item:

```
/* Create an Image View to replace the Title View */
UIImageView *imageView =
[[UIImageView alloc]
 initWithFrame:CGRectMake(0.0f, 0.0f, 100.0f, 40.0f)];

/* Load an image. Be careful, this image will be cached */
UIImage     *image = [UIImage imageNamed:@"FullSizeLogo.png"];

/* Set the image of the Image View */
[imageView setImage:image];

/* Set the Title View */
self.navigationItem.titleView = imageView;

/* Release the image view now */
[imageView release];
```

The preceding code must be executed in a view controller that has a navigation controller.

Discussion

The navigation item of every view controller can display a title for the view controller to which it is assigned in two different ways:

- By displaying simple text
- By displaying a view

If you want to use text, as we discussed before, you can use the `title` property of the navigation item. However, if you want more control over the title or if you simply want to display an image or whatnot up on the navigation bar, you can use the `titleView` property of the navigation item of a view controller. You can assign any object that is an explicit or implicit subclass of the `UIView` class. In the example, we are creating an image view and assigning an image to it. Then we are displaying it as the title of the current view controller on the navigation controller.

2.13 Creating and Managing Buttons on a Navigation Bar

Problem

You want to create buttons on a navigation bar.

Solution

Use the current view controller's navigation item to display buttons on the right or the left side of the navigation bar. Remember that each button has to be of type `UIBarButtonItem`:

```
- (void) performLeft:(id)paramSender{
  /* Peform an action here */
}

- (void) performRight:(id)paramSender{
  /* Perform another action here */
}

- (void)viewDidLoad {
  [super viewDidLoad];

  /* Create the button that appears on the Right Side */
  UIBarButtonItem *rightButton = [[UIBarButtonItem alloc]
                             initWithTitle:@"Right"
                             style:UIBarButtonItemStylePlain
                             target:self
```

```
                               action:@selector(performRight:)];

    /* Create the button that appears on the Left Side */
    UIBarButtonItem *leftButton = [[UIBarButtonItem alloc]
                                   initWithTitle:@"Left"
                                   style:UIBarButtonItemStyleDone
                                   target:self
                                   action:@selector(performLeft:)];

    /* Assign the buttons to the Navigation Item's properties */
    self.navigationItem.rightBarButtonItem = rightButton;
    self.navigationItem.leftBarButtonItem = leftButton;

    [rightButton release];
    [leftButton release];
}

- (void) viewDidUnload{
    [super viewDidUnload];

    self.navigationItem.rightBarButtonItem = nil;
    self.navigationItem.leftBarButtonItem = nil;
}
```

The results are depicted in Figure 2-15.

Figure 2-15. iPhone4 Simulator displaying two buttons on a navigation controller

In iOS, using UIBarButtonItem, we can create different types of buttons by simply passing one of these values or the style parameter of the initWithTitle:style:tar get:action: instance method of UIBarButtonItem:

UIBarButtonItemStylePlain

This is a simple button and is the default style for bar button items.

`UIBarButtonItemStyleBordered`

> This is again a simple button like the aforementioned plain button. No visual difference between the two can be recognized when the application runs on a device.

`UIBarButtonItemStyleDone`

> This is another simple button, but with a major difference in color contrast to highlight the fact that pressing this button will finish the current mode (such as an editing mode).

A quotation from the iPhone Human Interface Guidelines cites the fact that all items on the navigation bar are bordered:

> ...buttons in a navigation bar include a bezel around them. In iPhone OS, this style is called the bordered style. All controls in a navigation bar should use the bordered style. In fact, if you place a plain (borderless) control in a navigation bar, it will automatically convert to the bordered style.

In addition to simple bar button items, you can create a bar button item that derives its properties from the items that iOS creates on various internal applications such as Mail. To do this, you can use the `initWithBarButtonSystemItem:target:action:` instance method of the `UIBarButtonItem` class, like so:

```
- (void)viewDidLoad {
  [super viewDidLoad];

  UIBarButtonItem *rightButton =
  [[UIBarButtonItem alloc]
   initWithBarButtonSystemItem:UIBarButtonSystemItemPlay
   target:self
   action:@selector(performRight:)];

  UIBarButtonItem *leftButton =
  [[UIBarButtonItem alloc]
   initWithBarButtonSystemItem:UIBarButtonSystemItemCamera
   target:self
   action:@selector(performLeft:)];

  [self.navigationItem setRightBarButtonItem:rightButton
                                    animated:YES];

  [self.navigationItem setLeftBarButtonItem:leftButton
                                   animated:YES];

  /* Release the buttons */
  [rightButton release];
  [leftButton release];

}

- (void) viewDidUnload{
  [super viewDidUnload];

  self.navigationItem.rightBarButtonItem = nil;
```

```
        self.navigationItem.leftBarButtonItem = nil;
}
```

Your results will look similar to Figure 2-16.

Figure 2-16. System bar button items shown on a navigation bar

Discussion

The navigation item of every view controller has various properties that deal with buttons it might need to show on the current navigation bar. These properties are:

backBarButtonItem
The back button used to represent the current view controller on the navigation bar

leftBarButtonItem
The button that is displayed on the left side of the navigation bar when the current view controller is the top item on the stack

rightBarButtonItem
The button that is displayed on the right side of the navigation bar when the current view controller is the top item on the stack

Each button must be of type UIBarButtonItem.

The only item that could be a bit confusing is the backBarButtonItem. This button is the back button *associated* with the current view controller, not the back button shown when the current view controller is displayed. For instance, let's assume you have two view controllers, namely Controller1 and Controller2. In Controller1, you create an instance of the UIBarButtonItem class and assign it to the backBarButtonItem property of the navigationItem property of Controller1. Now if you push Controller2 onto the stack, the back button that you created in Controller1 will be shown on the navigation bar because this is the back button associated with Controller1—or in other words, the button that will take the user back to Controller1 when pressed.

 The target that you assign to the backBarButtonItem of the navigation item must be nil. Even if you do assign a target to your custom back button, iOS will simply ignore it.

Any of the buttons on a navigation bar, whether on the left or the right side, can only display one button at a time unless you initialize the UIBarButtonItem with a custom view, like so:

```
- (void)viewDidLoad {
[super viewDidLoad];

NSArray *items = [NSArray arrayWithObjects:@"Up", @"Down", nil];

UISegmentedControl *segmentedButton =
[[UISegmentedControl alloc]
  initWithItems:items];

/* Make the buttons not stay in the On state when they are tapped */
segmentedButton.momentary = YES;

[segmentedButton setSegmentedControlStyle:UISegmentedControlStyleBar];

UIBarButtonItem *rightButton = [[UIBarButtonItem alloc]
                        initWithCustomView:segmentedButton];

[self.navigationItem setRightBarButtonItem:rightButton
                        animated:YES];

[rightButton release];
[segmentedButton release];

}

- (void) viewDidUnload{
[super viewDidUnload];

self.navigationItem.rightBarButtonItem = nil;
self.navigationItem.leftBarButtonItem = nil;
}
```

When you run this code, you will get results similar to Figure 2-17.

 You can pass an array of instances of NSString or UIImage to the items property of an instance of UISegmentedControl. In this example, we used NSString to merely display text. But you can use images as well if you want to.

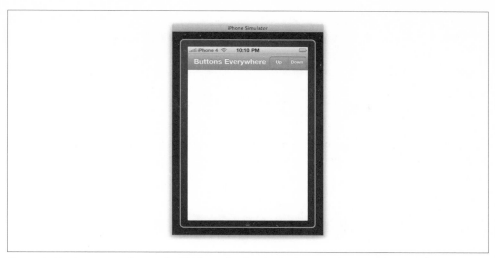

Figure 2-17. A segmented control added as a bar button item on a navigation bar

See Also

Recipe 2.9

2.14 Removing a View from a Navigation Controller

Problem

You would like to remove (pop) a view controller from the hierarchy of a navigation controller.

Solution

Use the `popViewControllerAnimated:` method of the `UINavigationController` class:

```
/* This code is run inside a View Controller
 with a Navigation Controller */
[self.navigationController popViewControllerAnimated:YES];
```

This code will run inside a view controller with a valid navigation controller.

Discussion

The `popViewControllerAnimated:` method accepts a `BOOL` parameter that determines whether the top-of-the-stack view controller's pop transition has to be animated or not.

The last view controller on the stack (created first), which is also called the root view controller, cannot be popped from the hierarchy.

The return value of this method is an object of type UIViewController, which is the view controller that has been popped from the navigation controller's hierarchy of the view controller; in other words, the topmost view controller.

See Also

Recipe 2.15

2.15 Manipulating a Navigation Controller's Array of View Controllers

Problem

You would like to directly manipulate the array of view controllers associated with a specific navigation controller.

Solution

Use the viewControllers property of the UINavigationController class to access and modify the array of view controllers associated with a navigation controller:

```
- (void) goBack{
    /* Get the current array of View Controllers */
    NSArray *currentControllers = self.navigationController.viewControllers;

    /* Create a mutable array out of this array */
    NSMutableArray *newControllers = [NSMutableArray
                                arrayWithArray:currentControllers];

    /* Remove the last object from the array */
    [newControllers removeLastObject];

    /* Assign this array to the Navigation Controller */
    self.navigationController.viewControllers = newControllers
}
```

You can call this method inside any view controller in order to pop the last view controller from the hierarchy of the navigation controller associated with the current view controller.

Discussion

An instance of the UINavigationController class retains an array of UIViewController objects that it holds at any instance. After retrieving this array, you can manipulate it in any way. For instance, you can remove a view controller from an arbitrary place in the array.

Manipulating the view controllers of a navigation controller directly by assigning an array to the viewControllers property of the navigation controller will commit the operation without a transition. If you wish this operation to be animated, use the set ViewControllers:animated: method of the UINavigationController, as shown in the following snippet:

```
- (void) goBack{
    /* Get the current array of View Controllers */
    NSArray *currentControllers = self.navigationController.viewControllers;

    /* Create a mutable array out of this array */
    NSMutableArray *newControllers = [NSMutableArray
                                      arrayWithArray:currentControllers];

    /* Remove the last object from the array */
    [newControllers removeLastObject];

    /* Assign this array to the Navigation Controller with animation */
    [self.navigationController setViewControllers:newControllers
                                         animated:YES];
}
```

See Also

Recipe 2.14

2.16 Incorporating a Tab Bar into Your Application

Problem

You would like to use a tab bar in your application.

Solution

Use the UITabBarController class in, for instance, your application's delegate. Here is the .h file of the application delegate with a tab bar controller:

```
#import <UIKit/UIKit.h>

@interface ViewsAndVCAppDelegate : NSObject <UIApplicationDelegate> {
    UIWindow                    *window;
    UITabBarController          *tabBarController;
}
```

```
@property (nonatomic, retain) IBOutlet UIWindow    *window;
@property (nonatomic, retain) UITabBarController   *tabBarController;

@end
```

Here is the implementation (.*m*) file of the application delegate with a tab bar:

```
#import "ViewsAndVCAppDelegate.h"
#import "FirstViewController.h"
#import "SecondViewController.h"

@implementation ViewsAndVCAppDelegate

@synthesize window;
@synthesize tabBarController;

- (BOOL)             application:(UIApplication *)application
  didFinishLaunchingWithOptions:(NSDictionary *)launchOptions {

  /* Create the first View Controller */
  FirstViewController *firstController =
  [[FirstViewController alloc]
   initWithNibName:@"FirstViewController"
   bundle:nil];

  /* Now create the second View Controller */
  SecondViewController *secondController =
  [[SecondViewController alloc]
   initWithNibName:@"SecondViewController"
   bundle:nil];

  /* Stack up all the View Controllers into an array */
  NSArray *arrayofViewControllers = [NSArray arrayWithObjects:
                                     firstController,
                                     secondController,
                                     nil];

  [firstController release];
  [secondController release];

  /* Instantiate your Tab Bar Controller */
  UITabBarController *theTabBarController =
  [[UITabBarController alloc] init];

  self.tabBarController = theTabBarController;
  [theTabBarController release];

  /* Set the array of View Controllers of the Tab Bar Controller */
  [self.tabBarController setViewControllers:arrayofViewControllers
                          animated:YES];

  /* Show the View of the Tab Bar Controller */
  [window addSubview:self.tabBarController.view];

  [window makeKeyAndVisible];
```

```
        return YES;

}

- (void)dealloc {
    [tabBarController release];
    [window release];
    [super dealloc];
}

@end
```

Discussion

A tab bar is a special control that appears at the bottom of the screen of an iOS application (see Figure 2-18 for an example). Using a tab bar, you can present different types of data to the users of your application in separate *sections*. These *sections* are called tab bar items. The Clock and Phone applications included in iOS provide examples of tab bars (the Phone application is exclusive to the iPhone).

We can create tab bars by instantiating an object of type UITabBarController and adding it to the main window of our application, as shown in the code in this recipe's Solution.

Figure 2-18. A simple tab bar at the bottom of the screen displaying two tab bar items (labeled "First" and "Second," from left to right)

Instances of the UITabBarController retain a property called viewControllers of type NSArray. This property is the list of all the instances of UIViewController that the tab bar must display on the screen. Please bear in mind that UINavigationController is also a subclass of UIViewController, and therefore can be added to this array. By setting this property to an array, the new set of view controllers will get displayed in the tab bar (if the tab bar itself has already been added to the current window), without an animation. However, to enable the animation, you must call the setViewControllers:animated: instance method of UITabBarController.

UIViewController instances are able to change their tab bar item's behavior by accessing a property named tabBarItem. View controllers can then change the image (for instance) that gets displayed on their corresponding tab bar item by assigning an instance of UIImage to the image property of tabBarItem:

```
- (void)viewDidLoad {
    [super viewDidLoad];
```

```
    UIImage *tabImage = [UIImage imageNamed:@"FirstTabImage.png"];
    self.tabBarItem.image = tabImage;
}

- (void) viewDidUnload{
    [super viewDidUnload];

    self.tabBarItem.image = nil;
}
```

2.17 Pop Up Additional Information over iPad UI Elements

Problem

You want to display data to iPad users, without blocking the main screen's contents.

Solution

Use *popovers* on the iPad:

```
#import "ViewsAndVCAppDelegate.h"
#import "LeftViewController.h"
#import "RightViewController.h"

@implementation ViewsAndVCAppDelegate

@synthesize window;
@synthesize splitViewController;
@synthesize rightViewController;
@synthesize leftViewController;

- (void)splitViewController:(UISplitViewController*)svc
      willHideViewController:(UIViewController *)aViewController
           withBarButtonItem:(UIBarButtonItem*)barButtonItem
        forPopoverController:(UIPopoverController*)pc{

    barButtonItem.title = NSLocalizedString(@"Left Item", nil);
    [self.rightViewController.navigationItem
     setLeftBarButtonItem:barButtonItem
     animated:YES];

}

Rest of application delegate's implementation...
```

Discussion

iOS applications running on the iPad can take advantage of popovers. An example is the New Element popover shown in Figure 2-19.

Popovers are an efficient way to display data temporarily to the user. They are commonly used in iPad applications that use split view controllers that support both

Figure 2-19. The New Element popover, which is displayed when the user selects the + button

landscape and portrait orientations. For more information about split view controllers, please refer to Recipe 2.6.

In portrait mode, a split view controller hides its left (master) view controller to give all the space on the screen to the right (detail) view controller. However, because users still need access to the hidden master controller, all it takes is a press of a button on the right view controller to display the contents of the left view controller in a popover. Popovers are different from modal view controllers because popovers use limited space on the main window and still allow interactivity with the screens beneath them; a modal view controller prevents interaction with other screens while the modal view controller is still displaying.

Popovers are created using the `UIPopoverViewController`. Popovers can be managed and displayed in two ways:

- Use a split view controller and listen to its delegate messages that can automatically create popovers for you.
- Manually create instances of `UIPopoverViewController` and present them to the user using various instance methods available in the aforementioned class.

Let's look at the first method. In this example, we will build on the same example code in Recipe 2.6. Here is what we want to accomplish:

- If the user opens the application in landscape mode:
  — The user will see one view controller on the left (master) side and another view controller on the right (detail) side.
  — When the user picks an item on the left (master) side, the same item gets displayed as the title of the right (detail) side.
- Now if the user rotates the device to portrait mode:
  — We want to display a button on the top left of the navigation bar of the right (detail) side.
  — Once the user taps on that button, we want the contents of the left (master) side to get displayed in a popover on the right (detail) side.

So, how do we go about doing this? Simple! We do this one step at a time:

1. In the application delegate (you might choose a different place to do this processing—for instance, from another view controller that you want to use to push a split view controller into the stack), instantiate the split view controller and retain the left (master) and right (detail) view controllers. Also, make sure the application delegate conforms to the UISplitViewControllerDelegate. We will use these delegate methods to easily manage popovers to represent the left (master) view controller in portrait mode.

```objc
#import <UIKit/UIKit.h>

@interface ViewsAndVCAppDelegate : NSObject
           <UIApplicationDelegate, UISplitViewControllerDelegate> {

  UIWindow                    *window;
  UISplitViewController       *splitViewController;
  UIViewController            *rightViewController;
  UIViewController            *leftViewController;
}

@property (nonatomic, retain)
IBOutlet UIWindow *window;

@property (nonatomic, retain)
UISplitViewController *splitViewController;

@property (nonatomic, retain)
UIViewController      *rightViewController;

@property (nonatomic, retain)
UIViewController      *leftViewController;

@end
```

2. When the application launches, assign the left and right view controllers to their respective properties in the delegate object:

```objc
- (BOOL) isiPad{

  BOOL result = NO;

  NSString *classAsString =
  NSStringFromClass([UISplitViewController class]);

  if (classAsString == nil ||
      [classAsString length] == 0){
    return(NO);
  }

  UIDevice *device = [UIDevice currentDevice];

  if ([device respondsToSelector:
       @selector(userInterfaceIdiom)] == NO){
    return(NO);
  }
```

```objc
  if ([device userInterfaceIdiom] != UIUserInterfaceIdiomPad){
    return(NO);
  }

  /* Do extra checking based on screen size
   for instance if you want */
  result = YES;

  return(result);

}

- (BOOL)              application:(UIApplication *)application
  didFinishLaunchingWithOptions:(NSDictionary *)launchOptions {

  /* First make sure this is an iPad running this application */

  if ([self isiPad] == YES){
    /* Create the View Controller that is shown on the left side */
    LeftViewController *leftVC =
    [[LeftViewController alloc]
     initWithNibName:@"LeftViewController"
     bundle:nil];

    self.leftViewController = leftVC;

    /* Create a Navigation Controller for
     the View Controller on the left */
    UINavigationController *leftNC =
    [[UINavigationController alloc]
     initWithRootViewController:leftVC];

    [leftVC release];

    /* Create the right-side View Controller now */
    RightViewController *rightVC =
    [[RightViewController alloc]
     initWithNibName:@"RightViewController"
     bundle:nil];

    self.rightViewController = rightVC;

    leftVC.delegate = rightVC;

    /* And a Navigation Controller for the
     View Controller on the right */
    UINavigationController *rightNC =
    [[UINavigationController alloc]
     initWithRootViewController:rightVC];

    [rightVC release];

    /* Put all the Navigation Controllers in one array to
     be passed to the Split View Controller */
```

```
        NSArray *navigationControllers =
          [NSArray arrayWithObjects:leftNC, rightNC, nil];

        [leftNC release];
        [rightNC release];

        /* Create the Split View Controller now. */
        UISplitViewController *splitController =
        [[UISplitViewController alloc] init];

        self.splitViewController = splitController;
        self.splitViewController.delegate = self;
        [splitController release];

        /* Place the Navigation Controllers
         (which are linked to our View Controllers),
         into the Split View Controller */
        [self.splitViewController
         setViewControllers:navigationControllers];

        /* Show the View of the Split View Controller which is
         now the mixture of the left and right View Controllers */
        [window addSubview:self.splitViewController.view];
    } else {
    /* Choose another interface path if the device is not an iPad */
    }

    [window makeKeyAndVisible];

    return YES;
}
```

As you can see, we have chosen the application delegate object to become the delegate of our split view controller. One of the split view controller delegate methods that we are interested in is `splitViewController:willHideViewControl` `ler:withBarButtonItem:forPopoverController:`. This delegate method gets called when the split view controller is displayed in portrait mode. This method is very intelligent indeed. The `withBarButtonItem` parameter contains an object of type `UIBarButtonItem` that the framework has already created for you. All you have to do is to set the `title` property of this button and assign it to your toolbar or navigation bar. When the user presses this button, the left (master) view controller will be displayed on the right (detail) view controller as a popover, so you don't have to create the popover manually. The `title` property is empty, so you must assign a title to your button; otherwise, you will not be able to see it.

3. Assign this button to the left bar button of the right (detail) view controller's navigation bar, like so:

```
- (void)splitViewController:(UISplitViewController*)svc
      willHideViewController:(UIViewController *)aViewController
           withBarButtonItem:(UIBarButtonItem*)barButtonItem
        forPopoverController:(UIPopoverController*)pc{
```

```
barButtonItem.title = NSLocalizedString(@"Left Item", nil);

[self.rightViewController.navigationItem
 setLeftBarButtonItem:barButtonItem
 animated:YES];

}
```

4. We are almost done! There is still one thing left to do. Assume that the user is in portrait mode and she presses the bar button item that we assigned to the right (detail) view controller's navigation bar in the preceding step. What happens if she now rotates the device and the orientation changes to landscape while the popover is showing? Well, the split view controller's delegate object must also implement the `splitViewController:willShowViewController:invalidatingBarButtonItem:` delegate method that gets called when the left (master) view controller is about to get displayed (in landscape mode, obviously). This is where the delegate object must remove the bar button item that we used in the preceding step.

```
- (void)splitViewController:(UISplitViewController*)svc
    willShowViewController:(UIViewController *)aViewController
 invalidatingBarButtonItem:(UIBarButtonItem *)barButtonItem{

[self.rightViewController.navigationItem setLeftBarButtonItem:nil
                                                     animated:YES];

}
```

Let's give it a go. First we will launch the application in landscape mode, as shown in Figure 2-20.

When the user rotates the device to portrait mode, the left (master) view controller will automatically get hidden by the split view controller and the `splitViewController:willHideViewController:withBarButtonItem:forPopoverController:` split view controller delegate method will get called. We will then place the bar button item on the navigation bar of the right (detail) view controller, as shown in Figure 2-21.

When the user selects the Left Item button, the split view controller will automatically present the left (master) view controller in a popover that gets displayed on the Left Item button, as shown in Figure 2-22.

Now that you've seen how easy it is to manage popovers on split view controllers, it's time to explore another way to manage and display popovers.

This second method, which is a bit more difficult than the first, involves manually creating instances of `UIPopoverViewController` and displaying them on various UI components such as bar buttons. (Bar buttons are instances of `UIBarButtonItem`. Please refer to Recipe 2.13 for more information about bar buttons.)

For this method, we will create a simple application consisting of three main components:

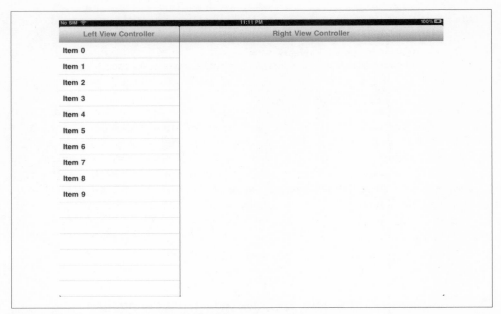

Figure 2-20. The left (master) and the right (detail) view controllers displayed on a split view controller

Figure 2-21. The bar button item representing the left (master) view controller, appearing on the navigation bar

Application delegate

In this example, the application delegate will simply create a navigation controller and will push the root view controller into it (see Recipe 2.10 for more information). The navigation controller will then get displayed on the main window as usual.

RootViewController

This view controller will create an instance of `UIBarButtonItem` and will add this button to the top-righthand side of its navigation bar. Pressing this button will display the `AddNewViewController`, which is explained next.

AddNewViewController

This view controller will get displayed in the popover that gets shown on the root view controller when the user presses the bar button that we display in the root view controller. For this example, we will set the size of the view of this view controller to 200 × 168 pixels. We will also put some dummy components, such as three instances of `UIButton` objects, on this view to make sure our popover works correctly.

Figure 2-23 shows how our popover will look when the bar button is pressed on the root view controller.

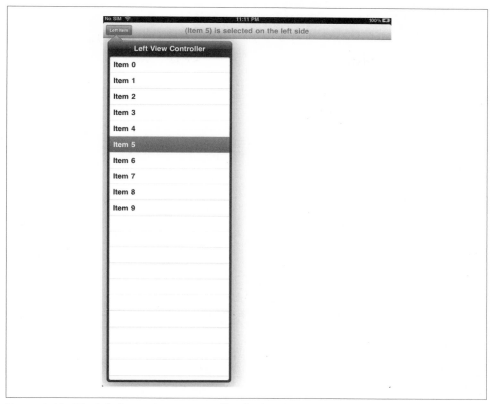

Figure 2-22. The left (master) view controller displayed in a popover on the right (detail) view controller

Figure 2-23. Contents of a view controller displayed in a popover

Let's start with our application delegate's declaration (.h) file:

```
#import <UIKit/UIKit.h>
```

```
@interface ViewsAndVCAppDelegate : NSObject
            <UIApplicationDelegate> {
@protected
  UIWindow *window;
  UINavigationController  *navigationController;
}

@property (nonatomic, retain)
IBOutlet UIWindow *window;

@property (nonatomic, retain)
UINavigationController  *navigationController;

@end
```

When the application launches, we want to show the root view controller inside a navigation controller:

```
#import "ViewsAndVCAppDelegate.h"
#import "RootViewController.h"

@implementation ViewsAndVCAppDelegate

@synthesize window;
@synthesize navigationController;

- (BOOL)          application:(UIApplication *)application
didFinishLaunchingWithOptions:(NSDictionary *)launchOptions {

  // Override point for customization after application launch.

  RootViewController *rootVC = [[RootViewController alloc]
                               initWithNibName:@"RootViewController"
                               bundle:nil];

  UINavigationController *newNC = [[UINavigationController alloc]
                                  initWithRootViewController:rootVC];

  [rootVC release];

  self.navigationController = newNC;

  [newNC release];

  [self.window addSubview:self.navigationController.view];
  [self.window makeKeyAndVisible];

  return YES;
}

- (void)dealloc {
  [navigationController release];
  [window release];
  [super dealloc];
}
```

```
@end
```

The root view controller's job is of utmost importance in this example application. This view controller has to display the contents of another view controller inside a popover when a bar button is pressed on the navigation bar, so let's go ahead and declare the root view controller:

```objc
#import <UIKit/UIKit.h>

@interface RootViewController : UIViewController{
@protected
  UIBarButtonItem     *barButtonAdd;
  UIPopoverController *popoverController;
}

@property (nonatomic, retain)
UIBarButtonItem    *barButtonAdd;

@property (nonatomic, retain)
UIPopoverController *popoverController;

@end
```

When the root view controller is initialized, we will allocate and initialize the bar button and the popover controller and retain them, like so:

```objc
- (id)initWithNibName:(NSString *)nibNameOrNil
              bundle:(NSBundle *)nibBundleOrNil {

    self = [super initWithNibName:nibNameOrNil
                            bundle:nibBundleOrNil];

    if (self != nil){

      self.title = @"Scrap Book";

      /* Allocate and initialize the add button first */
      UIBarButtonItem *newAddButton =
      [[UIBarButtonItem alloc]
       initWithBarButtonSystemItem:UIBarButtonSystemItemAdd
       target:self
       action:@selector(addNew:)];

      barButtonAdd = [newAddButton retain];
      [newAddButton release];

      /* This will be the View Controller that will get
       displayed in the Popover Controller */
      AddNewViewController *addNewViewController =
      [[AddNewViewController alloc]
       initWithNibName:@"AddNewViewController"
       bundle:nil];

      /* Create the popover using the View Controller */
```

```
    UIPopoverController *newPopoverController =
    [[UIPopoverController alloc]
     initWithContentViewController:addNewViewController];

    [addNewViewController release];

    popoverController = [newPopoverController retain];

    [newPopoverController release];

  }
  return self;
}
```

The bar button will fire the addNew: instance method (which we will implement shortly) whenever it is pressed. Once the bar button is pressed, we would like to display the popover controller as shown in Figure 2-23. The implementation of the addNew: method will be as simple as this:

```
- (void) addNew:(id)paramSender{

  if (self.popoverController.popoverVisible == YES){

    [self.popoverController dismissPopoverAnimated:YES];

  } else {

    [self.popoverController
     presentPopoverFromBarButtonItem:self.barButtonAdd
     permittedArrowDirections:UIPopoverArrowDirectionUp
     animated:YES];

  }

}
```

Before I forget, to control the size of the popover that gets displayed on a popover controller, you need to set the contentSizeForViewInPopover property of the view controller that gets displayed in a popover. In this example, the instance of the AddNewView Controller class is the view controller whose contentSizeForViewInPopover property has to be set according to its view size. Assuming that the view in this view controller is 200 × 168 pixels, we can implement the viewDidLoad method of this view controller like so:

```
- (void)viewDidLoad {
  [super viewDidLoad];

  self.contentSizeForViewInPopover = CGSizeMake(200.0f,
                                                168.0f);
}
```

This forces the AddNewViewController to fit its contents, when displayed in a popover controller, to the rectangular space of 200 × 168 pixels.

In this example, we used the `presentPopoverFromBarButtonItem:permittedArrowDirec` `tions:animated:` instance method of our popover controller to display the popover on a bar button. Sometimes you may need to display a popover on a certain point on a certain view. In those cases, you need to use the `presentPopoverFromRect:inView:per` `mittedArrowDirections:animated:` instance method of the popover controller where the `presentPopoverFromRect` parameter will specify the rectangular area from where the popover has to be displayed.

When a popover is displayed, it will be automatically dismissed whenever a view that is not the popover's view is tapped on by the user. If you want the popover to be displayed as a modal popover—where the popover has to manually be dismissed by an action on its contents, such as tapping on a button—you can set the `modalInPopover` property of the view controller that is displayed in the popover to `YES`. If we set this property to `YES` in the `AddNewViewController`'s `viewDidLoad` method, as shown here, the popover that gets displayed on the root view controller will not be dismissed if the user taps on any other view on the screen:

```
- (void)viewDidLoad {
    [super viewDidLoad];

    self.modalInPopover = YES;
    self.contentSizeForViewInPopover = CGSizeMake(200.0f,
                                                  168.0f);
}
```

See Also

Recipe 2.10

CHAPTER 3

Constructing and Using Table Views

3.0 Introduction

A table view is simply a scrolling view that is separated into sections, each of which is further separated into rows. Each row is an instance of the `UITableViewCell` class, and you can create custom table view rows by *subclassing* this class.

Using table views is an ideal way to present a list of items to users. You can embed images, text, and whatnot into your table view cells; you can customize their height, shape, grouping, and much more. The simplicity of the structure of table views is what makes them highly customizable.

A table view can be fed with data using a table view data source, and you can receive various events and control the physical appearance of table views using a table view delegate object. These are defined, respectively, in the `UITableViewDataSource` and `UITableViewDelegate` protocols.

Although an instance of `UITableView` subclasses `UIScrollView`, table views can only scroll vertically. This is more a feature than a limitation. In this chapter, we will discuss the different ways of creating, managing, and customizing table views.

3.1 Creating a Table View Using Interface Builder

Problem

You want to create a table view using Interface Builder.

Solution

Drop an instance of the `UITableView` from the Library onto a container, such as an instance of `UIView` (see Figure 3-1).

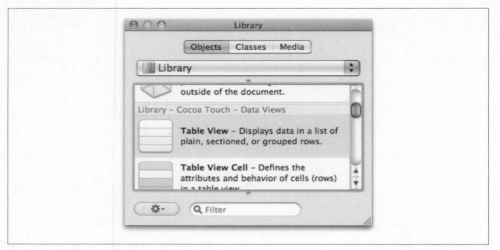

Figure 3-1. The Library panel of Interface Builder

If you can't see it on the screen, you can enable the Library panel in Interface Builder, using the Tools→Library menu.

Discussion

You can drag and drop an instance of the UITableView to any object that directly or indirectly inherits from UIView. A table view uses an object conforming to the UITableViewDataSource protocol as its data source and an object conforming to the UITableViewDelegate protocol as its delegate object. The data source's job is to provide data to the table view, and the delegate object receives various events from the table view. These two objects can be assigned to the delegate and the dataSource properties of an instance of the UITableView, respectively, in this way:

```
id<UITableViewDelegate> tableViewDelegate = /* Refer to an object */;
id<UITableViewDataSource> tableViewDataSource = /* Refer to an object */;

UITableView *myTableView = /* Refer to a table view */;
myTableView.delegate = tableViewDelegate;
myTableView.dataSource = tableViewDataSource;
```

The data source object and the delegate of a table view can be two separate objects.

See Also

Recipe 3.2

3.2 Creating a Table View Using Code

Problem

You would like to programmatically create a table view in your application.

Solution

Create an instance of the UITableView class and add it to another view:

```
- (void)viewDidLoad {
  [super viewDidLoad];

  /* We want a full-screen Table View which is as
   big as the View which is attached to the current
   View Controller */
  CGRect tableViewFrame = self.view.bounds;

  /* Create the Table View First */
  UITableView *tableView = [[UITableView alloc]
                            initWithFrame:tableViewFrame
                            style:UITableViewStylePlain];

  /* Add this Table View to our View */
  [self.view addSubview:tableView];

  /* Release the allocated Table View. It is now retained
   by this View Controller's View */
  [tableView release];

}
```

This code will create a table view the size of the view on the current view controller that owns the code. In other words, the table view will fill the available space on the screen, outside of fixed elements such as the status bar.

Discussion

It may be useful to create a table view at runtime in order to gain more control over when and how your table view is created. For instance, if on a certain view controller you require a table view to be created only when the user presses a button, creating the table view in code might be more advantageous than having the table view existing as a hidden control on the view, as the latter case will consume memory for the table view even though the table view is not visible to the user. However, if you are working with simple and straightforward interfaces, as we normally do, you are better off using Interface Builder, which allows you to control the contents of your GUI in a WYSIWYG manner.

In order to display a table view to the users of your application, you need to add it as a subview of an instance of the UIView class (UIScrollView, UIButton, etc.). Your table view can have a rectangular dimension specified by the left border (x), the top border

(y), the width, and the height, which form a `CGRect` structure. You can create a `CGRect` structure using the `CGRectMake` macro.

See Also

Recipe 3.1

3.3 Assigning an Event Handler to a Table View Using Interface Builder

Problem

You want to connect a table view to its delegate object using Interface Builder.

Solution

Follow these steps:

1. Open the XIB file that contains your table view.
2. Select your table view object. If it is dropped on any other object, such as a view (Figure 3-2), make sure you select the table view by first selecting the top-level view (e.g., the view of a view controller first) and then selecting the table view normally in order to make sure the focus is 100% on the table view.
3. Open the Connections Inspector in Interface Builder by selecting Tools→Connections Inspector, as shown in Figure 3-3.
4. Find the `delegate` property of the table view object and drag and drop the circle to its right to the object in your XIB file that wants to become the delegate of the table view.

Make sure the object you supplied as the delegate of your table view conforms to the `UITableViewDelegate` protocol.

Discussion

Like all objects in the iPhone user interface, a table view requires a delegate object to notify it of all relevant events.

The object you choose as the `delegate` property of your table view must conform to the `UITableViewDelegate` protocol. This object will then be responsible for responding to various selectors defined by this protocol. One of the important delegate messages the delegate object of a table view receives is the `tableView:didSelectRowAtIndexPath:` selector, which lets the delegate object know that a row has been selected by the user in a given table view.

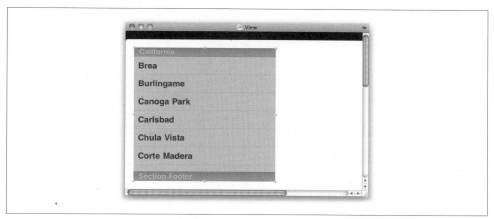

Figure 3-2. A table view object on an instance of UIView in Interface Builder

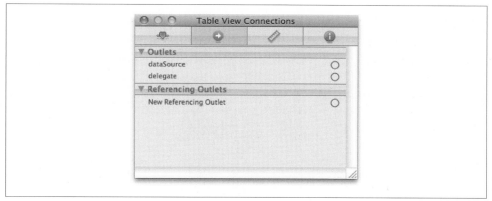

Figure 3-3. The Connections Inspector panel of Interface Builder

Some of the most important messages that a table view delegate object can respond to and interact with are:

- An event fired by the table view before it displays any rows or cells. You can use this event to customize your cells just before they are displayed.
- Events that you can use to return a value to your table view to specify the height of each cell inside your table view.
- Events that let you know when a cell is selected or deselected.

It is worth noting that the object you specify as the delegate of a `UITableView` doesn't necessarily have to implement all the selectors defined in the `UITableViewDelegate` protocol, only those that are mandatory and marked with the `@required` directive, if any.

See Also

Recipe 3.4; Recipe 3.6

3.4 Assigning an Event Handler to a Table View Using Xcode

Problem

You have decided to assign a delegate to a table view without the help of Interface Builder.

Solution

Assign an object that conforms to the UITableViewDelegate protocol to the delegate property of your table view object:

```
- (void)viewDidLoad {
  [super viewDidLoad];

  /* We want a full-screen Table View which is as
   big as the View which is attached to the current
   View Controller */
  CGRect tableViewFrame = self.view.bounds;

  /* Create the Table View First */
  UITableView *tableView = [[UITableView alloc]
                            initWithFrame:tableViewFrame
                            style:UITableViewStylePlain];

  self.myTableView = tableView;
  self.myTableView.delegate = self;

  /* Release the allocated Table View */
  [tableView release];

  /* Add this Table View to our View */
  [self.view addSubview:tableView];

}
```

The highlighted statement assigns the current object as the delegate of the table view. myTableView is a property of type UITableView belonging to the calling object. The statement is embedded in the viewDidLoad method because the calling object here is an instance of UIViewController, and this method is the right place to put the statement so that the association is made just once.

Discussion

The UITableView class defines a property called delegate. The table view should assign to this property an object that conforms to the UITableViewDelegate protocol. In other words, this delegate must *promise* to reply to the messages defined in this protocol, which are sent to the delegate object by the table view itself. Think of the delegate of a table view as an object that listens to various events sent by the table view, such as when a cell is selected or when the table view wants to figure out the height of each of its cells. We can modify the visual appearance of a table and its cells, to some extent, using Interface Builder, too. Just open Interface Builder and select a table view that you created previously, and then select Tools→Size Inspector. In the Size Inspector panel, you can modify the visual appearance of the table view by changing values such as the height of the table view's cells.

To make the delegate object that you choose for a table view conform to the UITable ViewDelegate protocol, you need to add that protocol to that object's interface declaration in this way:

```objc
#import <UIKit/UIKit.h>

@interface RootViewController : UIViewController
                               <UITableViewDelegate> {
@private
  UITableView    *myTableView;
}

@property (nonatomic, retain) UITableView    *myTableView;

@end
```

 It is mandatory for the delegate object to respond to messages that are marked as @required by the UITableViewDelegate protocol. Responding to other messages is optional, but the delegate must respond to any messages you want to affect the table view.

Messages sent to the delegate object of a table view carry a parameter that tells the delegate object which table view has fired that message in its delegate. This is very important to make note of because you might, under certain circumstances, require more than one table view to be placed on one object (usually a view). Because of this, it is highly recommended that you make your decisions based on which table view has actually sent that specific message to your delegate object, like so:

```objc
- (CGFloat)    tableView:(UITableView *)tableView
  heightForRowAtIndexPath:(NSIndexPath *)indexPath{

  CGFloat result = 20.0f;

  if ([tableView isEqual:self.myTableView]){
    result = 40.0f;
  }

  return(result);

}
```

It is worth noting that the location of a cell in a table view is represented by its index path. An index path is the combination of the section and the row index, where the section index is the zero-based index specifying which grouping or section each cell belongs to, and the cell index is the zero-based index of that particular cell in its section.

See Also

Recipe 3.3; Recipe 3.6

3.5 Populating a Table View with Data

Problem

You would like to populate your table view with data.

Solution

Create an object that conforms to the UITableViewDataSource protocol and assign it to a table view instance. Then, by responding to the data source messages, provide information to your table view. For this example, let's go ahead and declare the *.h* file of our view controller, which will later create a table view on its own view, in code:

```
#import <UIKit/UIKit.h>

@interface RootViewController : UIViewController
                             <UITableViewDataSource> {
@public
  UITableView      *myTableView;
}

@property (nonatomic, retain) UITableView      *myTableView;

@end
```

Now let's implement the view controller in this way (in the *.m* file):

```
#import "RootViewController.h"

@implementation RootViewController

@synthesize myTableView;

- (NSInteger)tableView:(UITableView *)table
 numberOfRowsInSection:(NSInteger)section{

  NSInteger result = 0;

  /* For each section, we will return a specific number of
   rows. This number can/will be different depending on the
   requirements of the project */

  if ([table isEqual:self.myTableView] == YES){
```

```
    switch (section){

      case 0:{
        /* Let's have 3 rows for the first section */
        result = 3;
        break;
      }

      case 1:{
        /* And 5 rows for the second */
        result = 5;
        break;
      }

      default:{
        /* And last but not least, let's create
         8 rows for the third section */
        result = 8;
      }

    }
  }

  return(result);

}


- (UITableViewCell *)tableView:(UITableView *)tableView
        cellForRowAtIndexPath:(NSIndexPath *)indexPath{

  UITableViewCell *result = nil;

  if ([tableView isEqual:self.myTableView] == YES){

    static NSString *MyCellIdentifier = @"SimpleCell";

    /* We will try to retrieve an existing
     cell with the given identifier */
    result = [tableView
            dequeueReusableCellWithIdentifier:MyCellIdentifier];

    if (result == nil){
      /* If a cell with the given identifier does
       not exist, we will create the cell with the
       identifier and hand it to the table view */

      result = [[[UITableViewCell alloc]
              initWithStyle:UITableViewCellStyleDefault
              reuseIdentifier:MyCellIdentifier] autorelease];
    }

    result.textLabel.text = [NSString stringWithFormat:
                            @"Section %lu, Cell %lu",
                            (unsigned long)indexPath.section,
                            (unsigned long)indexPath.row];
```

```
    } /* if ([tableView isEqual:self.myTableView] == YES){ */

    return(result);

}


- (NSInteger)numberOfSectionsInTableView:(UITableView *)tableView{
    NSInteger result = 0;

    if ([tableView isEqual:self.myTableView] == YES){
        /* For now, let's just create 3 sections in our Table
         View for the sake of simplicity in this example. We
         can increase or decrease this value. Each one of the
         sections will have a number of rows in it, determined with
         the data source's [tableView:numberOfRowsInSection:] method */
        result = 3;
    }

    return(result);
}

- (void)viewDidLoad {
    [super viewDidLoad];

    /* Our Table View, in this example, will fill the area
     of the whole view */
    CGRect tableViewRect = self.view.bounds;

    UITableView *tableView = [[UITableView alloc]
                              initWithFrame:tableViewRect
                              style:UITableViewStylePlain];

    self.myTableView = tableView;

    [tableView release];

    /* Make sure our Table View adjusts its width/height/left/top
     spaces whenever its container view changes its
     orientation or size */
    self.myTableView.autoresizingMask =
      UIViewAutoresizingFlexibleHeight |
      UIViewAutoresizingFlexibleWidth;

    /* Add our Table View to our view and also make the current
     View Controller the data source of the Table View.
     The data source could be any other object, though, and
     doesn't necessarily have to be a View Controller */

    [self.view addSubview:self.myTableView];

    self.myTableView.dataSource = self;

}
```

```
- (void) viewDidUnload{
  [super viewDidUnload];

  self.myTableView = nil;

}

- (BOOL)shouldAutorotateToInterfaceOrientation:
  (UIInterfaceOrientation)interfaceOrientation {
  return YES;
}

- (void)dealloc {
  [myTableView release];
  [super dealloc];
}

@end
```

Because this view controller implements the `UITableViewDataSource` protocol, it acts as its own data source. This might not be the case in your application, but it makes it easy for this example to generate dummy data with which to populate the table view. Similarly, the view controller implements the `UITableViewDelegate` protocol, so it can act as its own delegate.

The result of this view controller running as the root view controller of a plain application is shown in Figure 3-4.

Discussion

The `UITableView` class defines a property called `dataSource`. This is an untyped object that must conform to the `UITableViewDataSource` protocol. Every time a table view is refreshed and reloaded using the `reloadData` method, the table view will call various methods in its data source to find out about the data you intend to populate it with. A table view data source can implement three important methods, two of which are mandatory for every data source:

`numberOfSectionsInTableView:`
> This method allows the data source to inform the table view of the number of sections that must be loaded into the table.

`tableView:numberOfRowsInSection:`
> This method tells the view controller how many cells or rows have to be loaded for each section. The section number is passed to the data source in the `numberOfRows InSection` parameter. The implementation of this method is mandatory in the data source object.

`tableView:cellForRowAtIndexPath:`
> This method is responsible for returning instances of the `UITableViewCell` class as rows that have to be populated into the table view. The implementation of this method is mandatory in the data source object.

Figure 3-4. A table view populated with data

When a table view is reloaded or refreshed, it queries its data source through the `UITableViewDataSource` protocol, asking for various bits of information. Among the important methods mentioned previously, the table view will first ask for the number of sections. Each section is responsible for holding rows or cells. After the data source specifies the number of sections, the table view will ask for the number of rows that have to be loaded into each section. The data source gets the zero-based index of each section and, based on this, can decide how many cells have to be loaded into each section.

The table view, after determining the number of cells in the sections, will continue to ask the data source about the view that will represent each cell in each section. You can allocate instances of the `UITableViewCell` class and return them to the table view. There are, of course, properties that can be set for each cell, including the title, subtitle, and color of each cell, among other properties. For more information about this, please refer to Recipe 3.9.

3.6 Receiving and Handling Table View Events

Problem

You would like to respond to various events that a table view can generate.

Solution

Provide your table view with a delegate object.

Here is an excerpt of the *.h* file of a view controller with a table view:

```
#import <UIKit/UIKit.h>

@interface RootViewController : UIViewController
                                <UITableViewDataSource,
                                 UITableViewDelegate> {
@public
  UITableView      *myTableView;
}

@property (nonatomic, retain) UITableView      *myTableView;

@end
```

The *.m* file of the same view controller implements a method defined in the UITable
ViewDelegate protocol:

```
- (void)          tableView:(UITableView *)tableView
  didSelectRowAtIndexPath:(NSIndexPath *)indexPath{

  if ([tableView isEqual:self.myTableView] == YES){
    self.title =
    [NSString stringWithFormat:@"Cell %lu in Section %lu is selected",
     (unsigned long)indexPath.row, (unsigned long)indexPath.section];
  }

}

- (void)viewDidLoad {
  [super viewDidLoad];

  /* Our Table View, in this example, will fill the area
   of the whole view */
  CGRect tableViewRect = self.view.bounds;

  UITableView *tableView = [[UITableView alloc]
                            initWithFrame:tableViewRect
                            style:UITableViewStylePlain];

  self.myTableView = tableView;

  [tableView release];
```

```
/* Make sure our Table View adjusts its width/height/left/top
 spaces whenever its container view changes
 its orientation or size */
self.myTableView.autoresizingMask =
  UIViewAutoresizingFlexibleHeight |
  UIViewAutoresizingFlexibleWidth;

/* Add our Table View to our view and also make the current
 View Controller the data source of the Table View.
 The data source could be any other object, though,
 and doesn't necessarily have to be a View Controller */

[self.view addSubview:self.myTableView];

self.myTableView.dataSource = self;
self.myTableView.delegate = self;

}
```

Discussion

While a data source is responsible for providing data to the table view, the table view consults the delegate whenever an event occurs, or if the table view needs further information before it can complete a task. For instance, the table view invokes a delegate's method:

- When and before a cell is selected or deselected
- When the table view needs to find the height of each cell
- When the table view needs to construct the header and footer of every section

As you can see in the example code in this recipe's Solution, the current object is set as the delegate of a table view. The delegate implements the `tableView:didSelectRowAtIndexPath:` selector in order to get notified when the user selects a cell or a row on a table view. The documentation for the `UITableViewDelegate` protocol in the SDK shows you all the methods that the delegate can define and the view can invoke.

See Also

"UITableViewDelegate Protocol Reference," *http://developer.apple.com/library/ios/ #documentation/uikit/reference/UITableViewDelegate_Protocol/Reference/Reference .html*

3.7 Using Different Types of Accessories in a Table View

Problem

You want to grab users' attention in a table view by displaying accessories, and offer different ways to interact with each cell in your table view.

Solution

Use the `accessoryType` of the `UITableViewCell`, which you provide to your table view in its data source object:

```
- (UITableViewCell *)tableView:(UITableView *)tableView
        cellForRowAtIndexPath:(NSIndexPath *)indexPath{

  UITableViewCell* result = nil;

  if ([tableView isEqual:self.myTableView] == YES){

    static NSString *MyCellIdentifier = @"SimpleCell";

    /* We will try to retrieve an existing cell
     with the given identifier */
    result = [tableView
            dequeueReusableCellWithIdentifier:MyCellIdentifier];

    if (result == nil){
      /* If a cell with the given identifier does not
       exist, we will create the cell with the identifier
       and hand it to the table view */

      result = [[[UITableViewCell alloc]
              initWithStyle:UITableViewCellStyleDefault
              reuseIdentifier:MyCellIdentifier] autorelease];
    }

    result.textLabel.text =
    [NSString stringWithFormat:@"Section %lu, Cell %lu",
     indexPath.section,
     indexPath.row];

    result.accessoryType =
    UITableViewCellAccessoryDetailDisclosureButton;

  } /* if ([tableView isEqual:self.myTableView] == YES){ */

  return(result);

}
```

Discussion

You can assign any of the values defined in the `UITableViewCellAccessoryType` enumeration to the `accessoryType` property of an instance of the `UITableViewCell` class. Two very useful accessories are the *disclosure indicator* and the *detail disclosure button*. They both display a chevron indicating to users that if they tap on the associated table view cell, a new view or view controller will be displayed. In other words, the users will be taken to a new screen with further information about their current selector. The difference between these two accessories is that the disclosure indicator produces no event, whereas the detail disclosure button fires an event to the delegate when pressed. In other words, pressing the button has a different effect from pressing the cell itself. Thus, the detail disclosure button allows the user to perform two separate but related actions on the same row.

Figure 3-5 shows these two different accessories on a table view. The first row has a disclosure indicator and the second row has a detail disclosure button.

Figure 3-5. A table view with two different accessory views

If you tap any detail disclosure button assigned to a table view cell, you will immediately realize that it truly is a separate button. Now the question is: how does the table view know when the user taps this button?

Table views, as explained before, fire events on their delegate object. The detail disclosure button on a table view cell also fires an event that can be captured by the delegate object of a table view:

```
- (void)                         tableView:(UITableView *)tableView
  accessoryButtonTappedForRowWithIndexPath:(NSIndexPath *)indexPath{

  if ([tableView isEqual:self.myTableView] == YES){

    UITableViewCell *ownerCell =
    [tableView cellForRowAtIndexPath:indexPath];

    /* Do something with the cell which owns the button */
    NSLog(@"%@", ownerCell.textLabel.text);
  }

}
```

This code finds the table view cell whose detail disclosure button has been tapped and prints the contents of the text label of that cell into the console screen. As a reminder, you can display the console screen in Xcode by selecting Run→Console.

3.8 Creating Custom Table View Accessories

Problem

The accessories provided to you by the iOS SDK are not sufficient, and you would like to create your own accessories.

Solution

Assign an instance of the `UIView` class to the `accessoryView` property of any instance of the `UITableViewCell` class:

```
- (UITableViewCell *)tableView:(UITableView *)tableView
        cellForRowAtIndexPath:(NSIndexPath *)indexPath{

    UITableViewCell* result = nil;

    if ([tableView isEqual:self.myTableView] == YES){

        static NSString *MyCellIdentifier = @"SimpleCell";

        /* We will try to retrieve an existing cell
          with the given identifier */
        result = [tableView
                dequeueReusableCellWithIdentifier:MyCellIdentifier];

        if (result == nil){
          /* If a cell with the given identifier does not
            exist, we will create the cell with the
            identifier and hand it to the table view */

          result = [[[UITableViewCell alloc]
                  initWithStyle:UITableViewCellStyleDefault
                  reuseIdentifier:MyCellIdentifier] autorelease];
        }

        result.textLabel.text =
        [NSString stringWithFormat:@"Section %lu, Cell %lu",
         (unsigned long)indexPath.section,
         (unsigned long)indexPath.row];

        UIButton *button =
        [UIButton buttonWithType:UIButtonTypeRoundedRect];
        button.frame = CGRectMake(0.0, 0.0, 150.0, 25.0);

        [button setTitle:@"Expand"
                forState:UIControlStateNormal];
```

```
    [button addTarget:self
               action:@selector(performExpand:)
      forControlEvents:UIControlEventTouchUpInside];

    result.accessoryView = button;

  } /* if ([tableView isEqual:self.myTableView] == YES){ */

  return(result);

}
```

As you can see, this code uses the `performExpand:` method as the selector for each button. Here is the definition of this method:

```
- (void) performExpand:(id)paramSender{
  /* Take an action here */
}
```

This example code snippet assigns a custom button to the accessory view of every row in the targeted table. The result is shown in Figure 3-6.

Figure 3-6. Table view cells with custom accessory views

Discussion

An object of type `UITableViewCell` retains a property named `accessoryView`. This is the view you can assign a value to if you are not completely happy with the built-in iOS SDK table view cell accessories. After this property is set, Cocoa Touch will ignore the value of the `accessoryType` property and will use the view assigned to the `accessory View` property as the accessory assigned to the cell.

The code listed in this recipe's Solution creates buttons for all the cells populated into the table view. When a button is pressed in any cell, the performExpand: method gets called, and if you are like me, you have probably already started thinking about how you can determine which cell the sender button belongs to. So, now we have to somehow link our buttons with the cells to which they belong.

One way to handle this situation is to retrieve the superview of the button that fires the event. Since the accessory view of the cells of a table view adds the cells' accessory views as their subviews, retrieving the superview of the button will return the table view cell that owns the button as its accessory view:

```
- (void) performExpand:(UIButton *)paramSender{

    UITableViewCell *ownerCell =
    (UITableViewCell*)paramSender.superview;

    if (ownerCell != nil){

      /* Now we will retrieve the index path of the cell
         which contains the section and the row of the cell */

      NSIndexPath *ownerCellIndexPath =
        [self.myTableView indexPathForCell:ownerCell];

      /* Now we can use these two values to truly determine that
         the accessory button of which cell was the sender of this event:

         OwnerCellIndexPath.section
         OwnerCellIndexPath.row

      */

      if (ownerCellIndexPath.section == 0 &&
          ownerCellIndexPath.row == 1){
        /* This is the second row in the first section */
      }

      /* And so forth with the other checks ... */

    }

}
```

3.9 Customizing the Appearance of a Table View's Contents

Problem

The UITableViewCell class doesn't provide the visual functionality you are looking for.

Solution

Subclass the `UITableViewCell` class and feed your table views with your custom table view cells.

Discussion

The `UITableViewCell` class retains a property of type `UIView` called `contentView`. The content view is the view that represents the table view cell, and of course, it is a read-only property. However, you can add subviews to the content view in order to add your own components to each cell. For instance, the following code examples allow you to add a label to your table view cells that positions itself on the top-right side of the cell and allows you to show a date as a string.

Here is the *.h* file of the custom table view cell:

```
#import <UIKit/UIKit.h>

@interface CustomTableViewCell : UITableViewCell {
@public
  UILabel *labelDate;
}

@property (nonatomic, retain) UILabel *labelDate;

- (BOOL) displayCurrentDateWithDate:(NSDate*)ParamDate;

@end
```

Here is the *.m* file of the custom table view cell:

```
#import "CustomTableViewCell.h"

@implementation CustomTableViewCell

@synthesize labelDate;

- (void) layoutSubviews{

  /* Position the subviews here */
  [super layoutSubviews];

  if (self.labelDate != nil){
    [self.contentView bringSubviewToFront:self.labelDate];

    /* The width of our label is always 100 pixels */
    CGRect frameforLabel;
    frameforLabel.size.width = 100.0f;
    /* The height of our label is also constant */
    frameforLabel.size.height = 25.0f;
    /* The x position of the label is the width of this cell minus the
     width of the label itself */
    frameforLabel.origin.x = self.contentView.frame.size.width -
    frameforLabel.size.width;
```

```
    if (self.accessoryView != nil){
      /* If there is an accessory view set for this cell, then we
       shift our label farther to the left */
      frameforLabel.origin.x -= self.accessoryView.frame.size.width;
    }

    /* The y position of the label is also constant */
    frameforLabel.origin.y = 5.0;
    self.labelDate.frame = frameforLabel;
  }

}

- (BOOL) displayCurrentDateWithDate:(NSDate*)ParamDate{
  BOOL result = NO;

  if (self.labelDate == nil){
    return(NO);
  }

  if (ParamDate == nil){
    return(NO);
  }

  /* Format the given date and display it in the label */
  NSDateFormatter *formatter = [[NSDateFormatter alloc] init];
  [formatter setDateFormat:@"dd/MM/YYYY"];
  self.labelDate.text = [formatter stringFromDate:ParamDate];
  [formatter release];

  return(result);
}

- (id)initWithStyle:(UITableViewCellStyle)style
    reuseIdentifier:(NSString *)reuseIdentifier {

  if ((self = [super initWithStyle:style
                   reuseIdentifier:reuseIdentifier])) {

    /* Create our label here. We are not retaining the label so we
     will add it to the content view, remove it and then release it */
    UILabel *label = [[UILabel alloc] initWithFrame:CGRectZero];
    labelDate = [label retain];
    [label release];

    labelDate.textAlignment = UITextAlignmentLeft;
    labelDate.textColor = [UIColor blackColor];
    labelDate.opaque = NO;
    labelDate.backgroundColor = [UIColor clearColor];
    labelDate.font = [UIFont boldSystemFontOfSize:16.0f];
    [self.contentView addSubview:labelDate];

  }
```

```
    return self;
}

- (void)dealloc {
  [labelDate release];
  [super dealloc];
}

@end
```

As you can see, the custom table view cell that we created implements a method named `displayCurrentDateWithDate:`, which we can use in our table view data source to provide a date to the cell as an `NSDate` object. Our custom cell is intelligent enough to format this date and displays it in the relevant label. Figure 3-7 shows how our custom table view cell looks.

Figure 3-7. A custom table view cell with the ability to display formatted dates

A few things must be noted about our custom table view cell's code:

- The `contentView` property of a table view cell can contain various default components of a cell, such as its label and its detail text. For this reason, when adding a new view such as a label to a content view, make sure you bring that view on top of the subview stack of the content view, or your view might never get shown to the user. The location of a view on a stack depends on when you add your views to the content view. But if you decide to add your views in the `initWithStyle` method as we did in this example, you must bring your views to the top in the stack of subviews of the content view of your cell, or you will never see the views on the screen because the main label of the cell will come on top of your views.

- We have overridden the `layoutSubviews` method of the `UITableViewCell` class in order to change the position of the date label whenever the cell's properties, such as its width or height, change. Since we need to first see how the cell itself reacts to these changes, we call the same method in the superclass of our current class and then reposition the custom views that we have created.

Here is how we can insert these custom cells into a table view through the table view's data source object:

```
- (UITableViewCell *)tableView:(UITableView *)tableView
        cellForRowAtIndexPath:(NSIndexPath *)indexPath{

    CustomTableViewCell* result = nil;

    /* Reusable identifier of our cells */
    static NSString *MyCellIdentifier = @"SimpleCells";

    if ([tableView isEqual:self.myTableView] == YES){

        /* Can we reuse the cells? */
        result = (CustomTableViewCell *)
        [tableView dequeueReusableCellWithIdentifier:MyCellIdentifier];

        if (result == nil){
            /* Could not get a reusable cell. Create one */
            result = [[[CustomTableViewCell alloc]
                    initWithStyle:UITableViewCellStyleDefault
                    reuseIdentifier:MyCellIdentifier] autorelease];
        } /* if (result == nil){ */

        result.textLabel.text =
        [NSString stringWithFormat:@"Section %lu, Cell %lu",
         indexPath.section,
         indexPath.row];

        [result displayCurrentDateWithDate:[NSDate date]];

        result.accessoryType =
        UITableViewCellAccessoryDetailDisclosureButton;

    } /* if ([tableView isEqual:self.myTableView] == YES){ */

    return(result);

}
```

See Also

Recipe 3.8

3.10 Displaying Hierarchical Data

Problem

You want to be able to display hierarchical data in a table view.

Solution

Use the indentation functionality of table view cells:

```objc
- (UITableViewCell *)tableView:(UITableView *)tableView
        cellForRowAtIndexPath:(NSIndexPath *)indexPath{

  UITableViewCell* result = nil;

  /* Reusable identifier of our cells */
  static NSString *MyCellIdentifier = @"SimpleCells";

  if ([tableView isEqual:self.myTableView] == YES){

    /* Can we reuse the cells? */
    result = (UITableViewCell *)
    [tableView dequeueReusableCellWithIdentifier:MyCellIdentifier];

    if (result == nil){
      /* Could not get a reusable cell. Create one */
      result = [[[UITableViewCell alloc]
                  initWithStyle:UITableViewCellStyleDefault
                  reuseIdentifier:MyCellIdentifier] autorelease];
    } /* if (result == nil){ */

    result.textLabel.text =
    [NSString stringWithFormat:@"Section %lu, Cell %lu",
     (unsigned long)indexPath.section,
     (unsigned long)indexPath.row];

    result.indentationLevel = indexPath.row;
    result.indentationWidth = 10.0f;

  } /* if ([tableView isEqual:self.myTableView] == YES){ */

  return(result);

}
```

The indentation level is simply multiplied by the indentation width in order to give a margin to the content view of each cell. Figure 3-8 depicts how these cells look when displayed inside a table view.

Figure 3-8. Table view cells with indentation

Discussion

Although you might rarely find it useful, you can apply indentation to table view cells in the iOS SDK. Each cell can have two properties related to indentation: *indentation level* and *indentation width*. The indentation level is simply multiplied by the indentation width, and the resultant value is the offset by which the table view cell's content is shifted to the right or left.

For instance, if the indentation level of a cell is set to 2 and its indentation width is set to 3, the resultant value is 6. This means the content view of the cell is shifted to the right by six pixels when displayed in a table view.

> The indentation level is defined as a signed integer value, making it possible for you to assign negative values to it. This will obviously shift the content view of your cells to the left.

The indentation level assigned to table view cells enables programmers to present hierarchical data, and it is up to the programmer to determine the indentation level and the indentation width of each cell.

3.11 Effectively Managing Memory with Table Views

Problem

You want to use the best trade-off of memory use and performance when displaying table view cells.

Solution

If you are looking for higher performance and can live with the memory being kept busy, reuse your table view cells. If you want to make sure you are not using too much memory, refrain from having reusable cells.

Discussion

There are two distinct ways to manage memory when dealing with table view cells:

- Creating reusable cells
- Allocating cells on the fly

First let's see how reusable cells work. Think of each reusable cell as a template that a table view can reuse whenever it is needed. You will assign an *identifier* to each type of *cell* that you think you are going to need to reuse in the future. The table view will then find out if another cell has already been created with the given identifier. If so, the table view will try to reuse that cell, which essentially means the cell will not get allocated again; it will simply get displayed again, and it is your job to change its contents.

Just to make things simpler, let's go through this list of actions that we have to take in order to be able to use reusable table view cells effectively:

1. For every *type* of cell that you are using in a table view, create a unique identifier of type `NSString`.

2. When the data source is required to return an instance of the `UITableViewCell` to your table view, use the `dequeueReusableCellWithIdentifier:` method of your table view to determine whether a specific cell identifier has already been created and can be used. The return value of this method is an instance of the `UITableView Cell` object, or `nil` if the cell with the given identifier has not been created before.

3. Depending on whether the cell could be reused or not, given the identifier you passed to the `dequeueReusableCellWithIdentifier:` method, you need to allocate a new reusable cell if the returned value of this method is `nil`. Make sure your newly allocated cells are marked as `autorelease`.

4. Ignore whether your table view cells are already allocated or not. Whenever you are asked to return a new cell, you must set the cell's properties to the values that you need your users to see. This is because when you call the `dequeueReusableCell WithIdentifier:` method and your table view returns an instance of a reusable cell,

---

it contains the contents you previously set on this instance, and they will get displayed again unless you change them.

Here is the correct way to reuse table view cells in a table view data source:

```objc
- (UITableViewCell *)tableView:(UITableView *)tableView
        cellForRowAtIndexPath:(NSIndexPath *)indexPath{

    UITableViewCell* result = nil;

    if ([tableView isEqual:self.myTableView] == YES){

        static NSString *MyCellIdentifier = @"SimpleCell";

        /* We will try to retrieve an existing cell with the given identifier */
        result = [tableView
                dequeueReusableCellWithIdentifier:MyCellIdentifier];

        if (result == nil){
            /* If a cell with the given identifier does not exist,
            we will create the cell with the identifier and hand
            it to the table view */

            result = [[[UITableViewCell alloc]
                    initWithStyle:UITableViewCellStyleDefault
                    reuseIdentifier:MyCellIdentifier] autorelease];
        }

        result.textLabel.text =
        [NSString stringWithFormat:@"Section %lu, Cell %lu",
         (unsigned long)indexPath.section,
         (unsigned long)indexPath.row];

    } /* if ([tableView isEqual:self.myTableView] == YES){ */

    return(result);

}
```

You can see that the value of the textLabel property of the cells is being set every time we are asked to return an instance of the UITableViewCell to a table view.

Now for the second way to allocate cells. Imagine you want to create a table view that displays all the contacts in your users' address books with the contacts' pictures on each cell. Now let's say one of the users of your application has 300 contacts, and each contact's picture fetched out of the address book is roughly 60 KB in size. This is 18,000 KB or 17.57 MB worth of pictures. If you decide to use reusable table view cells for your contents, none of this data will get deallocated unless the system decides to do it or you deallocate your table view from memory, which allows its contents to get deallocated, too.

In such memory-intensive situations, you may want to manage memory better and only allocate as much memory as is needed to display the contents of the cells that are on the screen. You can do so by allocating table view cells on the fly and deallocating them as soon as they scroll out of view, instead of storing and reusing them. Here is how to do this:

1. In the data source object, when you need to return an instance of the UITableView Cell to a table view, instead of calling the dequeueReusableCellWithIdentifier: method, call the cellForRowAtIndexPath: method with an index path for the required cell.

   If the return value of the cellForRowAtIndexPath: method is nil, allocate and initialize an autorelease instance of the UITableViewCell class and set its properties, such as its picture, label text, and so on. By checking for a nil return value, you create a cell whenever it is needed (each time it reappears on the screen).

2. Make sure that when you initialize your cells, you pass nil as their reusable identifier.

Here is an example:

```
- (UITableViewCell *)tableView:(UITableView *)tableView
        cellForRowAtIndexPath:(NSIndexPath *)indexPath{

  UITableViewCell* result = nil;

  if ([tableView isEqual:self.myTableView] == YES){

    result = [tableView cellForRowAtIndexPath:indexPath];

    if (result == nil){

      result = [[[UITableViewCell alloc]
                  initWithStyle:UITableViewCellStyleDefault
                  reuseIdentifier:nil] autorelease];

      result.textLabel.text =
      [NSString stringWithFormat:@"Section %lu, Cell %lu",
       (unsigned long)indexPath.section,
       (unsigned long)indexPath.row];

    }

  } /* if ([tableView isEqual:self.myTableView] == YES){ */

  return(result);

}
```

If you follow these steps, the only instances of table view cells that you create are those that are currently being displayed on the screen. Cells that were created and used previously, but that have now scrolled off the screen, are deallocated. Using this method, you will save memory but you will use more processing power since you are allocating and deallocating cells whenever they are displayed and scrolled off the visible portion of the table view that owns them.

To conclude, you can choose which of the aforementioned methods to use for your current implementation of your table view. If performance is your number-one goal, reuse your cells. If you are concerned about the amount of memory you are using, simply manage your cells manually and do not use reusable cells.

3.12 Editing and Moving Data in a Table View

Problem

You want to allow your users to add, remove, and move table view cells at runtime.

Solution

You can use the `setEditing` method of the `UITableView` class:

```
[self.myTableView setEditing:YES
                     animated:YES];
```

Discussion

A table view can go into editing mode when its `editing` Boolean property is set to `YES`. When the editing property is modified manually by the programmer, the transition from normal mode to editing mode is not animated. However, to animate into editing mode, you can use the `setEditing:animated:` method with the `animated` parameter set to `YES`.

Different editing modes can be used on a table view. These modes are:

- Moving rows
- Deleting rows
- Inserting rows

Let's start with moving rows. Imagine you have a table view whose rows you want your users to be able to move. In other words, you would like to allow your users to pick a cell from a row in a section and drop it into an arbitrary row in the same or a different section. Follow these steps to allow your users to move rows in a table view:

1. In your GUI, create an Edit button that indicates to your users that they can edit the contents of the current screen, with the current screen having the table view as its main component.

2. When the Edit button is pressed, set the value of the editing property of your table view to `YES`.

3. After the Edit button is pressed, you need to remove the button and replace it with a button that is commonly referred to as a Done button. Of course, if you are using a navigation bar, you can use instances of the `UIBarButtonItem` class to represent the Edit and Done buttons.

4. In the data source object of your table view, to allow reordering of cells, implement the `tableView:moveRowAtIndexPath:toIndexPath:` method. This method gets called whenever the user drags and drops a cell from one row to another. You get objects of type `NSIndexPath` as parameters to this method. The `moveRowAtIndexPath` parameter indicates where the row originated and the `toIndexPath` parameter indicates where the row is moved to.

5. Optionally, you can implement the `tableView:canMoveRowAtIndexPath:` method in the data source object of your table view to tell your table view whether it is allowed to move a specific row from one place to another.

Here is an example of a view controller with a single table view on it. An array is prepopulated with simple `NSString` values and these values are fed to our table view's cells as their title. Here we make use of a navigation bar with Edit and Done buttons dynamically created. All the user is allowed to do is to press the Edit button, move rows around, and then press the Done button.

This is the *.h* file of the view controller:

```
#import <UIKit/UIKit.h>

@interface RootViewController : UIViewController
                                <UITableViewDataSource,
                                 UITableViewDelegate> {
@public
  UITableView     *myTableView;
  NSMutableArray  *arrayOfRows;
  UIBarButtonItem *editButton;
  UIBarButtonItem *doneButton;
}

@property (nonatomic, retain) UITableView     *myTableView;
@property (nonatomic, retain) NSMutableArray  *arrayOfRows;
@property (nonatomic, retain) UIBarButtonItem *editButton;
@property (nonatomic, retain) UIBarButtonItem *doneButton;

@end
```

Now we can move on to implement our view controller. Let's begin by writing the code in the `viewDidLoad` method of the view controller to create our table view, set its delegate and data source, and so on. In this method, we are also populating our array of rows (defined as a property named `arrayOfRows`) with instances of the `NSString` class that can later be given to our table view rows:

```
- (void) initializeNavigationBarButtons{

  /* Give the user an Edit Button on the Navigation Bar */
  UIBarButtonItem *newEditButton =
  [[UIBarButtonItem alloc]
   initWithBarButtonSystemItem:UIBarButtonSystemItemEdit
   target:self
   action:@selector(performEdit:)];
```

```objc
self.editButton = newEditButton;
[newEditButton release];

self.navigationItem.rightBarButtonItem = self.editButton;


UIBarButtonItem *newDoneButton =
[[UIBarButtonItem alloc]
 initWithBarButtonSystemItem:UIBarButtonSystemItemDone
 target:self
 action:@selector(performDone:)];

self.doneButton = newDoneButton;

[newDoneButton release];

}

- (void) initializeArrayOfRows{

/* Put strings in an array that will later be feeding our table */
NSMutableArray *mutableArray = [[NSMutableArray alloc] init];
self.arrayOfRows = mutableArray;
[mutableArray release];

NSUInteger rowCounter = 0;
for (rowCounter = 0; rowCounter < 100; rowCounter++){
  NSString *rowText = [NSString stringWithFormat:@"Row %lu",
                        (unsigned long)rowCounter];
  [self.arrayOfRows addObject:rowText];
}

}

- (void)viewDidLoad {
[super viewDidLoad];

[self initializeArrayOfRows];

[self initializeNavigationBarButtons];

UITableView *tableView = [[UITableView alloc]
                          initWithFrame:self.view.bounds
                          style:UITableViewStylePlain];

self.myTableView = tableView;

[tableView release];

self.myTableView.autoresizingMask =
  UIViewAutoresizingFlexibleWidth |
  UIViewAutoresizingFlexibleHeight;

[self.view addSubview:self.myTableView];
```

```
    /* This View Controller is both the Data Source
     and the Delegate of the Table View */
    self.myTableView.dataSource = self;
    self.myTableView.delegate = self;

}

- (void) viewDidUnload{
    [super viewDidUnload];

    self.myTableView = nil;
    self.arrayOfRows = nil;
    self.editButton = nil;
    self.doneButton = nil;

}
```

Obviously, we are also creating an Edit button and assigning it to the right button of our navigation bar. This button will fire up the performEdit: method. We will use this method to:

1. Put the table view into editing mode.
2. Change the Edit button to a Done button.

This is the actual implementation of our performEdit: method:

```
- (void) performDone:(id)paramSender{

    [self.myTableView setEditing:NO
                     animated:YES];

    [self.navigationItem setRightBarButtonItem:self.editButton
                                    animated:YES];

}

- (void) performEdit:(id)paramSender{

    [self.myTableView setEditing:YES
                     animated:YES];

    [self.navigationItem setRightBarButtonItem:self.doneButton
                                    animated:YES];

}
```

After the Edit button puts the table view into editing mode, it asks its delegate what mode of editing it has to enter into. Subsequently, it calls the tableView:editingStyle ForRowAtIndexPath: method of the delegate to get an answer to this question. Since we just want to allow users to move the rows around and we don't want them to be able to delete or add rows, we can implement this method in this way:

```
- (UITableViewCellEditingStyle)tableView:(UITableView *)tableView
         editingStyleForRowAtIndexPath:(NSIndexPath *)indexPath{
```

```
/* When we are moving the cells around, we do not want to have
 any deletion or insertion styles for our cells */

UITableViewCellEditingStyle result = UITableViewCellEditingStyleNone;

if ([tableView isEqual:self.myTableView] == YES){
  /* You can customize this value here. We return
   UITableViewCellEditingStyleNone for this table
   view like other table views that we might be managing
   at this time */
}

return(result);

}
```

At this stage, our table view can go into editing mode, but it will not allow the user to move rows around. This is because we have not yet implemented the `tableView:moveRowAtIndexPath:toIndexPath:` method in the data source. Even an empty implementation of this method in the data source of a table view indicates that the table view can now allow users to move rows around. However, you must handle this movement of rows in the source of the data (in our example, in the `arrayOfRows` array) like so:

```
- (void)       tableView:(UITableView *)tableView
    moveRowAtIndexPath:(NSIndexPath *)sourceIndexPath
           toIndexPath:(NSIndexPath *)destinationIndexPath{

  if ([tableView isEqual:self.myTableView] == YES){

    /* First get the row's title out of our array */
    NSString *sourceObject =
    [[self.arrayOfRows objectAtIndex:sourceIndexPath.row] retain];

    /* Now remove the string from where it was before. This is
     the source index path */
    [self.arrayOfRows removeObjectAtIndex:sourceIndexPath.row];

    /* And insert the same string into the destination row */
    [self.arrayOfRows insertObject:sourceObject
                    atIndex:destinationIndexPath.row];

    /* Make sure we release the retained string */
    [sourceObject release];

  } /* if ([tableView isEqual:self.myTableView] == YES){ */

}
```

Our table view is populated with only one section in this application; hence, we can simply move our row around in one array. However, if you would like to have more than one section in your table view and allow users to move rows between sections, you can simply do so using the `NSIndexPath` objects that are passed to the `tableView:moveRowAtIndexPath:toIndexPath:` method of a table view's data source.

Having written all these methods, we can now simply provide data to our table view using the array of **NSString** objects that we created earlier:

```
- (NSInteger)tableView:(UITableView *)table
 numberOfRowsInSection:(NSInteger)section{

  NSInteger result = 0;

  if ([table isEqual:self.myTableView] == YES){
    if (self.arrayOfRows != nil){
      /* Number of Rows that we have in total */
      result = [self.arrayOfRows count];
    }
  }

  return(result);

}

- (UITableViewCell *)tableView:(UITableView *)tableView
        cellForRowAtIndexPath:(NSIndexPath *)indexPath{

  UITableViewCell* result = nil;

  /* Reusable identifier of our cells */
  static NSString *MyCellIdentifier = @"SimpleCells";

  if ([tableView isEqual:self.myTableView] == YES){

    /* Can we reuse the cells? */
    result = [tableView dequeueReusableCellWithIdentifier:MyCellIdentifier];

    if (result == nil){
      /* Could not get a reusable cell. Create one */
      result = [[[UITableViewCell alloc]
                 initWithStyle:UITableViewCellStyleDefault
                 reuseIdentifier:MyCellIdentifier] autorelease];
    } /* if (result == nil){ */

    /* Set the main text of the cell, grab the text
     out of the array that we've already constructed */
    if (self.arrayOfRows != nil &&
        indexPath.row < [self.arrayOfRows count]){
      result.textLabel.text =
      [self.arrayOfRows objectAtIndex:indexPath.row];
    }

  } /* if ([tableView isEqual:self.myTableView] == YES){ */

  return(result);

}

- (NSInteger)numberOfSectionsInTableView:(UITableView *)tableView{
```

```
    NSInteger result = 0;

    if ([tableView isEqual:self.myTableView] == YES){
      /* We just have one section */
      result = 1;
    }

    return(result);

}
```

Figure 3-9 shows how this table view looks before editing begins.

Figure 3-9. An editable table view with contents being moved from one row to the other

Now that we know how to allow users to move cells around, we can allow them to
delete the cells, too. In the previous example, UITableViewCellEditingStyleNone was
the return value of the tableView:editingStyleForRowAtIndexPath: method. If we want
to allow our users to be able to delete rows when the table view is in editing mode, we
have to return the UITableViewCellEditingStyleDelete value in this method:

```
- (UITableViewCellEditingStyle)tableView:(UITableView *)tableView
         editingStyleForRowAtIndexPath:(NSIndexPath *)indexPath{

    UITableViewCellEditingStyle result = UITableViewCellEditingStyleNone;

    if ([tableView isEqual:self.myTableView] == YES){
      result = UITableViewCellEditingStyleDelete;
    }

    return(result);

}
```

Now when the table view goes into editing mode, our users will be able to delete each row by pressing on a small red button on the lefthand side of the cell and confirming the action by pressing the Delete button, which will then appear on the righthand side of the cell.

After the user has pressed the Delete button in the editing mode of a table view, the table view will call the `tableView:commitEditingStyle:forRowAtIndexPath:` method in its data source, and the implementation of this method will then be responsible for removing the target cell from the table view and the source from which the table is being fed. Figure 3-10 depicts the table view in deletion editing style.

Figure 3-10. A table view in editing mode with deletion editing style

```
- (void) tableView:(UITableView *)tableView
  commitEditingStyle:(UITableViewCellEditingStyle)editingStyle
  forRowAtIndexPath:(NSIndexPath *)indexPath{

  if ([tableView isEqual:self.myTableView] == YES){
    if (editingStyle == UITableViewCellEditingStyleDelete){
      if (self.arrayOfRows != nil &&
          indexPath.row < [self.arrayOfRows count]){
        /* First remove this object from the source */
        [self.arrayOfRows removeObjectAtIndex:indexPath.row];
        /* Then remove the associated cell from the Table View */
        [tableView deleteRowsAtIndexPaths:[NSArray arrayWithObject:indexPath]
                        withRowAnimation:UITableViewRowAnimationLeft];
      }
    }
  }

}
```

The Delete button will not delete the cell in the table view; it is your responsibility to do so.

You can specify the type of animation that is used when your cells are deleted, in the `deleteRowsAtIndexPaths:withRowAnimation:` method of a table view.

There is one more editing mode: insertion. This editing style is similar to the deletion editing style, except that you have to create data in the data source instead of deleting it, and insert rows into the table view instead of deleting rows:

```
- (UITableViewCellEditingStyle)tableView:(UITableView *)tableView
          editingStyleForRowAtIndexPath:(NSIndexPath *)indexPath{

  UITableViewCellEditingStyle result = UITableViewCellEditingStyleNone;

  if ([tableView isEqual:self.myTableView] == YES){
    result = UITableViewCellEditingStyleInsert;
  }

  return(result);

}
```

Now we will handle the insertion event:

```
- (void)  tableView:(UITableView *)tableView
 commitEditingStyle:(UITableViewCellEditingStyle)editingStyle
  forRowAtIndexPath:(NSIndexPath *)indexPath{

  if ([tableView isEqual:self.myTableView] == YES){
    if (editingStyle == UITableViewCellEditingStyleInsert){

      /* We want to insert a new cell into the row after the
       cell that has fired this event */
      indexPath = [NSIndexPath indexPathForRow:indexPath.row + 1
                              inSection:indexPath.section];

      /* Insert a new value into the source */
      [self.arrayOfRows insertObject:@"New Cell"
                      atIndex:indexPath.row];

      /* And eventually insert a new row into the Table View */
      [tableView insertRowsAtIndexPaths:[NSArray arrayWithObject:indexPath]
              withRowAnimation:UITableViewRowAnimationMiddle];

    }
  }

}
```

Imagine this scenario: if the user taps on the plus button (for insertion) on the first row, which has the zero-based row index of 0, the `NSIndexPath` object passed to the `tableView:commitEditingStyle:forRowAtIndexPath:` method will contain the value 0, the index of the cell that fired this event. It is up to you and your design whether you want to allow users to enter rows after or before the cell that fires this event. The code just shown inserts a cell after the selected cell. The results are shown in Figure 3-11.

Figure 3-11. A table view in editing mode with insertion editing style

See Also

"UITableViewDelegate Protocol Reference," *http://developer.apple.com/library/ios/ #documentation/uikit/reference/UITableViewDelegate_Protocol/Reference/Reference .html*

"UITableViewDataSource Protocol Reference," *http://developer.apple.com/library/ios/ #documentation/UIKit/Reference/UITableViewDataSource_Protocol/Reference/Refer ence.html*

3.13 Enabling Swipe Deletion

Problem

You want your application users to be able to delete rows from a table view easily.

Solution

Implement the `tableView:editingStyleForRowAtIndexPath:` selector in the delegate and the `tableView:commitEditingStyle:forRowAtIndexPath:` selector in the data source of your table view:

```objc
- (void)          tableView:(UITableView*)tableView
    didEndEditingRowAtIndexPath:(NSIndexPath *)indexPath{

  if ([tableView isEqual:self.myTableView] == YES){
    if (self.myTableView.editing == NO){
      [self performDone:nil];
    }

  }

}


- (void)          tableView:(UITableView*)tableView
  willBeginEditingRowAtIndexPath:(NSIndexPath *)indexPath{

  if ([tableView isEqual:self.myTableView] == YES){
    if (self.myTableView.editing == NO){
      [self performEdit:nil];
    }

  }

}

- (UITableViewCellEditingStyle)tableView:(UITableView *)tableView
        editingStyleForRowAtIndexPath:(NSIndexPath *)indexPath{
  UITableViewCellEditingStyle result = UITableViewCellEditingStyleNone;

  if ([tableView isEqual:self.myTableView] == YES){
    result = UITableViewCellEditingStyleDelete;
  }

  return(result);

}

- (void)  tableView:(UITableView *)tableView
  commitEditingStyle:(UITableViewCellEditingStyle)editingStyle
  forRowAtIndexPath:(NSIndexPath *)indexPath{

  if ([tableView isEqual:self.myTableView] == YES){
    if (editingStyle == UITableViewCellEditingStyleDelete){
      if (self.arrayOfRows != nil &&
          indexPath.row < [self.arrayOfRows count]){
        /* First remove this object from the source */
        [self.arrayOfRows removeObjectAtIndex:indexPath.row];
        /* Then remove the associated cell from the Table View */
        [tableView
         deleteRowsAtIndexPaths:[NSArray arrayWithObject:indexPath]
         withRowAnimation:UITableViewRowAnimationLeft];
      }
    }
  }

}
```

```
- (void) performDone:(id)paramSender{

  [self.myTableView setEditing:NO
                    animated:YES];

  [self.navigationItem setRightBarButtonItem:self.editButton
                                   animated:YES];

}

- (void) performEdit:(id)paramSender{

  [self.myTableView setEditing:YES
                    animated:YES];

  [self.navigationItem setRightBarButtonItem:self.doneButton
                                   animated:YES];

}
```

The `tableView:editingStyleForRowAtIndexPath:` method can enable deletions. It is called by the table view and its return value determines what the table view allows the user to do (insertion, deletion, etc.). The `tableView:commitEditingStyle:forRowAtIndexPath:` method carries out the user's requested deletion. The latter method is defined in the delegate, but its functionality is a bit overloaded: not only do you use the method to delete data, but you also have to delete rows from the table here.

Discussion

In Recipe 3.12, we saw how to use editing mode to allow a user to delete rows (as well as to move and add rows). This recipe shows another way to allow insertions and deletions. Here we enable the conventional behavior where the user swipes his finger from right to left or left to right on a row. The table view responds to the swipe by showing an additional Delete button on the right side of the targeted row (Figure 3-12). The table view is *not* in editing mode, but the button allows the user to delete the row.

This mode is enabled by the `tableView:editingStyleForRowAtIndexPath:` method, whose return value indicates whether the table should allow insertions, deletions, both, or neither. By implementing the `tableView:commitEditingStyle:forRowAtIndexPath:` method in the data source of a table view, you can then get notified if a user has performed an insertion or deletion.

The second parameter of the `deleteRowsAtIndexPaths:withRowAnimation:` method allows you to specify an animation method that will be performed when rows are deleted from a table view. Our example specifies that we want rows to disappear when moving from right to left when deleted.

Figure 3-12. Delete button on the right side of a row, displayed after user has swiped his finger on the row from right to the left

3.14 Grouping Data

Problem

You want to organize your data in a table view into categories.

Solution

Implement the following methods inside the data source object of your table view:

`tableView:titleForHeaderInSection:`
Use this method to return an `NSString` for the header of each section inside a table view.

`tableView:titleForFooterInSection:`
Use this method to return an `NSString` for the footer of each section inside a table view.

You can implement the following methods inside the delegate object of your table view if you prefer to provide your customized views as the header and footer for each section of your table view:

`tableView:viewForHeaderInSection:`
Implement this method to return an instance of `UIView` that will be displayed as the header of each section.

`tableView:viewForFooterInSection:`
> Implement this method to return an instance of `UIView` that will be displayed as the footer of each section.

For example code, please refer to this recipe's Discussion below.

Discussion

Each section of a table view can optionally have a header and a footer, which can simply be made up of a string that you return to your table view *or* that could be constructed manually by you as an instance of `UIView`. You can specify the number of sections and the number of rows in each section using various methods provided to you in the `UITableViewDataSource` protocol. Bear in mind that you cannot change the style of a table view after you have created it, as the style property of an instance of `UITable View` is read-only and can be set only when the property is initialized.

You can populate a sectioned table view in different ways. One way is to create an array of arrays. The root arrays will be your sections, each of which can hold an array of `NSString` objects representing rows inside each section, like so:

```
- (void)viewDidLoad {
[super viewDidLoad];

/* Put strings in array that will later be feeding our table */
/* Create the array of sections */

NSMutableArray *mutableArray = [[NSMutableArray alloc] init];
self.arrayOfSections = mutableArray;
[mutableArray release];

/* The first section with all its values */
NSMutableArray *section1 = [[NSMutableArray alloc] init];
[section1 addObject:@"Section 1 Cell 1"];
[section1 addObject:@"Section 1 Cell 2"];
[self.arrayOfSections addObject:section1];
[section1 release];

/* The second section with all its values */
NSMutableArray *section2 = [[NSMutableArray alloc] init];
[section2 addObject:@"Section 2 Cell 1"];
[section2 addObject:@"Section 2 Cell 2"];
[self.arrayOfSections addObject:section2];
[section2 release];


UITableView *tableView = [[UITableView alloc]
                            initWithFrame:self.view.frame
                            style:UITableViewStyleGrouped];

self.myTableView = tableView;
[tableView release];

self.myTableView.autoresizingMask =
```

```
                UIViewAutoresizingFlexibleWidth |
                UIViewAutoresizingFlexibleHeight;

                [self.view addSubview:self.myTableView];
                /* This View Controller is both the Data Source
                 and the Delegate of the Table View */
                self.myTableView.dataSource = self;
                self.myTableView.delegate = self;

        }

        - (void) viewDidUnload{
                [super viewDidUnload];

                self.myTableView = nil;
                self.arrayOfSections = nil;

        }
```

After having created this array, you can provide it to the data source of your table view and populate your table view using the NSString instances in each array. First you need to provide the number of sections in your table view:

```
        - (NSInteger)numberOfSectionsInTableView:(UITableView *)tableView{

                NSInteger result = 0;

                if ([tableView isEqual:self.myTableView] == YES){
                  /* We just have one section */
                  if (self.arrayOfSections != nil){
                    result = [self.arrayOfSections count];
                  }
                }

                return(result);

        }
```

The arrayOfSections property is an instance of the NSMutableArray class:

```
        #import <UIKit/UIKit.h>

        @interface RootViewController : UIViewController <UITableViewDataSource,
                                                           UITableViewDelegate> {
        @public
          UITableView     *myTableView;
          NSMutableArray  *arrayOfSections;
        }

        @property (nonatomic, retain) UITableView    *myTableView;
        @property (nonatomic, retain) NSMutableArray *arrayOfSections;

        @end
```

Along with returning the number of sections you created in your array in numberOfSectionsInTableView:, your table view will call the tableView:numberOfRowsIn

Section: method in its data source, for each section, to return the number of rows that particular section will contain:

```
- (NSInteger)tableView:(UITableView *)table
  numberOfRowsInSection:(NSInteger)section{

  NSInteger result = 0;

  if ([table isEqual:self.myTableView] == YES){

    if (self.arrayOfSections != nil &&
        [self.arrayOfSections count] > section){

      NSMutableArray *thisSection =
      [self.arrayOfSections objectAtIndex:section];

      result = [thisSection count];

    }

  } /* if ([table isEqual:self.myTableView] == YES){ */

  return(result);

}
```

Before it is displayed, you need to provide your table view with values that can be displayed on the screen for each cell:

```
- (UITableViewCell *)tableView:(UITableView *)tableView
        cellForRowAtIndexPath:(NSIndexPath *)indexPath{

  UITableViewCell* result = nil;

  /* Reusable identifier of our cells */
  static NSString *MyCellIdentifier = @"SimpleCells";

  if ([tableView isEqual:self.myTableView] == YES){

    /* Can we reuse the cells? */
    result =
    [tableView
     dequeueReusableCellWithIdentifier:MyCellIdentifier];

    if (result == nil){
      /* Could not get a reusable cell. Create one */
      result = [[[UITableViewCell alloc]
                  initWithStyle:UITableViewCellStyleDefault
                  reuseIdentifier:MyCellIdentifier] autorelease];
    }

    /* Does this array have values? */
    if (self.arrayOfSections != nil &&
        [self.arrayOfSections count] > indexPath.section){

      /* Get the section */
```

```
    NSMutableArray *thisSection =
    [self.arrayOfSections objectAtIndex:indexPath.section];

    /* Now get the row */
    if ([thisSection count] > indexPath.row){
      result.textLabel.text =
      [thisSection objectAtIndex:indexPath.row];
    }
  }

  } /* if ([tableView isEqual:self.myTableView] == YES){ */

  return(result);

}
```

Figure 3-13 gives you a good idea how the results will look.

Figure 3-13. A table view with sections

There is more to sectioned table views than what we have done so far. We can also display headers and footers for each section. If we want to use the same architecture we used before and store our sections and their rows in instances of NSMutableArray, we have a choice of places to store the values for the header and footer of each section, if any. One way is to create instances of NSMutableDictionary for each section and fill our root NSMutableArray with NSMutableDictionary objects instead of more NSMutableArray objects. The inner arrays of NSMutableArray objects still exist, but are stored in the NSMutableDictionary for their section:

```
/* Create the array of sections */
NSMutableArray *mutableArray = [[NSMutableArray alloc] init];
self.arrayOfSections = mutableArray;
[mutableArray release];

/* The first section with all its values */
NSMutableDictionary *section1 = [[NSMutableDictionary alloc] init];
NSMutableArray *section1Values = [[NSMutableArray alloc] init];
[section1Values addObject:@"Section 1 Cell 1"];
[section1Values addObject:@"Section 1 Cell 2"];
[section1 setValue:@"Section 1 Header" forKey:@"Header"];
[section1 setValue:@"Section 1 Footer" forKey:@"Footer"];
[section1 setValue:section1Values forKey:@"Values"];
[self.arrayOfSections addObject:section1];
[section1Values release];
[section1 release];

NSMutableDictionary *section2 = [[NSMutableDictionary alloc] init];
NSMutableArray *section2Values = [[NSMutableArray alloc] init];
[section2Values addObject:@"Section 2 Cell 1"];
[section2Values addObject:@"Section 2 Cell 2"];
[section2 setValue:@"Section 2 Header" forKey:@"Header"];
[section2 setValue:@"Section 2 Footer" forKey:@"Footer"];
[section2 setValue:section2Values forKey:@"Values"];
[self.arrayOfSections addObject:section2];
[section2Values release];
[section2 release];
```

Finding out the number of sections and returning it in the data source of our table view is quite easy now. All we have to do is to count the number of NSMutableDictionary objects in our array of sections, as we did before:

```
- (NSInteger)numberOfSectionsInTableView:(UITableView *)tableView{

    NSInteger result = 0;

    if ([tableView isEqual:self.myTableView] == YES){
      /* We just have one section */
      if (self.arrayOfSections != nil){
        result = [self.arrayOfSections count];
      }
    }

    return(result);

}
```

Now, to determine the number of *rows* in each *section*, we have to go through these steps:

1. Get the NSMutableDictionary instance that corresponds to our current section from the array of sections.

2. Get the instance of the NSMutableArray object inside this dictionary. This array holds our rows.

3. Return the number of rows in this array:

```objc
- (NSInteger)tableView:(UITableView *)table
  numberOfRowsInSection:(NSInteger)section{

    NSInteger result = 0;

    if ([table isEqual:self.myTableView] == YES){

      if (self.arrayOfSections != nil &&
          [self.arrayOfSections count] > section){

        /* Get the section */
        NSMutableDictionary *thisSection =
        [self.arrayOfSections objectAtIndex:section];

        NSMutableArray *thisSectionValues =
        [thisSection valueForKey:@"Values"];

        result = [thisSectionValues count];

      }

    } /* if ([table isEqual:self.myTableView] == YES){ */

    return(result);

}
```

After returning the number of sections and the number of rows in each section, we need to construct our cells:

```objc
- (UITableViewCell *)tableView:(UITableView *)tableView
        cellForRowAtIndexPath:(NSIndexPath *)indexPath{

    UITableViewCell* result = nil;

    /* Reusable identifier of our cells */
    static NSString *MyCellIdentifier = @"SimpleCells";

    if ([tableView isEqual:self.myTableView] == YES){

      /* Can we reuse the cells? */
      result =
      [tableView
       dequeueReusableCellWithIdentifier:MyCellIdentifier];

      if (result == nil){
        /* Could not get a reusable cell. Create one */
        result = [[[UITableViewCell alloc]
                 initWithStyle:UITableViewCellStyleDefault
                 reuseIdentifier:MyCellIdentifier] autorelease];
      }

      /* Does this array have values? */
      if (self.arrayOfSections != nil &&
```

```
        [self.arrayOfSections count] > indexPath.section){

      /* Get the section */
      NSMutableDictionary *thisSection =
      [self.arrayOfSections objectAtIndex:indexPath.section];

      /* Get the array of values out of the dictionary */
      NSMutableArray *thisSectionValues =
      [thisSection valueForKey:@"Values"];

      if ([thisSectionValues count] > indexPath.row){
        result.textLabel.text =
        [thisSectionValues objectAtIndex:indexPath.row];
      }

    }

  } /* if ([tableView isEqual:self.myTableView] == YES){ */

  return(result);

}
```

The code is long but basically pretty simple. We created an array of dictionaries, with each dictionary representing the data we want to populate into a section, including the cells, the header, and the footer. After constructing the cells, we can start returning the value for each section's header. Each value must be of type NSString.

You can construct headers and footers in two ways. One is by simply returning instances of the NSString class, which obviously can hold string values. The other way, which allows you to use any pictures or visual indicators of your choice, is to construct instances of the UIView class and return them as custom headers and footers. However, for simple headers and footers, NSString will suffice. We can return the header value using the tableView:titleForHeaderInSection: method, which is defined in the UITableViewDataSource protocol:

```
- (NSString *)  tableView:(UITableView *)tableView
  titleForHeaderInSection:(NSInteger)section{

  NSString *result = nil;

  if ([tableView isEqual:self.myTableView] == YES){

    if (self.arrayOfSections != nil &&
        [self.arrayOfSections count] > section){

      NSMutableDictionary *thisSection =
      [self.arrayOfSections objectAtIndex:section];

      if (thisSection != nil){
        result = [thisSection objectForKey:@"Header"];
      }

    }
```

```
  } /* if ([tableView isEqual:self.myTableView] == YES){ */

  return(result);

}
```

The method that we can use to return the string value for the footer of each section is `tableView:titleForFooterInSection:` and is defined in the `UITableViewDataSource` protocol as well:

```
- (NSString *) tableView:(UITableView *)tableView
  titleForFooterInSection:(NSInteger)section{

  NSString *result = nil;

  if ([tableView isEqual:self.myTableView] == YES){

    if (self.arrayOfSections != nil &&
        [self.arrayOfSections count] > section){

      NSMutableDictionary *thisSection =
      [self.arrayOfSections objectAtIndex:section];

      if (thisSection != nil){
        result = [thisSection objectForKey:@"Footer"];
      }

    } /* if (self.arrayofSections != nil && ... */

  } /* if ([tableView isEqual:self.myTableView] == YES){ */

  return(result);

}
```

Figure 3-14 depicts the output of our program. You can see that our headers are left-aligned and the footers are center-aligned. If you are like me and you want more control over how your headers and footers are rendered, you probably would like to find out how you can create custom headers and footers. To do this, use views by replacing the `tableView:titleForHeaderInSection:` and `tableView:titleForFooterInSection:` methods from the previous example with the `tableView:viewForHeaderInSection:` and `tableView:viewForFooterInSection:` methods defined in the `UITableViewDelegate` protocol.

If you don't specifically tell the table view about the height of each custom view, the table view will use a default height. So, the `UITableViewDelegate` protocol also defines the `tableView:heightForHeaderInSection:` and `tableView:heightForFooterInSection:` methods, returning floating-point values as the heights of your headers and footers.

Figure 3-14. A table view with headers and footers set as simple text

Overall, these are the tasks you need to take care of before you can create custom headers and footers for sections of a table view:

1. Return a floating-point value for the `tableView:heightForHeaderInSection:` method and specify the height of each custom header you wish to create for the sections of each cell. Also return a floating-point value for the `tableView:heightForFooterInSection:` method as the height of the footer for each section of your table view.

2. Use the `tableView:viewForHeaderInSection:` method to create a custom `UIView` for your sections' headers. You can also use the `tableView:viewForFooterInSection:` method to return custom `UIView` instances for any footer in your table view.

```
- (CGFloat)      tableView:(UITableView *)tableView
 heightForHeaderInSection:(NSInteger)section{

  CGFloat result = 0;

  if ([tableView isEqual:self.myTableView] == YES){
    result = 40.0f;
  }

  return(result);

}

- (CGFloat)      tableView:(UITableView *)tableView
 heightForFooterInSection:(NSInteger)section{

  CGFloat result = 0;

  if ([tableView isEqual:self.myTableView] == YES){
```

```
        result = 40.0f;
    }

    return(result);

}
```

Now begin returning a custom `UIView` for each section header:

```
- (UIView *)      tableView:(UITableView *)tableView
  viewForHeaderInSection:(NSInteger)section{

  UIView *result = nil;

  if ([tableView isEqual:self.myTableView] == YES){

    /* Do we have as many sections as specified? */
    if (self.arrayOfSections != nil &&
        [self.arrayOfSections count] > section){

      /* Get the dictionary for our section */
      NSMutableDictionary *thisSection =
      [self.arrayOfSections objectAtIndex:section];

      /* We can't insert nil into an array, so if we can get the dictionary
         out of our array of sections, we don't have to do any check to see if
         it is nil. */

      /* Get the value to display for our header */
      NSString *headerValue = [thisSection objectForKey:@"Header"];

      /* Construct a label for our header. Size is ignored here */
      UILabel *myLabel = [[[UILabel alloc]
                          initWithFrame:CGRectZero] autorelease];

      myLabel.backgroundColor = [UIColor clearColor];
      myLabel.textAlignment = UITextAlignmentCenter;
      myLabel.text = headerValue;
      result = myLabel;

    } /* if (self.arrayofSections != nil && ... */

  } /* if ([tableView isEqual:self.myTableView] == YES){ */

  return(result);

}
```

We are simply passing `CGRectZero` for the size of our label. This is because the `UITableView` object will ignore that parameter. The only thing it cares about is the custom height that we specified through the `tableView:heightForHeaderInSection:` and `tableView:heightForFooterInSection:` methods. The width, *x*, and *y* positions will be determined by the table view itself.

In the same way, we can construct our footers just by changing the `NSString` value of `"Header"` in our example to `"Footer"`, like so:

```
- (UIView *)  tableView:(UITableView *)tableView
  viewForFooterInSection:(NSInteger)section{

  UIView *result = nil;

  if ([tableView isEqual:self.myTableView] == YES){

    if (self.arrayOfSections != nil &&
        [self.arrayOfSections count] > section){

      NSMutableDictionary *thisSection =
      [self.arrayOfSections objectAtIndex:section];

      NSString *footer = [thisSection objectForKey:@"Footer"];

      UILabel *myLabel =
      [[[UILabel alloc] initWithFrame:CGRectZero] autorelease];

      myLabel.backgroundColor = [UIColor clearColor];
      myLabel.textAlignment = UITextAlignmentCenter;
      myLabel.font = [UIFont boldSystemFontOfSize:14.0f];
      myLabel.text = footer;
      result = myLabel;

    } /* if (self.arrayOfSections != nil && ... */

  } /* if ([tableView isEqual:self.myTableView] == YES){ */

  return(result);

}
```

We can celebrate our success by looking at Figure 3-15.

Figure 3-15. A grouped table view with custom headers and footers for each section

CHAPTER 4

Core Location and Maps

4.0 Introduction

The Core Location and Map Kit frameworks can be used to create location-aware and map-based applications. The Core Location framework uses the internal device's hardware to determine the current location of the device. The Map Kit framework enables your application to display maps to your users, put custom annotations on the maps, and so on. The availability of location services depends on the availability of hardware on the device; if the hardware is there, it must be enabled and switched on for the Map Kit framework to work.

To use the Core Location and Map Kit frameworks, you need to first add them to your project and make sure appropriate header files are imported. Follow these steps to add these two frameworks to your project:

1. Right-click on the Frameworks item in Xcode's project explorer (on the lefthand side of the main Xcode window).
2. Select Add→Existing Frameworks, as shown in Figure 4-1.
3. To select both frameworks in one go, hold down the Command key on your keyboard and select CoreLocation.framework and MapKit.framework, as shown in Figure 4-2.
4. Click Add.

After adding these two frameworks, you will need to add two header files to your *.m* or *.h* file (in your *.h* file if you are referring to any entity that is included in either of the two aforementioned frameworks):

```
#import <CoreLocation/CoreLocation.h>
#import <MapKit/MapKit.h>
```

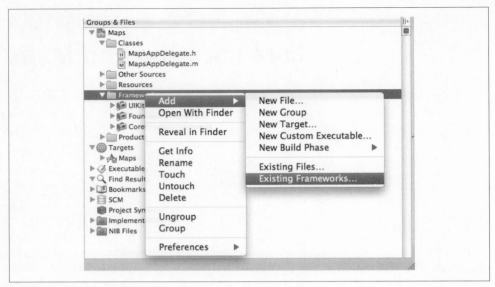

Figure 4-1. Adding a new framework to an Xcode project

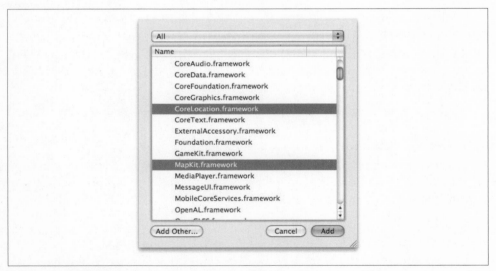

Figure 4-2. Adding the Core Location and Map Kit frameworks to an Xcode project

4.1 Creating a Map Using Interface Builder

Problem

You would like to use Interface Builder to create a map control.

Solution

Use Interface Builder to add a map view to one of your application's objects (usually a view). By adding it to an XIB file, you create an MKMapView object that can be accessed by your program.

1. Choose an XIB file to host the map. Normally, we use a View XIB file, which comes with a view attached to the XIB file by default. If you already have an XIB file, skip to step 2. To create a new XIB file, in Xcode choose File→New File. In the New File dialog, on the left side, choose User Interface and then, based on your need, select whichever type of XIB file you would like.
2. Once Interface Builder has opened the XIB file, select Tools→Library to open the Library pane.
3. In the Library pane, find the map view and drag and drop it into your XIB file.

Discussion

Now that you have an instance of the MKMapView object in your XIB file (as shown in this recipe's Solution), you can navigate to the Connections Inspector pane in Interface Builder by selecting Tools→Connections Inspector. Select your map object and you will see that the Connections Inspector allows you to see the *delegate* and the *new referencing outlet* connections of your map view. You need to drop the *new referencing outlet* to an IBOutlet of type MKMapView in your header files and drop the delegate connection to an object in your XIB file that implements the MKMapViewDelegate protocol.

See Also

Recipe 4.2

4.2 Creating a Map Using Code

Problem

You want to instantiate and display a map using code.

Solution

Create an instance of the MKMapView class and add it to a view or assign it as a subview of your view controller. Here is the sample *.h* file of a view controller that creates an instance of MKMapView and displays it full-screen on its view:

```
#import <UIKit/UIKit.h>
#import <MapKit/MapKit.h>

@interface RootViewController : UIViewController{
@public
  MKMapView          *myMapView;
```

```
}

@property (nonatomic, retain) MKMapView    *myMapView;

@end
```

This is a simple root view controller with a variable of type MKMapView. Later in the implementation of this view controller (.m file), we will initialize the map and set its type to Satellite, like so:

```
#import "RootViewController.h"

@implementation RootViewController

@synthesize myMapView;

- (void)viewDidLoad {
  [super viewDidLoad];

  /* Create a map as big as our view */

  MKMapView *mapView = [[MKMapView alloc]
                        initWithFrame:self.view.bounds];
  self.myMapView = mapView;
  [mapView release];

  /* Set the map type to Satellite */
  self.myMapView.mapType = MKMapTypeSatellite;

  self.myMapView.autoresizingMask =
    UIViewAutoresizingFlexibleWidth |
    UIViewAutoresizingFlexibleHeight;

  /* Add it to our view */
  [self.view addSubview:self.myMapView];

}

- (void) viewDidUnload{
  [super viewDidUnload];

  self.myMapView = nil;

}

- (BOOL)shouldAutorotateToInterfaceOrientation:
  (UIInterfaceOrientation)interfaceOrientation {
  /* Support all orientations */
  return YES;
}

- (void)dealloc {

  [myMapView release];
  [super dealloc];
```

```
}

@end
```

Discussion

Creating an instance of the `MKMapView` class is quite straightforward. We can simply assign a frame to it using its constructor, and after the map is created, we will add it as a subview of the view on the screen just so that we can see it.

 `MKMapView` is a subclass of `UIView`, so you can manipulate any map view the way you manipulate an instance of `UIView`.

If you haven't already noticed, the `MKMapView` class has a property called `mapType` that can be set to satellite, standard, or hybrid. In this example, we are using the satellite map type (see Figure 4-3).

Figure 4-3. A satellite map view

We can change the visual representation type of a map view using the `mapType` property of an instance of `MKMapView`. Here are the different values we can use for this property:

`MKMapTypeStandard`
Use this map type to display a standard map (this is the default).

`MKMapTypeSatellite`
Use this map type to display a satellite image map (as depicted in Figure 4-3).

MKMapTypeHybrid
Use this map type to display a standard map overlaid on a satellite image map.

See Also

Recipe 4.1

4.3 Handling the Events of a Map

Problem

You want to handle various events that a map view can send to its delegate.

Solution

Assign a delegate object, which conforms to the MKMapViewDelegate protocol, to the delegate property of an instance of the MKMapView class:

```
/* Create a map as big as our view */
MKMapView *mapView = [[MKMapView alloc]
                         initWithFrame:self.view.bounds];
self.myMapView = mapView;
[mapView release];

/* Set the map type to Satellite */
self.myMapView.mapType = MKMapTypeSatellite;

self.myMapView.delegate = self;

self.myMapView.autoresizingMask =
  UIViewAutoresizingFlexibleWidth |
  UIViewAutoresizingFlexibleHeight;

/* Add it to our view */
[self.view addSubview:self.myMapView];
```

This code can easily run in the viewDidLoad method of a view controller object that has a property named MapView of type MKMapView:

```
#import <UIKit/UIKit.h>
#import <MapKit/MapKit.h>

@interface RootViewController : UIViewController
                                <MKMapViewDelegate>{
@public
  MKMapView          *myMapView;
}

@property (nonatomic, retain) MKMapView    *myMapView;

@end
```

Discussion

The delegate object of an instance of the `MKMapView` class must implement the methods defined in the `MKMapViewDelegate` protocol in order to receive various messages from the map view and, as we will see later, to be able to provide information to the map view. Various methods are defined in the `MKMapViewDelegate` protocol, such as the `mapViewWillStartLoadingMap:` method that will get called in the delegate object whenever the map loading process starts. Bear in mind that a delegate for a map view is not a required object, meaning that you can create map views without assigning delegates to them; these views simply won't respond to user manipulation.

Here is a list of some of the methods declared in the `MKMapViewDelegate` protocol and what they are meant to report to the delegate object of an instance of `MKMapView`:

`mapViewWillStartLoadingMap:`
> This method is called on the delegate object whenever the map view starts to load the data that visually represents the map to the user.

`mapView:viewForAnnotation:`
> This method is called on the delegate object whenever the map view is asking for an instance of `MKAnnotationView` to visually represent an annotation on the map. For more information about this, please refer to Recipe 4.5.

`mapViewWillStartLocatingUser:`
> This method, as its name implies, gets called on the delegate object whenever the map view starts to detect the user's location. For information about finding a user's location, please refer to Recipe 4.4.

`mapView:regionDidChangeAnimated:`
> This method gets called on the delegate object whenever the region displayed by the map is changed.

4.4 Pinpointing a Device's Location

Problem

You want to find the latitude and longitude of a device.

Solution

Use the `CLLocationManager` class:

```
CLLocationManager *newLocationManager =
[[CLLocationManager alloc] init];

self.locationManager = newLocationManager;

[newLocationManager release];
```

```
    self.locationManager.delegate = self;

    self.locationManager.purpose =
    NSLocalizedString(@"To provide functionality based on\
                      user's current location.", nil);

    [self.locationManager startUpdatingLocation];
```

In this code, locationManager is a property of type CLLocationManager. The current class is also the delegate of the location manager in this sample code.

Discussion

The Core Location framework in the SDK provides functionality for programmers to be able to detect the current spatial location of an iOS device. Because in iOS, the user is allowed to disable location services using the Settings, before instantiating an object of type CLLocationManager, it is best to first determine whether the location services are enabled on the device.

 The delegate object of an instance of CLLocationManager must conform to the CLLocationManagerDelegate protocol.

This is how we will declare our location manager object in the *.h* file of a view controller (the object creating an instance of CLLocationManager does not necessarily have to be a view controller):

```
#import <UIKit/UIKit.h>
#import <CoreLocation/CoreLocation.h>

@interface RootViewController : UIViewController
                                <CLLocationManagerDelegate>{
@public
  CLLocationManager   *locationManager;
}

@property (nonatomic, retain) CLLocationManager *locationManager;

@end
```

The implementation of our view controller is as follows:

```
#import "RootViewController.h"
#import <objc/runtime.h>

@implementation RootViewController

@synthesize locationManager;

- (void)locationManager:(CLLocationManager *)manager
    didUpdateToLocation:(CLLocation *)newLocation
```

```
            fromLocation:(CLLocation *)oldLocation{

  /* We received the new location */

  NSLog(@"Latitude = %f", newLocation.coordinate.latitude);
  NSLog(@"Longitude = %f", newLocation.coordinate.longitude);

}

- (void)locationManager:(CLLocationManager *)manager
       didFailWithError:(NSError *)error{

  /* Failed to receive user's location */

}

- (void)viewDidLoad {
  [super viewDidLoad];

  BOOL locationServicesAreEnabled = NO;

  Method requiredClassMethod =
  class_getClassMethod([CLLocationManager class],
                       @selector(locationServicesEnabled));

  if (requiredClassMethod != nil){

    locationServicesAreEnabled =
    [CLLocationManager locationServicesEnabled];

  } else {

    CLLocationManager *DummyManager =
    [[CLLocationManager alloc] init];

    locationServicesAreEnabled = [DummyManager locationServicesEnabled];
    [DummyManager release];

  }

  if (locationServicesAreEnabled == YES){

    CLLocationManager *newLocationManager =
    [[CLLocationManager alloc] init];

    self.locationManager = newLocationManager;

    [newLocationManager release];

    self.locationManager.delegate = self;

    self.locationManager.purpose =
      NSLocalizedString(@"To provide functionality based on\
                        user's current location.", nil);
```

```
      [self.locationManager startUpdatingLocation];

  } else {

    /* Location services are not enabled.
     Take appropriate action: for instance, prompt the
     user to enable the location services */
    NSLog(@"Location services are not enabled");

  }

}

- (void) viewDidUnload{
  [super viewDidUnload];

  if (self.locationManager != nil){
    [self.locationManager stopUpdatingLocation];
  }

  self.locationManager = nil;
}

- (BOOL)shouldAutorotateToInterfaceOrientation:
  (UIInterfaceOrientation)interfaceOrientation {
  /* Support all orientations */
  return YES;
}

- (void)dealloc {
  [locationManager stopUpdatingLocation];
  [locationManager release];
  [super dealloc];
}

@end
```

The startUpdateLocation instance method of CLLocationManager reports the success or failure of retrieving the user's location to its delegate through the location Manager:didUpdateToLocation:fromLocation: and locationManager:didFailWithError: methods of its delegate object, in that order.

The locationServicesEnabled class method of CLLocationManager is available in SDK 4.0 and later. For this reason, we must detect the availability of this class method before invoking it if we are targeting devices with earlier versions of iOS installed.

The CLLocationManager class implements a property named purpose. This property allows us to customize the message that is shown to the users of our application, asking for their permission to allow location services for our application using Core Location functionalities. A good practice is to use localized strings for the value of this property.

4.5 Displaying Built-in Pins on a Map View

Problem

You want to point out a specific location on a map to the user.

Solution

Use built-in map view annotations.

Follow these steps:

1. Create a new class and call it `MyAnnotation`.
2. Make sure this class conforms to the `MKAnnotation` protocol.
3. Define a property for this class of type `CLLocationCoordinate2D` and name it `coor dinate`. Also make sure you set it as a `readonly` property since the `coordinate` property is defined as `readonly` in the `MKAnnotation` protocol.
4. Optionally, define two properties of type `NSString`, namely `title` and `subtitle`, which will be able to carry the title and the subtitle information for your annotation view.
5. Create an initializer method for your class that will accept a parameter of type `CLLocationCoordinate2D`. In this method, assign the passed location parameter to the property that we defined in step 3. Since this property is `readonly`, it cannot be assigned by code outside the scope of this class. Therefore, the initializer of this class acts as a bridge here and allows us to indirectly assign a value to this property.
6. Instantiate the `MyAnnotation` class and add it to your map using the `add Annotation:` method of the `MKMapView` class.

Discussion

As explained in this recipe's Solution, we must create an object that conforms to the `MKAnnotation` protocol and later instantiate this object and pass it to the map to be displayed. We will write the *.h* file of this object like so:

```
#import <Foundation/Foundation.h>
#import <MapKit/MapKit.h>

@interface MyAnnotation : NSObject <MKAnnotation> {
@private
  CLLocationCoordinate2D coordinate;
  NSString *title;
  NSString *subtitle;
}

@property (nonatomic, assign, readonly) CLLocationCoordinate2D coordinate;
@property (nonatomic, copy) NSString *title;
@property (nonatomic, copy) NSString *subtitle;
```

```
- (id) initWithCoordinates:(CLLocationCoordinate2D)paramCoordinates
                     title:(NSString *)paramTitle
                  subTitle:(NSString *)paramSubTitle;

@end
```

The *.m* file of the MyAnnotation class sets up the class to display location information as follows:

```
#import "MyAnnotation.h"

@implementation MyAnnotation

@synthesize coordinate, title, subtitle;

- (id) initWithCoordinates:(CLLocationCoordinate2D)paramCoordinates
                     title:(NSString *)paramTitle
                  subTitle:(NSString *)paramSubTitle{

  self = [super init];

  if (self != nil){
    coordinate = paramCoordinates;
    title = [paramTitle copy];
    subtitle = [paramSubTitle copy];
  }

  return(self);

}

- (void) dealloc {
  [title release];
  [subtitle release];
  [super dealloc];
}

@end
```

Later we will instantiate this class and add it to our map, for instance, in the *.m* file of a view controller that creates and displays a map view:

```
#import "RootViewController.h"
#import "MyAnnotation.h"

@implementation RootViewController

@synthesize myMapView;

- (void)viewDidLoad {
  [super viewDidLoad];

  /* Create a map as big as our view */
  MKMapView *newMapView = [[MKMapView alloc]
                            initWithFrame:self.view.bounds];
  self.myMapView = newMapView;
```

```
    [newMapView release];

    self.myMapView.delegate = self;

    /* Set the map type to Standard */
    self.myMapView.mapType = MKMapTypeStandard;

    self.myMapView.autoresizingMask =
      UIViewAutoresizingFlexibleWidth |
      UIViewAutoresizingFlexibleHeight;

    /* Add it to our view */
    [self.view addSubview:self.myMapView];

    /* This is just a sample location */
    CLLocationCoordinate2D location;
    location.latitude = 50.82191692907181;
    location.longitude = -0.13811767101287842;

    /* Create the annotation using the location */
    MyAnnotation *annotation =
    [[MyAnnotation alloc] initWithCoordinates:location
                                  title:@"My Title"
                               subTitle:@"My Sub Title"];

    /* And eventually add it to the map */
    [self.myMapView addAnnotation:annotation];

    [annotation release];

}

- (void) viewDidUnload{
    [super viewDidUnload];
    self.myMapView = nil;
}

- (BOOL)shouldAutorotateToInterfaceOrientation:
    (UIInterfaceOrientation)interfaceOrientation {
    /* Support all orientations */
    return YES;
}

- (void)dealloc {
    [myMapView release];
    [super dealloc];
}

@end
```

Figure 4-4 depicts the output of the program when run in iPhone Simulator.

See Also

Recipe 4.6; Recipe 4.7

Figure 4-4. A built-in pin dropped on a map

4.6 Displaying Pins with Different Colors on a Map View

Problem

The default color for pins dropped on a map view is red. You want to be able to display pins in different colors in addition to the default red pin.

Solution

Return instances of `MKPinAnnotationView` to your map view through the `mapView:view ForAnnotation:` delegate method.

Every annotation that is added to an instance of `MKMapView` has a corresponding view that gets displayed on the map view. These views are called *annotation views*. An annotation view is an object of type `MKAnnotationView`, which is a subclass of `UIView`. If the delegate object of a map view implements the `mapView:viewForAnnotation:` delegate method, the delegate object will have to return instances of the `MKAnnotationView` class to represent and, optionally, customize the annotation views to be displayed on a map view.

Discussion

To set up our program so that we can customize the color (choosing from the default SDK pin colors) of the annotation view that gets dropped on a map view, representing an annotation, we must return an instance of the `MKPinAnnotationView` class instead of an instance of `MKAnnotationView` in the `mapView:viewForAnnotation:` delegate method.

Bear in mind that the `MKPinAnnotationView` class is a subclass of the `MKAnnotationView` class.

```
- (MKAnnotationView *)mapView:(MKMapView *)mapView
          viewForAnnotation:(id <MKAnnotation>)annotation{

  MKAnnotationView *result = nil;

  if ([annotation isKindOfClass:[MyAnnotation class]] == NO){
    return(result);
  }

  if ([mapView isEqual:self.myMapView] == NO){
    /* We want to process this event only for the Map View
       that we have created previously */
    return(result);
  }

  /* First typecast the annotation for which the Map View has
     fired this delegate message */
  MyAnnotation *senderAnnotation = (MyAnnotation *)annotation;

  /* Using the class method we have defined in our custom
     annotation class, we will attempt to get a reusable
     identifier for the pin we are about to create */
  NSString *pinReusableIdentifier =
    [MyAnnotation
     reusableIdentifierforPinColor:senderAnnotation.pinColor];

  /* Using the identifier we retrieved above, we will
     attempt to reuse a pin in the sender Map View */
  MKPinAnnotationView *annotationView = (MKPinAnnotationView *)
    [mapView
     dequeueReusableAnnotationViewWithIdentifier:
     pinReusableIdentifier];

  if (annotationView == nil){
    /* If we fail to reuse a pin, then we will create one */
    annotationView =
    [[[MKPinAnnotationView alloc]
      initWithAnnotation:senderAnnotation
      reuseIdentifier:pinReusableIdentifier] autorelease];

    /* Make sure we can see the callouts on top of
       each pin in case we have assigned title and/or
       subtitle to each pin */
    [annotationView setCanShowCallout:YES];
  }

  /* Now make sure, whether we have reused a pin or not, that
     the color of the pin matches the color of the annotation */
  annotationView.pinColor = senderAnnotation.pinColor;

  result = annotationView;
```

```
          return(result);
     }
```

An annotation view must be reused by giving it an identifier (an NSString). By determining which type of pin you would like to display on a map view and setting a unique
identifier for each type of pin (e.g., blue pins can be treated as one type of pin
and red pins as another), you must reuse the proper type of pin using the
dequeueReusableAnnotationViewWithIdentifier: instance method of MKMapView as demonstrated in the code.

We have set the mechanism of retrieving the unique identifiers of each pin in our custom
MyAnnotation class. Here is the *.h* file of the MyAnnotation class:

```
#import <Foundation/Foundation.h>
#import <MapKit/MapKit.h>

/* These are the standard SDK pin colors. We are setting
 unique identifiers per color for each pin so that later we
 can reuse the pins that have already been created with the same
 color */

#define REUSABLE_PIN_RED     @"Red"
#define REUSABLE_PIN_GREEN   @"Green"
#define REUSABLE_PIN_PURPLE  @"Purple"

@interface MyAnnotation : NSObject <MKAnnotation> {
@private
  CLLocationCoordinate2D  coordinate;
  NSString              *title;
  NSString              *subtitle;
  MKPinAnnotationColor   pinColor;
}

@property (nonatomic, assign, readonly) CLLocationCoordinate2D coordinate;
@property (nonatomic, copy) NSString  *title;
@property (nonatomic, copy) NSString  *subtitle;
@property (nonatomic, assign) MKPinAnnotationColor  pinColor;

- (id) initWithCoordinates:(CLLocationCoordinate2D)paramCoordinates
                     title:(NSString*)paramTitle
                  subTitle:(NSString*)paramSubTitle;

+ (NSString *)   reusableIdentifierforPinColor
                  :(MKPinAnnotationColor)paramColor;

@end
```

Annotations are not the same as annotation views. An annotation is the location that
you want to show on a map and an annotation view is the view that represents
that annotation on the map. The MyAnnotation class is the annotation, not the annotation view. When we create an annotation by instantiating the MyAnnotation class, we
can assign a color to it using the pinColor property that we have defined and implemented. When the time comes for a map view to display an annotation, the map view

will call the `mapView:viewForAnnotation:` delegate method and ask its delegate for an annotation view. The `forAnnotation` parameter of this method passes the annotation that needs to be displayed. By getting a reference to the annotation, we can type-case the annotation to an instance of `MyAnnotation`, retrieve its `pinColor` property, and based on that, create an instance of `MKPinAnnotationView` with the given pin color and return it to the map view.

This is the *.m* file of `MyAnnotation`:

```
#import "MyAnnotation.h"

@implementation MyAnnotation

@synthesize coordinate, title, subtitle, pinColor;

+ (NSString *)    reusableIdentifierforPinColor
                  :(MKPinAnnotationColor)paramColor{

  NSString *result = nil;

  switch (paramColor){
    case MKPinAnnotationColorRed:{
      result = REUSABLE_PIN_RED;
      break;
    }
    case MKPinAnnotationColorGreen:{
      result = REUSABLE_PIN_GREEN;
      break;
    }
    case MKPinAnnotationColorPurple:{
      result = REUSABLE_PIN_PURPLE;
      break;
    }
  }

  return(result);
}

- (id) initWithCoordinates:(CLLocationCoordinate2D)paramCoordinates
                     title:(NSString*)paramTitle
                  subTitle:(NSString*)paramSubTitle{

  self = [super init];

  if (self != nil){
    coordinate = paramCoordinates;
    title = [paramTitle copy];
    subtitle = [paramSubTitle copy];
    pinColor = MKPinAnnotationColorGreen;
  }

  return(self);

}
```

```
- (void) dealloc {
  [title release];
  [subtitle release];
  [super dealloc];


}

@end
```

After implementing the MyAnnotation class, it's time to use it in our application (in this example, we will use it in a view controller). Here is the .h file of the view controller:

```
#import <UIKit/UIKit.h>
#import <MapKit/MapKit.h>
#import <CoreLocation/CoreLocation.h>

@interface RootViewController : UIViewController <MKMapViewDelegate> {
@public
  MKMapView          *myMapView;
}

@property (nonatomic, retain) MKMapView    *myMapView;

@end
```

The implementation is in the .m file like so:

```
#import "RootViewController.h"
#import "MyAnnotation.h"

@implementation RootViewController

@synthesize myMapView;

- (MKAnnotationView *)mapView:(MKMapView *)mapView
            viewForAnnotation:(id <MKAnnotation>)annotation{

  MKAnnotationView *result = nil;

  if ([annotation isKindOfClass:[MyAnnotation class]] == NO){
    return(result);
  }

  if ([mapView isEqual:self.myMapView] == NO){
    /* We want to process this event only for the Map View
     that we have created previously */
    return(result);
  }

  /* First typecast the annotation for which the Map View has
   fired this delegate message */
  MyAnnotation *senderAnnotation = (MyAnnotation *)annotation;

  /* Using the class method we have defined in our custom
   annotation class, we will attempt to get a reusable
```

```
  identifier for the pin we are about to create */
NSString *pinReusableIdentifier =
  [MyAnnotation
   reusableIdentifierforPinColor:senderAnnotation.pinColor];

/* Using the identifier we retrieved above, we will
   attempt to reuse a pin in the sender Map View */
MKPinAnnotationView *annotationView = (MKPinAnnotationView *)
  [mapView
   dequeueReusableAnnotationViewWithIdentifier:
   pinReusableIdentifier];

if (annotationView == nil){
  /* If we fail to reuse a pin, then we will create one */
  annotationView =
  [[[MKPinAnnotationView alloc]
    initWithAnnotation:senderAnnotation
    reuseIdentifier:pinReusableIdentifier] autorelease];

  /* Make sure we can see the callouts on top of
     each pin in case we have assigned title and/or
     subtitle to each pin */
  [annotationView setCanShowCallout:YES];
}

/* Now make sure, whether we have reused a pin or not, that
   the color of the pin matches the color of the annotation */
annotationView.pinColor = senderAnnotation.pinColor;

result = annotationView;

  return(result);
}

- (void)viewDidLoad {
  [super viewDidLoad];

  /* Create a map as big as our view */
  MKMapView *mapView = [[MKMapView alloc]
                         initWithFrame:self.view.bounds];
  self.myMapView = mapView;
  [mapView release];

  self.myMapView.delegate = self;

  /* Set the map type to Standard */
  self.myMapView.mapType = MKMapTypeStandard;

  self.myMapView.autoresizingMask =
    UIViewAutoresizingFlexibleWidth |
    UIViewAutoresizingFlexibleHeight;

  /* Add it to our view */
  [self.view addSubview:self.myMapView];
```

```
/* This is just a sample location */
CLLocationCoordinate2D location;
location.latitude = 50.82191692907181;
location.longitude = -0.13811767101287842;

/* Create the annotation using the location */
MyAnnotation *annotation =
[[MyAnnotation alloc] initWithCoordinates:location
                                    title:@"My Title"
                                 subTitle:@"My Sub Title"];

annotation.pinColor = MKPinAnnotationColorPurple;

/* And eventually add it to the map */
[self.myMapView addAnnotation:annotation];

[annotation release];
}


- (void) viewDidUnload{
  [super viewDidUnload];
  self.myMapView = nil;
}

- (BOOL)shouldAutorotateToInterfaceOrientation:
  (UIInterfaceOrientation)interfaceOrientation {
  /* Support all orientations */
  return YES;
}

- (void) didReceiveMemoryWarning{
  [super didReceiveMemoryWarning];
}

- (void)dealloc {
  /* Deallocate the map */
  [myMapView release];
  [super dealloc];
}

@end
```

4.7 Creating and Displaying Custom Pins on a Map View

Problem

Instead of the default iOS SDK pins, you would like to display your own images as pins on a map view.

Solution

Load an arbitrary image into an instance of the UIImage class and assign it to the image property of the MKAnnotationView instance that you return to your map view as a pin:

```
- (MKAnnotationView *)mapView:(MKMapView *)mapView
          viewForAnnotation:(id <MKAnnotation>)annotation{

  MKAnnotationView *result = nil;

  if ([annotation isKindOfClass:[MyAnnotation class]] == NO){
    return(result);
  }

  if ([mapView isEqual:self.myMapView] == NO){
    /* We want to process this event only for the Map View
       that we have created previously */
    return(result);
  }

  /* First typecast the annotation for which the Map View has
     fired this delegate message */
  MyAnnotation *senderAnnotation = (MyAnnotation *)annotation;

  /* Using the class method we have defined in our custom
     annotation class, we will attempt to get a reusable
     identifier for the pin we are about to create */
  NSString *pinReusableIdentifier =
  [MyAnnotation
   reusableIdentifierforPinColor:senderAnnotation.pinColor];

  /* Using the identifier we retrieved above, we will
     attempt to reuse a pin in the sender Map View */
  MKPinAnnotationView *annotationView = (MKPinAnnotationView *)
  [mapView
   dequeueReusableAnnotationViewWithIdentifier:
   pinReusableIdentifier];

  if (annotationView == nil){
    /* If we fail to reuse a pin, then we will create one */
    annotationView =
    [[[MKPinAnnotationView alloc]
      initWithAnnotation:senderAnnotation
      reuseIdentifier:pinReusableIdentifier] autorelease];

    /* Make sure we can see the callouts on top of
       each pin in case we have assigned title and/or
       subtitle to each pin */
    annotationView.canShowCallout = YES;

  }

  /* Now make sure, whether we have reused a pin or not, that
     the color of the pin matches the color of the annotation */
```

```
  annotationView.pinColor = senderAnnotation.pinColor;

  UIImage *pinImage = [UIImage imageNamed:@"BluePin.png"];
  if (pinImage != nil){
    annotationView.image = pinImage;
  }

  result = annotationView;

  return(result);
}
```

In this code, we are displaying an image named *BluePin.png* (in our application bundle) for any pin that is dropped on the map. For the definition and the implementation of the MyAnnotation class, refer to Recipe 4.6.

Discussion

The delegate object of an instance of the MKMapView class must conform to the MKMapViewDelegate protocol and implement the mapView:viewForAnnotation: method. The return value of this method is an instance of the MKAnnotationView class. Any object that subclasses the aforementioned class, by default, inherits a property called image. Assigning a value to this property will replace the default image provided by the Map Kit framework, as shown in Figure 4-5.

Figure 4-5. A custom image displayed on a map view

4.8 Retrieving Meaningful Addresses Using Spatial Coordinates

Problem

You have the latitude and longitude of a spatial location and you want to retrieve the address of this location.

Solution

The process of retrieving a meaningful address using spatial coordinates, x and y, is called *reverse geocoding*. Create and use an instance of the MKReverseGeocoder class and provide a delegate to this instance, making sure that the delegate object conforms to the MKReverseGeocoderDelegate protocol.

The *.h* file of a simple view controller for this purpose is defined like so:

```
#import <UIKit/UIKit.h>
#import <MapKit/MapKit.h>

@interface RootViewController : UIViewController <MKReverseGeocoderDelegate> {
@public
    MKReverseGeocoder *myReverseGeocoder;
}

@property (nonatomic, retain) MKReverseGeocoder *myReverseGeocoder;

@end
```

The *.m* file of this view controller is as follows:

```
#import "RootViewController.h"

@implementation RootViewController

@synthesize myReverseGeocoder;

- (void)reverseGeocoder:(MKReverseGeocoder *)geocoder
        didFindPlacemark:(MKPlacemark *)placemark{

    /* We received the results */
    NSLog(@"%@", placemark.country);
    NSLog(@"%@", placemark.postalCode);
    NSLog(@"%@", placemark.locality);

}

- (void)reverseGeocoder:(MKReverseGeocoder *)geocoder
        didFailWithError:(NSError *)error{

    /* An error has occurred. Use the [error] parameter to
       determine the cause of the issue */
    NSLog(@"An error occurred in the reverse geocoder");
```

```
}

- (void)viewDidLoad {
  [super viewDidLoad];

  CLLocationCoordinate2D location;
  location.latitude = +38.4112810;
  location.longitude = -122.8409780f;

  MKReverseGeocoder *reverseGeocoder =
  [[MKReverseGeocoder alloc] initWithCoordinate:location];

  self.myReverseGeocoder = reverseGeocoder;

  [reverseGeocoder release];

  self.myReverseGeocoder.delegate = self;
  [self.myReverseGeocoder start];

}

- (void) viewDidUnload{
  [super viewDidUnload];

  [self.myReverseGeocoder cancel];
  self.myReverseGeocoder = nil;

}

- (BOOL)shouldAutorotateToInterfaceOrientation:
  (UIInterfaceOrientation)interfaceOrientation {
  /* Support all orientations */
  return YES;
}

- (void)dealloc {
  [myReverseGeocoder cancel];
  [myReverseGeocoder release];
  [super dealloc];
}

@end
```

The NSLog methods in the preceding code write the results shown in Figure 4-6 in the console window for the given spatial location.

Discussion

Each application has a limit on the amount of reverse geocoding requests that it can make every day. To perform a reverse geocoding request, you must create an instance of the MKReverseGeocoder class. This class requires an active network connection in order to process requests successfully. The reverse geocoded values are reported to the

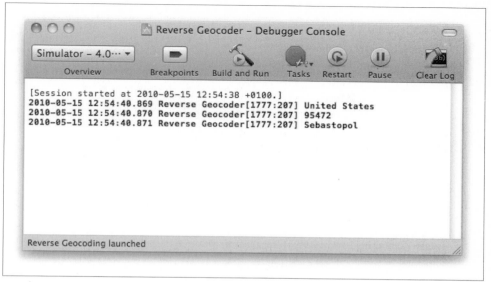

Figure 4-6. Reverse-geocoding console output

delegate object of this class. The delegate object assigned to an instance of the `MKReverseGeocoder` must conform to the `MKReverseGeocoderDelegate` protocol.

See Also

Recipe 4.9

4.9 Retrieving Spatial Coordinates Using Meaningful Addresses

Problem

You have an address of a location and you want to find the spatial location (*x*,*y*) of that address.

Solution

Use Google's publicly available Geocoder API available at *http://maps.google.com/ maps/geo*.

Before using this API, please make sure you read the terms of use for iPhone applications, at *http://code.google.com/apis/maps/iphone/terms.html*.

You can then call the Geocoder API in this way:

```
- (void)viewDidLoad {
  [super viewDidLoad];
```

```objc
/* We have our address */
NSString *oreillyAddress =
  @"1005 Gravenstein Highway North, Sebastopol, CA 95472, USA";

/* We will later insert the address and the format that we want
 our output in, into this API's URL */
NSString *geocodingURL =
@"http://maps.google.com/maps/geo?q=%@&output=%@";

/* Insert the address and the output format into the URL */
NSString *finalURL =
[NSString stringWithFormat:geocodingURL,
 oreillyAddress,
 GOOGLE_OUTPUT_FORMAT_CSV];

/* Now escape the URL using appropriate percentage marks */
finalURL =
  [finalURL
    stringByAddingPercentEscapesUsingEncoding:
    NSUTF8StringEncoding];

/* Create our URL */
NSURL *urlToCall = [NSURL URLWithString:finalURL];

/* And a request for the connection using the URL */
NSURLRequest *request = [NSURLRequest requestWithURL:urlToCall];

/* We will put all the connection's received data into this
 instance of the NSMutableData class */
NSMutableData *newMutableData = [[NSMutableData alloc] init];
self.connectionData = newMutableData;
[newMutableData release];

NSURLConnection *newConnection =
[[NSURLConnection alloc] initWithRequest:request
                                delegate:self];

/* Create the connection and start the downloading
 of geocoding results */
self.myConnection = newConnection;

[newConnection release];

}

- (void) viewDidUnload{
  [super viewDidUnload];

  [self.myConnection cancel];

  self.myConnection = nil;
  self.connectionData = nil;
}
```

The formats required for this API are defined in this way:

```
#define GOOGLE_OUTPUT_FORMAT_CSV  @"csv"
#define GOOGLE_OUTPUT_FORMAT_XML  @"xml"
```

This code is further explained in this recipe's Discussion.

Discussion

The reverse geocoding capabilities discussed in Recipe 4.8 are built into the iOS SDK's Map Kit framework. *Reverse geocoding* is the process of retrieving a meaningful address, city and country, and so on, using spatial locations (*x,y*). *Geocoding*, on the other hand, is the process of finding the spatial locations of a given address. Geocoding is not supported in the Map Kit framework and we need to call a third-party API, such as Google, for this. There are limitations on the number of geocoding requests that any client can send per day, so make sure you read the terms and conditions and the FAQs of Google's geocoding APIs before you begin using them:

- The Google Maps API Family FAQs page is located at *http://code.google.com/apis/maps/faq.html*.
- The Google Maps Terms of Service page for the iPhone is located at *http://code.google.com/apis/maps/iphone/terms.html*.
- The main URL of Google's Geocoder API is *http://maps.google.com/maps/geo*.

We can attach query and format parameters to this URL and call it as an API, synchronously or asynchronously. The query parameter can be provided using the q key and the format parameter using the output key. You can choose to have the output in CSV format or in XML by providing the value csv or xml, respectively, to the output parameter.

For instance, to find the spatial coordinates of Times Square in New York, we can construct the API URL in this way:

```
http://maps.google.com/maps/geo?q=Times%20Square&output=csv
```

 The value %20 in this URL represents a URL-encoded space character.

The output of this API call returns a value similar to this:

```
200,4,40.7590110,-73.9844722
```

where 200 is the status code, 4 is the accuracy, 40.7590110 is the latitude, and -73.9844722 is the longitude.

If you retrieve this value in a variable of type `NSString`, you can use the `components SeparatedByString:` method of the `NSString` class to split this string into four components separated with a comma character, as we will see shortly.

Now let's find the spatial location of O'Reilly's headquarters, located at this address:

1005 Gravenstein Highway North, Sebastopol, CA 95472, USA

The process through which we will find the spatial coordinates of this address is as follows:

1. Use the aforementioned Google Geocoder API URL, *http://maps.google.com/maps/ geo*, and attach the q parameter for the address to be queried and the `output` parameter for the required output. Here we will use the CSV output format.

2. Create an instance of `NSURL` from the string representing the URL constructed in step 1.

3. Create an instance of the `NSURLRequest` class using the `NSURL` instance created in step 2.

4. Instantiate a variable of type `NSURLConnection` and feed it with the `NSURLRequest` object created in step 3. Make sure you set the delegate property of the instance of the `NSURLConnection` class so that you get notified when the data is downloaded from the API.

In this example, we implement this functionality in a view controller and output the results into the console window using `NSLog`. The *.h* file of the view controller is defined in this way:

```
#import <UIKit/UIKit.h>
#import <MapKit/MapKit.h>

#define GOOGLE_OUTPUT_FORMAT_CSV  @"csv"
#define GOOGLE_OUTPUT_FORMAT_XML  @"xml"

@interface RootViewController : UIViewController {
@public
  NSURLConnection *myConnection;
  NSMutableData   *connectionData;
}

@property (nonatomic, retain) NSURLConnection *myConnection;
@property (nonatomic, retain) NSMutableData   *connectionData;

@end
```

We implement the view controller in this way in its *.m* file:

```
#import "RootViewController.h"

@implementation RootViewController

@synthesize myConnection, connectionData;
```

```objc
- (void)connection:(NSURLConnection *)connection
  didFailWithError:(NSError *)error{

  /* Handle the error here */
  NSLog(@"Connection error happened");

}

- (void) connection:(NSURLConnection *)connection
 didReceiveResponse:(NSURLResponse *)response{

  [self.connectionData setLength:0];

}

- (void)connection:(NSURLConnection *)connection
    didReceiveData:(NSData *)data{

  /* We received some data, let's append it to the end of the
   current mutable data that we have */
  [self.connectionData appendData:data];

}

- (void)connectionDidFinishLoading:(NSURLConnection *)connection{

  NSString *connectionString =
    [[NSString alloc] initWithData:self.connectionData
                          encoding:NSUTF8StringEncoding];

  if ([connectionString length] > 0){

    NSArray *components =
    [connectionString
     componentsSeparatedByString:@","];

    NSString *statusCode = nil,
             *accuracy = nil,
             *latitude = nil,
             *longitude = nil;

    if ([components count] == 4){

      statusCode = [components objectAtIndex:0];
      accuracy = [components objectAtIndex:1];
      latitude = [components objectAtIndex:2];
      longitude = [components objectAtIndex:3];

      NSLog(@"Status Code = %@", statusCode);
      NSLog(@"Accuracy = %@", accuracy);
      NSLog(@"Latitude = %@", latitude);
      NSLog(@"Longitude = %@", longitude);

    } else {
      /* Handle other situation where we have more or less
```

```
        than 4 values which we expect from this API */
    }

  } else {
    /* The string is empty, handle this problem here */
  }

  [connectionString release];
  connectionString = nil;

}

- (void)viewDidLoad {
  [super viewDidLoad];

  /* We have our address */
  NSString *oreillyAddress =
    @"1005 Gravenstein Highway North, Sebastopol, CA 95472, USA";

  /* We will later insert the address and the format that we want
   our output in, into this API's URL */
  NSString *geocodingURL =
  @"http://maps.google.com/maps/geo?q=%@&output=%@";

  /* Insert the address and the output format into the URL */
  NSString *finalURL =
  [NSString stringWithFormat:geocodingURL,
   oreillyAddress,
   GOOGLE_OUTPUT_FORMAT_CSV];

  /* Now escape the URL using appropriate percentage marks */
  finalURL =
    [finalURL
      stringByAddingPercentEscapesUsingEncoding:
      NSUTF8StringEncoding];

  /* Create our URL */
  NSURL *urlToCall = [NSURL URLWithString:finalURL];

  /* And a request for the connection using the URL */
  NSURLRequest *request = [NSURLRequest requestWithURL:urlToCall];

  /* We will put all the connection's received data into this
   instance of the NSMutableData class */
  NSMutableData *newMutableData = [[NSMutableData alloc] init];
  self.connectionData = newMutableData;
  [newMutableData release];

  NSURLConnection *newConnection =
  [[NSURLConnection alloc] initWithRequest:request
                                  delegate:self];

  /* Create the connection and start the downloading
   of geocoding results */
  self.myConnection = newConnection;
```

```
    [newConnection release];

}

- (void) viewDidUnload{
  [super viewDidUnload];

  [self.myConnection cancel];

  self.myConnection = nil;
  self.connectionData = nil;
}

- (BOOL)shouldAutorotateToInterfaceOrientation:
  (UIInterfaceOrientation)interfaceOrientation {
  /* Support all orientations */
  return YES;
}

- (void)dealloc {

  [myConnection cancel];
  [myConnection release];
  [connectionData release];

  [super dealloc];
}

@end
```

The output will get printed to the console, as shown in Figure 4-7.

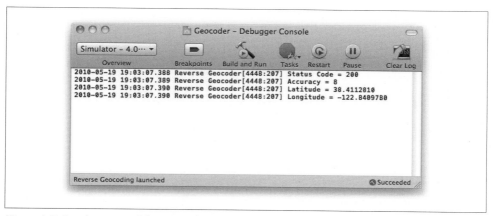

Figure 4-7. Results returned from Google's Geocoder API

See Also

Recipe 4.8

CHAPTER 5

Implementing Gesture Recognizers

5.0 Introduction

Prior to iPhone OS 3.2, developers had to write their own code to detect various touch events in their applications. For instance, the default iPhone Photo application allows the user to zoom into and out of a photo using two fingers while "pinching" the photo in and out. The mechanism to detect these gestures was never a concrete and reusable class, and developers had to create their own gesture recognizers. With the introduction of iOS 3.2, some of the most common gesture event detection code is encapsulated into reusable classes built into the SDK. These classes can be used to detect swipe, pinch, pan, tap, drag, long press, and rotation gestures.

Gesture recognizers must be added to instances of the UIView class. A single view can have more than one gesture recognizer. Once a view catches the gesture, that view will be responsible for passing down the same gesture to other views in the hierarchy if needed.

Some touch events required by an application might be complicated to process and might require the same event to be detectible in other views in the same application. This introduces the requirements for reusable gesture recognizers. There are six gesture recognizers in iOS SDK 3.2 and later:

- Swipe
- Rotation
- Pinch
- Pan
- Long press
- Tap

The basic framework for handling a gesture through a built-in gesture recognizer is as follows:

1. Create an object of the right data type for the gesture recognizer you want.

2. Add this object as a gesture recognizer to the view that will receive the gesture.

3. Write a method that is called when the gesture occurs and that takes the action you want.

The method associated as the target method of any gesture recognizer must follow these rules:

- It must return void.

- It must either accept no parameters, or accept a single parameter of type UIGestureRecognizer in which the system will pass the gesture recognizer that calls this method.

Here are two examples:

```
- (void) tapRecognizer:(UITapGestureRecognizer*)paramSender{
  /* */
}

- (void) tapRecognizer{
  /* */
}
```

Gesture recognizers are divided into two categories: *discrete* and *continuous*. Discrete gesture recognizers detect their gesture events and, once detected, call a method in their respective owners. Continuous gesture recognizers keep their owner objects informed of the events as they happen, and will call the method in their target object repeatedly as the event happens and until it ends.

For instance, a double-tap event is discrete. Even though it consists of two taps, the system recognizes that the taps occurred close enough together to be treated as a single event. The double-tap gesture recognizer calls the method in its target object once the double-tap event is detected.

An example of a continuous gesture recognizer is rotation. This gesture starts as soon as the user starts the rotation and only finishes when the user lifts his fingers off the screen. The method provided to the rotation gesture recognizer class gets called at short intervals until the event is finished.

Gesture recognizers can be added to any instance of the UIView class using the addGestureRecognizer: method of the view, and when needed, they can be removed from the view using the removeGestureRecognizer: method.

The UIGestureRecognizer class has a property named state. The state property represents the different states the gesture recognizer can have throughout the recognition process. Discrete and continuous gesture recognizers go through different sets of states.

Discrete gesture recognizers can pass through the following states:

1. UIGestureRecognizerStatePossible

2. UIGestureRecognizerStateRecognized

3. `UIGestureRecognizerStateFailed`

Depending on the situation, a discrete gesture recognizer might send the `UIGestureRe cognizerStateRecognized` state to its target, or it might send the `UIGestureRecognizer StateFailed` state if an error occurs during the recognition process.

Continuous gesture recognizers take a different path in the states they send to their targets:

1. `UIGestureRecognizerStatePossible`

2. `UIGestureRecognizerStateBegan`

3. `UIGestureRecognizerStateChanged`

4. `UIGestureRecognizerStateEnded`

5. `UIGestureRecognizerStateFailed`

Again, if the continuous gesture recognizer stumbles upon a situation that cannot be fixed internally, it will end with the `UIGestureRecognizerStateFailed` state instead of `UIGestureRecognizerStateEnded`.

5.1 Detecting Swipe Gestures

Problem

You want to be able to detect when the user performs a swipe gesture on a view—for instance, swiping a picture out of the window.

Solution

Instantiate an object of type `UISwipeGestureRecognizer` and add it to an instance of `UIView`:

```
- (void)viewDidLoad {
  [super viewDidLoad];

  /* Instantiate our object */
  UISwipeGestureRecognizer *newSwipeRecognizer =
  [[UISwipeGestureRecognizer alloc]
   initWithTarget:self
   action:@selector(handleSwipes:)];

  self.swipeGestureRecognizer = newSwipeRecognizer;

  [newSwipeRecognizer release];

  /* Swipes that are performed from right to
   left are to be detected */
  self.swipeGestureRecognizer.direction =
  UISwipeGestureRecognizerDirectionLeft;
```

```
/* Just one finger needed */
self.swipeGestureRecognizer.numberOfTouchesRequired = 1;

/* Add it to the view */
[self.view addGestureRecognizer:self.swipeGestureRecognizer];

}

- (void) viewDidUnload{
  [super viewDidUnload];

  self.swipeGestureRecognizer = nil;

}
```

A gesture recognizer could be created as a standalone object, but here, because we are using it just for one view, we have created it as a property of the view controller that will receive the gesture (self.swipeGestureRecognizer). This recipe's Discussion shows the handleSwipes: method used in this code as the target for the swipe gesture recognizer.

Discussion

The swipe gesture is one of the most straightforward gestures that can be recognized using built-in iOS SDK gesture recognizers. It is a simple movement of one or more fingers on a view from one direction to another. The UISwipeGestureRecognizer, like other gesture recognizers, inherits from the UIGestureRecognizer class and adds various functionalities to this class, such as properties that allow us to specify the direction in which the swipe gestures have to be performed in order to be detected, or how many fingers the user has to hold on the screen to be able to perform a swipe gesture. Please bear in mind that swipe gestures are discrete gestures.

The handleSwipes: method that we used for our gesture recognizer instance can be implemented in this way:

```
- (void) handleSwipes:(UISwipeGestureRecognizer*)paramSender{

  if (paramSender.direction & UISwipeGestureRecognizerDirectionDown){
    NSLog(@"Swiped Down.");
  }
  if (paramSender.direction & UISwipeGestureRecognizerDirectionLeft){
    NSLog(@"Swiped Left.");
  }
  if (paramSender.direction & UISwipeGestureRecognizerDirectionRight){
    NSLog(@"Swiped Right.");
  }
  if (paramSender.direction & UISwipeGestureRecognizerDirectionUp){
    NSLog(@"Swiped Up.");
  }

}
```

You can combine more than one direction in the direction property of an instance of the UISwipeGestureRecognizer class by using the bitwise OR operand. In Objective-C, this is done with the pipe (|) character. For instance, to detect diagonal swipes to the bottom left of the screen you can combine the UISwipeGestureRecognizerDirectionLeft and UISwipeGestureRecognizerDirectionDown values using the pipe character when constructing your swipe gesture recognizer. In our example, we are attempting to detect only swipes from the right side to the left.

Although swipe gestures are usually performed with one finger, the number of fingers required for the swipe gesture to be recognized can also be specified using the numberOfTouchesRequired property of the UISwipeGestureRecognizer class.

5.2 Reacting to Rotation Gestures

Problem

You want to detect when a user is attempting to rotate an element on the screen using her fingers.

Solution

Create an instance of the UIRotationGestureRecognizer class and attach it to your target view:

```
- (void)viewDidLoad {
[super viewDidLoad];

/* Create the gesture recognizer */
UIRotationGestureRecognizer *rotationRecognizer =
[[UIRotationGestureRecognizer alloc]
 initWithTarget:self
 action:@selector(handleRotations:)];

self.rotationGestureRecognizer = rotationRecognizer;

[rotationRecognizer release];

/* Add it to our view */
[self.view addGestureRecognizer:self.rotationGestureRecognizer];

}

- (void) viewDidUnload{
[super viewDidUnload];

self.helloWorldLabel = nil;
self.rotationGestureRecognizer = nil;
}
```

Discussion

The UIRotationGestureRecognizer, as its name implies, is the perfect candidate among gesture recognizers to detect rotation gestures and to help you build more intuitive graphical user interfaces. For instance, when the user encounters an image on the screen in your application in full-screen mode, it is quite intuitive for her to attempt to correct the orientation by rotating the image.

The UIRotationGestureRecognizer class implements a property named rotation that specifies the total amount and direction of rotation requested by the user's gesture, in radians. The rotation is determined from the fingers' initial position (UIGestureRecog nizerStateBegan) and final position (UIGestureRecognizerStateEnded).

To rotate UI elements that inherit from UIView class, you can pass the rotation property of the rotation gesture recognizer to the CGAffineTransformMakeRotation function to make an affine transform, as shown in the example.

The code in this recipe's Solution passes the current object, in this case a view controller, to the target of the rotation gesture recognizer. The target selector is specified as han dleRotations:, a method we have to implement. But before we do that, let's have a look at the header file of our view controller:

```
#import <UIKit/UIKit.h>

@interface RootViewController : UIViewController {
@protected
  UILabel                    *helloWorldLabel; ❶
  UIRotationGestureRecognizer *rotationGestureRecognizer; ❷
  CGFloat                    rotationAngleInRadians; ❸
}

@property (retain) UIRotationGestureRecognizer *rotationGestureRecognizer;
@property (retain) IBOutlet UILabel         *helloWorldLabel;
@property (assign) CGFloat                  rotationAngleInRadians;

@end
```

❶ helloWorldLabel is a label that we must create on the view of our view controller. Then we will write the code that will rotate this label whenever the user attempts to perform rotation gestures on the view that owns this label, in this case the view of the view controller.

❷ rotationGestureRecognizer is the instance of the rotation gesture recognizer that we will later allocate and initialize.

❸ rotationAngleInRadians is the value we will query as the exact rotation angle of our label. Initially we will set this to zero. Since the rotation angles reported by a rotation gesture recognizer are reset every time the rotation gesture is started again, we can keep the value of the rotation gesture recognizer whenever it goes into the UIGestureRecognizerStateEnded state. The next time the gesture is started, we will add the previous value to the new value to get an overall rotation angle.

Before we go any further, let's go into the XIB file of our view controller, create the corresponding Hello World label on its view, and link it to the IBOutlet that we declared in the header file of our view controller:

1. Open the corresponding XIB file in Interface Builder. You can double-click on the XIB file or press the Command + down arrow keys on the XIB file in Xcode.

2. In Interface Builder, use Tools→Library to bring up the Library pane.

3. Select the Label component from the Library pane, as shown in Figure 5-1, and drag and drop it into the view of the view controller whose XIB file we have opened, as shown in Figure 5-2.

4. Align the label in the center in the Attributes Inspector, which you can open using Tools→Attributes Inspector in Interface Builder (Figure 5-3).

5. Select the label that you dropped on the view. Now open the Connections Inspector in Interface Builder using Tools→Connections Inspector. Grab the New Reference Outlet of the label in the Connections Inspector and drop it into the instance of the label on the view, as shown in Figure 5-4.

Any component inheriting from UIView could be selected in the Library tab as our target component. All UIView descendants have the transform property that we are going to use in this section. For the sake of simplicity, I have chosen a label.

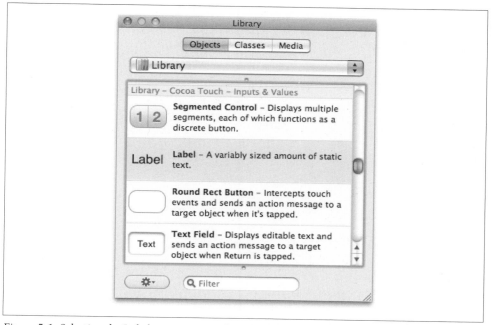

Figure 5-1. Selecting the Label component in the Library pane

Figure 5-2. Center-aligned label dropped on the view of a view controller with a navigation bar on top

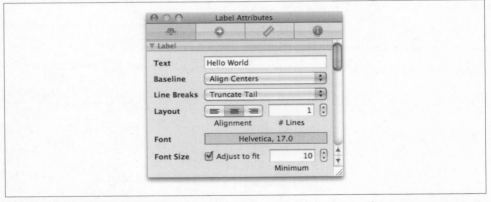

Figure 5-3. Center-aligning the label using the Layout option in the Attributes Inspector pane in Interface Builder

The size of the label does not matter much. Even the position of the label isn't that important, as we will only attempt to rotate the label around its center, no matter where on our view the label is positioned. The only important thing to remember is that in universal applications, the position of a label on a view controller used in different targets (devices) must be calculated dynamically using the size of its parent view. Otherwise, on different devices such as the iPad or the iPhone, it might appear in different places on the screen.

Using the Layout option, we will center-align the contents of our label. The rotation transformation that we will apply to this label rotates the label around its center, and left-aligned or right-aligned labels whose actual frame is bigger than the minimum frame that is required to hold their contents without truncation will appear to be rotating in an unnatural way and not on the center. If you are curious, just go ahead and left- or right-align the contents of the label and see what happens.

After connecting the New Referencing Outlet of the label to the File's Owner item as shown in Figure 5-4, you will be presented with a dialog asking which outlet you want to connect the component to, as shown in Figure 5-5.

Choose the helloWorldLabel outlet.

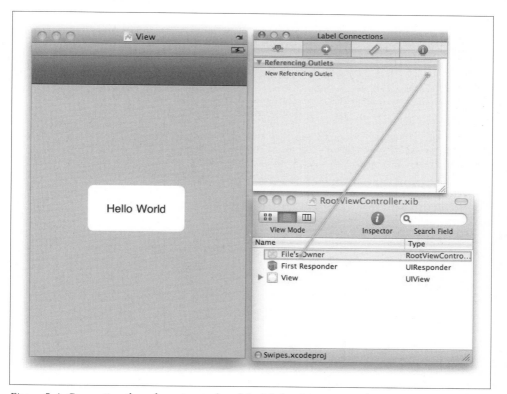

Figure 5-4. Connecting the referencing outlet of the label to its corresponding UILabel object in our view controller

Figure 5-5. Available outlets in our view controller's header file

Now we will implement our view controller in this way:

```objc
#import "RootViewController.h"

@implementation RootViewController

@synthesize rotationGestureRecognizer;
@synthesize helloWorldLabel;
@synthesize rotationAngleInRadians;

- (void) handleRotations:(UIRotationGestureRecognizer *)paramSender{

  if (self.helloWorldLabel == nil){
    return;
  }

  /* Take the previous rotation and add the current rotation to it */
  self.helloWorldLabel.transform =
  CGAffineTransformMakeRotation(self.rotationAngleInRadians +
                                paramSender.rotation);

  /* At the end of the rotation, keep the angle for later use */
  if (paramSender.state == UIGestureRecognizerStateEnded){
    self.rotationAngleInRadians += paramSender.rotation;
  }

}

- (void)viewDidLoad {
  [super viewDidLoad];

  /* Create the gesture recognizer */
  UIRotationGestureRecognizer *rotationRecognizer =
  [[UIRotationGestureRecognizer alloc]
   initWithTarget:self
   action:@selector(handleRotations:)];

  self.rotationGestureRecognizer = rotationRecognizer;

  [rotationRecognizer release];

  /* Add it to our view */
  [self.view addGestureRecognizer:self.rotationGestureRecognizer];

}

- (void) viewDidUnload{
  [super viewDidUnload];
```

```
    self.helloWorldLabel = nil;
    self.rotationGestureRecognizer = nil;
}

- (BOOL)shouldAutorotateToInterfaceOrientation
  :(UIInterfaceOrientation)interfaceOrientation {
    return YES;
}

- (void)dealloc {

    [rotationGestureRecognizer release];
    [helloWorldLabel release];

    [super dealloc];
}

@end
```

The way a rotation gesture recognizer sends us the rotation angles is very interesting. This gesture recognizer is continuous, which means it starts finding the angles as soon as the user begins her rotation gesture, and sends updates to the handler method at frequent intervals until the user finishes the gesture. Each message treats the starting angle as zero and reports the difference between the messages' starting point (which is the angle where the previous message left off) and its ending point. Thus, the complete effect of the gesture can be discovered only by summing the angles reported by the different events. Clockwise movement produces a positive angular value, whereas counterclockwise movement produces a negative value.

If you are using iPhone Simulator instead of a real device, you can still simulate the rotation gesture by holding down the Option key in the simulator. You will see two circles appearing on the simulator at the same distance from the center of the screen, representing two fingers. If you want to shift these fingers from the center to another location while holding down the Alt key, press the Shift key and point to somewhere else on the screen. Where you leave off your pointer will become the new center for these two fingers.

Now we will simply assign this angle to the rotation angle of our label. But can you imagine what will happen once the rotation is finished and another one starts? The second rotation gesture's angle will replace that of the first rotation in the rotation value reported to our handler. For this reason, whenever a rotation gesture is finished, we must keep the current rotation of our label. The value in each rotation gesture's angle must be added in turn, and we must assign the result to the label's rotation transformation as we saw before.

As we saw earlier, we used the CGAffineTransformMakeRotation function to create an affine transformation. Functions in the iOS SDK that start with "CG" refer to the Core

Graphics framework. For programs that use Core Graphics to compile and link successfully, you must make sure the Core Graphics framework is added to the list of frameworks. To do this, follow these steps:

1. Find and select your target output in Xcode, as shown in Figure 5-6.
2. Choose File→Get Info. Now you will see the Target Information page.
3. At the top of the Target Information page, choose the General tab.
4. At the bottom of the page, make sure Core Graphics is added to the Linked Libraries section. If it is not, click the + button and choose Core Graphics from the list. Finally, click the Add button, as shown in Figure 5-7.

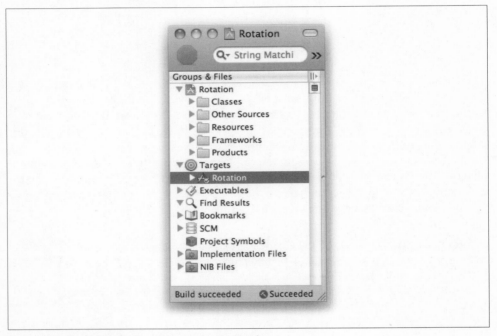

Figure 5-6. Finding and selecting our target in Xcode

 You must add the required frameworks for each target individually even if you have duplicated one target to create other targets. Adding a framework to the *Frameworks* folder directly will add the frameworks to the current target only.

Now that we are sure Core Graphics is added to our target, we can compile and run our application.

Figure 5-7. Adding the Core Graphics framework to our target

See Also

Recipe 5.6

5.3 Detecting Panning and Dragging Gestures

Problem

You want the users of your application to be able to move GUI elements around using their fingers.

Solution

Use the `UIPanGestureRecognizer` class:

```
- (void)viewDidLoad {
[super viewDidLoad];

/* Let's first create a label */
CGRect labelFrame = CGRectMake(0.0f,     /* X */
                               0.0f,     /* Y */
                               150.0f,   /* Width */
                               100.0f);  /* Height */

UILabel *newLabel = [[UILabel alloc] initWithFrame:labelFrame];
```

```
self.helloWorldLabel = newLabel;
[newLabel release];

self.helloWorldLabel.text = @"Hello World";
/* Contrasting colors to catch more attention */
self.helloWorldLabel.backgroundColor = [UIColor blackColor];
self.helloWorldLabel.textColor = [UIColor whiteColor];
/* Center align the text to make it look better */
self.helloWorldLabel.textAlignment = UITextAlignmentCenter;
/* Make sure to enable user interaction; otherwise, tap events
 won't be caught on this label */
self.helloWorldLabel.userInteractionEnabled = YES;
/* And now make sure this label gets displayed on our view */
[self.view addSubview:self.helloWorldLabel];


/* Create the Pan Gesture Recognizer */
UIPanGestureRecognizer *newPanGestureRecognizer =
[[UIPanGestureRecognizer alloc]
 initWithTarget:self
 action:@selector(handlePanGestures:)];

self.panGestureRecognizer = newPanGestureRecognizer;

[newPanGestureRecognizer release];

/* At least and at most we need only one finger to activate
 the pan gesture recognizer */
self.panGestureRecognizer.minimumNumberOfTouches = 1;
self.panGestureRecognizer.maximumNumberOfTouches = 1;

/* Add it to our view */
[self.helloWorldLabel addGestureRecognizer:self.panGestureRecognizer];

/* Get rid of the navigation bar for now as we don't need it */
self.navigationController.navigationBarHidden = YES;

}

- (void) viewDidUnload{

[super viewDidUnload];

self.panGestureRecognizer = nil;
self.helloWorldLabel = nil;

}
```

The pan gesture recognizer will call the handlePanGestures: method as its target method. This method is described in this recipe's Discussion.

Discussion

The UIPanGestureRecognizer, as its name implies, can detect *pan gestures*. Pan gestures are continuous movements of fingers on the screen; recall that swipe gestures were discrete gestures. This means the method set as the target method of a pan gesture recognizer gets called repeatedly from the beginning to the end of the recognition process. The pan gesture recognizer will go through the following states while recognizing the pan gesture:

1. `UIGestureRecognizerStateBegan`
2. `UIGestureRecognizerStateChanged`
3. `UIGestureRecognizerStateEnded`

We can implement our gesture recognizer target method as follows. The code will move the center of the label along with the user's finger continuously as `UIGestureRecognizerStateChanged` events are reported.

```
- (void) handlePanGestures:(UIPanGestureRecognizer*)paramSender{

  if (paramSender.state != UIGestureRecognizerStateEnded &&
      paramSender.state != UIGestureRecognizerStateFailed){

    CGPoint location =
      [paramSender locationInView:paramSender.view.superview];
    paramSender.view.center = location;

  }

}
```

 To be able to move the label on the view of our view controller, we need the position of the finger on the view, not the label. For this reason, we are calling the `locationInView:` method of the pan gesture recognizer and passing the superview of our label as the target view.

Use the `locationInView:` method of the pan gesture recognizer to find the point of the current panning finger(s). To detect the location of multiple fingers, use the `locationOfTouch:inView:` method. Using the `minimumNumberOfTouches` and `maximumNumberOfTouches` properties of the `UIPanGestureRecognizer`, you can detect more than one panning touch at a time. In our example, for the sake of simplicity, we are trying to detect only one finger.

 During the `UIGestureRecognizerStateEnded` state, the reported x and y values might not be a number; in other words, they could be equal to NAN. That is why we need to avoid using the reported values during this particular state.

5.4 Detecting Long Press Gestures

Problem

You want to be able to detect when the user taps and holds his finger on a view for a certain period of time.

Solution

Create an instance of the UILongPressGestureRecognizer class and add it to the view that has to detect long tap gestures. The *.h* file of our view controller is also defined in this way:

```
#import <UIKit/UIKit.h>

@interface RootViewController : UIViewController {
@public
  UILongPressGestureRecognizer  *longPressGestureRecognizer;
  UIButton                      *dummyButton;
}

@property (nonatomic, retain)
  UILongPressGestureRecognizer  *longPressGestureRecognizer;

@property (nonatomic, retain) IBOutlet UIButton *dummyButton;

@end
```

Before continuing any further, please open the XIB file associated with this view controller and place a simple button of type UIButton on the view. Now associate this button with the dummyButton outlet we defined in the *.h* file of our view controller. Here is the viewDidLoad instance method of our view controller that uses the long press gesture recognizer that we defined in the *.m* file:

```
- (void)viewDidLoad {
  [super viewDidLoad];

  /* First create the gesture recognizer */
  UILongPressGestureRecognizer *newLongPressGestureRecognizer =
  [[UILongPressGestureRecognizer alloc]
   initWithTarget:self
   action:@selector(handleLongPressGestures:)];

  self.longPressGestureRecognizer = newLongPressGestureRecognizer;
  [newLongPressGestureRecognizer release];

  /* The number of fingers that must be present on the screen */
  self.longPressGestureRecognizer.numberOfTouchesRequired = 2;

  /* Maximum 100 pixels of movement allowed before the gesture
   is recognized */
  self.longPressGestureRecognizer.allowableMovement = 100.0f;
```

```
    /* The user must press 2 fingers (numberOfTouchesRequired) for
      at least 1 second for the gesture to be recognized */
    self.longPressGestureRecognizer.minimumPressDuration = 1.0;

    /* Add this gesture recognizer to our view */
    [self.view addGestureRecognizer:self.longPressGestureRecognizer];

    /* Get rid of the navigation bar for now as we don't need it */
    self.navigationController.navigationBarHidden = YES;

}

- (void) viewDidUnload{
    [super viewDidUnload];

    self.longPressGestureRecognizer = nil;
    self.dummyButton = nil;

}
```

 Our code runs on a view controller with a property named longPress
GestureRecognizer of type UILongPressGestureRecognizer. For more information, refer to this recipe's Discussion.

Discussion

The iOS SDK comes with a long tap gesture recognizer class named UILongTap
GestureRecognizer. A long tap gesture is triggered when the user presses one or more
fingers (configurable by the programmer) on a UIView and holds the finger(s) for a
specific amount of time. Furthermore, you can narrow the detection of gestures down
to only those long tap gestures that are performed after a certain number of fingers are
tapped on a view for a certain number of times and are then kept on the view for a
specified number of seconds. Bear in mind that long taps are continuous events.

Four important properties can change the way the long tap gesture recognizer performs.
These are:

numberOfTapsRequired
    This is the number of taps the user has to perform on the target view, before the
    gesture can be triggered. Bear in mind that a tap is *not* a mere finger positioned on
    a screen. A tap is the movement of putting a finger down on the screen and lifting
    the finger off. The default value of this property is 0.

numberOfTouchesRequired
    This property specifies the number of fingers that are required to be touching the
    screen before the gesture can be recognized. You must specify the same number of
    fingers to detect the taps, if the numberOfTapsRequired property is set to a value
    larger than 0.

allowableMovement

    This is the maximum number of pixels that the fingers on the screen can be moved before the gesture recognition is aborted.

minimumPressDuration

    This property dictates how long, measured in seconds, the user must press her finger(s) on the screen before the gesture event can be detected.

In our example, these properties are set as follows:

- numberOfTapsRequired: default (we are not changing this value)
- numberOfTouchesRequired: 2
- allowableMovement: 100
- minimumPressDuration: 1

With these values, the long tap gesture will be recognized only if the user presses on the screen and holds both fingers for one second (minimumPressDuration) without moving her fingers more than 100 pixels around (allowableMovement).

Now when the gesture is recognized, it will call our handleLongPressGestures: method, which we can implement in this way:

```
- (void) handleLongPressGestures:(UILongPressGestureRecognizer*)paramSender{

    /* Here we want to find the midpoint of the two fingers
    that caused the long press gesture to be recognized. We configured
    this number using the numberOfTouchesRequired property of the
    UILongPressGestureRecognizer that we instantiated in the
    viewDidLoad instance method of this View Controller. If we
    find that another long press gesture recognizer is using this
    method as its target, we will ignore it */

    if ([paramSender isEqual:self.longPressGestureRecognizer] == YES){

      if (paramSender.numberOfTouchesRequired == 2){

        CGPoint touchPoint1 =
        [paramSender locationOfTouch:0
                            inView:paramSender.view];

        CGPoint touchPoint2 =
        [paramSender locationOfTouch:1
                            inView:paramSender.view];

        CGFloat midPointX = (touchPoint1.x + touchPoint2.x) / 2.0f;
        CGFloat midPointY = (touchPoint1.y + touchPoint2.y) / 2.0f;

        CGPoint midPoint = CGPointMake(midPointX, midPointY);

        self.dummyButton.center = midPoint;

      } else {
```

```
        /* This is a long press gesture recognizer with more
           or less than 2 fingers */

    }

  }

}
```

 One of the applications in iOS that uses long tap gesture recognizers is the Maps application. In this application, when you are looking at different locations, press your finger on a specific location and hold it for a while without lifting it off the screen. This will drop a pin on that specific location.

5.5 Responding to Tap Gestures

Problem

You want to be able to detect when users tap on a view.

Solution

Create an instance of the UITapGestureRecognizer class and add it to the target view, using the addGestureRecognizer: instance method of the UIView class. Let's have a look at the definition of our view controller (the *.h* file):

```
#import <UIKit/UIKit.h>

@interface RootViewController : UIViewController {
@public
  UITapGestureRecognizer  *tapGestureRecognizer;
}

@property (nonatomic, retain)
UITapGestureRecognizer  *tapGestureRecognizer;

@end
```

The implementation of the viewDidLoad instance method of our view controller is as follows:

```
- (void)viewDidLoad {
  [super viewDidLoad];

  /* Create the Tap Gesture Recognizer */
  UITapGestureRecognizer *newTapGestureRecognizer =
  [[UITapGestureRecognizer alloc]
   initWithTarget:self
   action:@selector(handleTaps:)];
```

```
self.tapGestureRecognizer = newTapGestureRecognizer;

[newTapGestureRecognizer release];

/* The number of fingers that must be on the screen */
self.tapGestureRecognizer.numberOfTouchesRequired = 2;

/* The total number of taps to be performed before the
 gesture is recognized */
self.tapGestureRecognizer.numberOfTapsRequired = 3;

/* Add this gesture recognizer to our view */
[self.view addGestureRecognizer:self.tapGestureRecognizer];

/* Get rid of the navigation bar for now as we don't need it */
self.navigationController.navigationBarHidden = YES;

}

- (void) viewDidUnload{
[super viewDidUnload];

self.tapGestureRecognizer = nil;
}
```

Discussion

The tap gesture recognizer is the best candidate among gesture recognizers to detect plain tap gestures. A tap event is the event triggered by the user touching and lifting his finger(s) off the screen. A tap gesture is a discrete gesture.

The locationInView: method of the UITapGestureRecognizer class can be used to detect the location of the tap event. If the tap gesture requires more than one touch, the locationOfTouch:inView: method of the UITapGestureRecognizer class can be called to determine individual touch points. In our code, we have set the numberOfTouches Required property of our tap gesture recognizer to 2. With this value set, our gesture recognizer will require two fingers to be on the screen on each tap event. The number of taps that are required for the gesture recognizer to recognize this gesture is set to 3 using the numberOfTapsRequired property. We have provided the handleTaps: method as the target method of our tap gesture recognizer:

```
- (void) handleTaps:(UITapGestureRecognizer*)paramSender{

NSUInteger touchCounter = 0;

for (touchCounter = 0;
     touchCounter < paramSender.numberOfTouchesRequired;
     touchCounter++){

  CGPoint touchPoint =
  [paramSender locationOfTouch:touchCounter
                        inView:paramSender.view];
```

```
NSLog(@"Touch #%lu: %@",
        (unsigned long)touchCounter+1,
        NSStringFromCGPoint(touchPoint));

    }

}
```

In this code, we are going through the number of touches that our tap gesture recognizer was asked to look for. Based on that number, we are finding the location of each tap. Figure 5-8 shows the result of tapping while this gesture recognizer is in effect.

 If you are using the simulator, you can simulate two touches at the same time by holding down the Option key and moving your mouse on the simulator's screen. You will now have two concentric touch points on the screen.

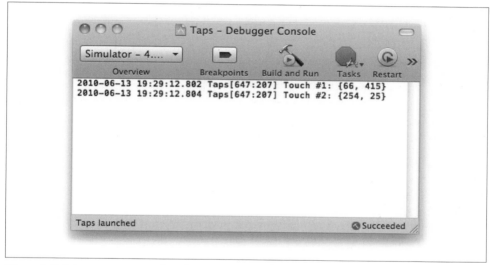

Figure 5-8. A tap gesture recognizer detecting two touch points on a view

One function that is worth noting is the NSStringFromCGPoint method, which, as its name implies, can convert a CGPoint structure to NSString. We use this function to convert the CGPoint of each touch on the screen to an NSString so that we can log it to the console window using NSLog. You can bring up the console window using Run→Console.

5.6 Responding to Pinch Gestures

Problem

You want your users to be able to perform a pinch gesture on a view.

Solution

Create an instance of the UIPinchGestureRecognizer class and add it to your target view, using the addGestureRecognizer: instance method of the UIView class:

```
- (void)viewDidLoad {
[super viewDidLoad];

CGRect labelRect = CGRectMake(0.0f,      /* X */
                             0.0f,       /* Y */
                             200.0f,     /* Width */
                             200.0f);    /* Height */

UILabel *newLabel =
[[UILabel alloc] initWithFrame:labelRect];

self.myBlackLabel = newLabel;

[newLabel release];

/* Put the label in the center of the view */
self.myBlackLabel.center = self.view.center;

/* The black color, of course */
self.myBlackLabel.backgroundColor = [UIColor blackColor];

/* Without this line, our pinch gesture recognizer
 will not work */
self.myBlackLabel.userInteractionEnabled = YES;

/* Add this label to our view */
[self.view addSubview:self.myBlackLabel];

/* Create the Pinch Gesture Recognizer */
self.pinchGestureRecognizer =
  [[UIPinchGestureRecognizer alloc]
    initWithTarget:self
           action:@selector(handlePinches:)];

/* Add this gesture recognizer to our view */
[self.myBlackLabel
 addGestureRecognizer:self.pinchGestureRecognizer];

/* Get rid of the navigation bar for now as we don't need it */
self.navigationController.navigationBarHidden = YES;

}
```

```
- (void) viewDidUnload{
  [super viewDidUnload];

  self.myBlackLabel = nil;
  self.pinchGestureRecognizer = nil;

}
```

The *.h* file of our view controller is defined in this way:

```
#import <UIKit/UIKit.h>

@interface RootViewController : UIViewController {
@public
  UIPinchGestureRecognizer   *pinchGestureRecognizer;
  UILabel *myBlackLabel;
  CGFloat currentScale;
}

@property (nonatomic, retain)
UIPinchGestureRecognizer   *pinchGestureRecognizer;

@property (nonatomic, retain) UILabel *myBlackLabel;
@property (nonatomic, assign) CGFloat currentScale;

@end
```

Discussion

Pinching allows users to scale GUI elements up and down easily. For instance, the Safari web browser on iOS allows users to pinch on a web page in order to zoom into the contents being displayed. Pinching works in two ways: scaling up and scaling down. It is a continuous gesture that must always be performed using two fingers on the screen.

The state of this gesture recognizer changes in this order:

1. UIGestureRecognizerStateBegan
2. UIGestureRecognizerStateChanged
3. UIGestureRecognizerStateEnded

Once the pinch gesture is recognized, the action method in the target object will be called and will continue to be called until the pinch gesture is ended. Inside the action method you can access two very important properties of the pinch gesture recognizer: scale and velocity. scale is the factor by which you should scale the x- and y-axes of a GUI element to reflect the size of the user's gesture. velocity is the velocity of the pinch in pixels per second. The velocity is a negative value if the touch points are getting closer to each other and a positive value if they are getting farther away from each other.

The value of the `scale` property can be provided to the `CGAffineTransformMakeScale` Core Graphics function in order to retrieve an affine transformation. This affine transformation can be applied to the transform property of any instance of the `UIView` class in order to change its transformation. We are using this function in this way:

```
- (void) handlePinches:(UIPinchGestureRecognizer*)paramSender{

    if (paramSender.state == UIGestureRecognizerStateEnded){
      self.currentScale = paramSender.scale;
    } else if (paramSender.state == UIGestureRecognizerStateBegan &&
               self.currentScale != 0.0f){
      paramSender.scale = self.currentScale;
    }

    if (paramSender.scale != NAN &&
        paramSender.scale != 0.0){
      paramSender.view.transform =
        CGAffineTransformMakeScale(paramSender.scale,
                                   paramSender.scale);
    }

}
```

Since the `scale` property of a pinch gesture recognizer is reset every time a new pinch gesture is recognized, we are storing the last value of this property in an instance property of our view controller called `currentScale`. The next time a new gesture is recognized, we start the scale factor from the previously reported scale factor, as demonstrated in the code.

CHAPTER 6

Networking and XML

6.0 Introduction

iOS applications can take advantage of iOS SDK networking and XML parsing classes in order to create richer and connected applications. Whether you are trying to download a new resource from the Internet into your application's bundle or make a call to a web service, you can find various classes and functionalities in the SDK that will help you make the process much easier than you thought. The iOS SDK provides the NSXMLParser class to parse XML and the NSURLConnection class (among others) for web services.

In this chapter, we will go through the implementation and usage of these classes, and then mix them to parse remote XML files. We will also discuss caching mechanisms and how to implement a custom mechanism for specific needs.

6.1 Opening and Reading a Local XML File

Problem

You want to read the contents of an XML file that has been read into memory, perhaps into an instance of the NSData class.

Solution

Create an instance of NSXMLParser and provide it with a delegate that conforms to the NSXMLParser protocol:

```
NSXMLParser *parser = [[NSXMLParser alloc] initWithData:nil];
[parser setDelegate:self];
[parser parse];
```

In this code, we are passing nil to the initWithData: method of the NSXMLParser. In this recipe's Discussion, you will see how to load the contents of an XML file from our application bundle into an instance of NSData and pass it to the NSXMLParser to parse.

Discussion

Instances of the `NSXMLParser` class are able to parse XML documents. `NSXMLParser` notifies its `delegate` of various pieces of data found in an XML file as it parses it. It produces data from each element as simple strings, giving you great flexibility to form your application's data model (objects) from the raw XML data.

Before we can continue with this receipt, we need to create an XML file in our project. To do so, follow these steps:

1. Locate a target directory in Xcode where you want your XML file to be created. The *Resources* folder is usually a good candidate, as shown in Figure 6-1.

2. In Xcode, choose File→New File, as shown in Figure 6-2.

3. In the New File dialog, choose Other and then Empty File, and click Next, as shown in Figure 6-3.

4. Give your file a name and make sure your project is selected as the default target. Once done, click Finish, as shown in Figure 6-4.

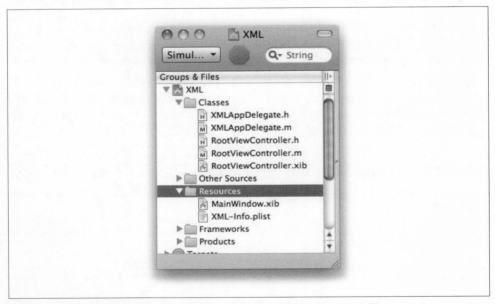

Figure 6-1. Choosing the Resources folder as the target folder of our XML file

In Xcode, the currently selected folder is the default target for files to be created.

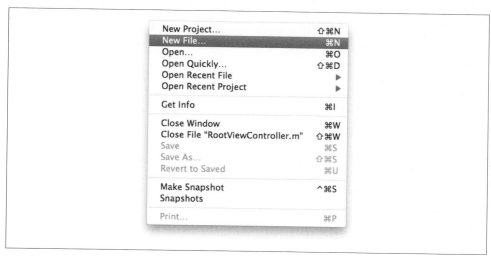

Figure 6-2. Selecting File→New File in Xcode in order to create a new XML file

Figure 6-3. Choosing Empty File in the Other category before assigning an XML extension to a filename

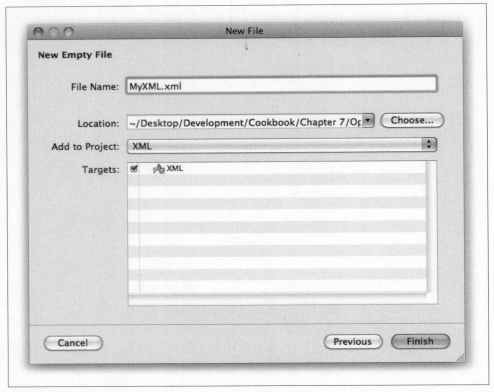

Figure 6-4. Assigning a name and extension to a file and making sure the current project is the selected target project

The *Resources* folder is a virtual folder, meaning that any files created under it will actually be created under the project's root folder. Don't be surprised to see that folder in the New File dialog.

Another way to create an XML file is to choose Resources as the template group in the New File dialog under the Mac OS X category, and then to choose Property List on the righthand side. This will create a traditional *.plist* file. However, to be able to create the XML from scratch and structure it according to your needs, I recommend that you use the Empty File item under the Other category.

We will use the name that we assign to this XML file later, in order to load its contents into an instance of NSData; to be able to follow the instructions in this recipe, keep this file's name as *MyXML.xml*.

Now we will use Xcode to enter a very simple XML document into the empty XML file we just created. Here is the sample content I entered:

```
<?xml version="1.0" encoding="UTF-8"?>
  <parent>
    <child name="James">
      <age>10</age>
      <sex>male</sex>
    </child>
  </parent>
</xml>
```

Now we will proceed to loading the contents of this XML file into an instance of NSData. We must then pass this data to an NSXMLParser and start the parsing process by implementing the required delegate methods defined under the NSXMLParserDelegate protocol. We will start with the .h file of a view controller.

> You do not have to parse XML files in a view controller. You can do that in any other object, as long as you are sure that object at some point gets allocated and initialized.

```
#import <UIKit/UIKit.h>

@interface RootViewController : UIViewController <NSXMLParserDelegate> {
@public
  NSXMLParser    *myXMLParser;
}

@property (nonatomic, retain)  NSXMLParser    *myXMLParser;

@end
```

As you can see in the .h file, we have defined a property named myXMLParser of type NSXMLParser. In our implementation file, we must allocate and initialize this property, choose it as the delegate of the current class, and start the parsing process:

```
- (void)viewDidLoad {
  [super viewDidLoad];

  NSBundle *mainBundle = [NSBundle mainBundle];

  /* Let's get the path of the XML file in the app bundle */
  NSString *xmlFilePath = [mainBundle pathForResource:@"MyXML"
                                               ofType:@"xml"];

  if ([xmlFilePath length] == 0){
    /* The file could not be found in the resources folder.
     Do something appropriate here */
    return;
  }

  /* Let's see if the XML file exists... */
  NSFileManager *fileManager = [[NSFileManager alloc] init];

  if ([fileManager fileExistsAtPath:xmlFilePath] == NO){
```

```
       /* The XML file doesn't exist, we double checked this just
        to make sure we don't leave any stones unturned. You can now
        throw an error/exception here or do other appropriate processing */
       NSLog(@"The file doesn't exist.");
       [fileManager release];
       return;
     }

     [fileManager release];

     /* Load the contents of the XML file into an NSData */
     NSData *xmlData = [NSData dataWithContentsOfFile:xmlFilePath];

     NSXMLParser *xmlParser = [[NSXMLParser alloc] initWithData:xmlData];
     self.myXMLParser = xmlParser;
     [xmlParser release];

     [self.myXMLParser setDelegate:self];

     if ([self.myXMLParser parse] == YES){
       /* Successfully started to parse */
     } else {
       /* Failed to parse. Do something with this error */
       NSError *parsingError = [self.myXMLParser parserError];
       NSLog(@"%@", parsingError);
     }

   }

   - (void) viewDidUnload{
     [super viewDidUnload];

     self.myXMLParser = nil;

   }
```

A very useful method defined in the NSXMLParserDelegate protocol will help the delegate object of our XML parser to know when a new element in the XML is being parsed:

```
   - (void)                    parser:(NSXMLParser *)parser
               didStartElement:(NSString *)elementName
                  namespaceURI:(NSString *)namespaceURI
                 qualifiedName:(NSString *)qName
                    attributes:(NSDictionary *)attributeDict{

     NSLog(@"Element Name = %@", elementName);
     NSLog(@"Number of attributes = %lu",
           (unsigned long)[attributeDict count]);

     if ([attributeDict count] > 0){
       NSLog(@"Attributes dictionary = %@", attributeDict);
     }

     NSLog(@"=========================");

   }
```

This method gets called whenever an XML tag gets parsed. The system passes the following as parameters:

parser
> This is the XML parser that has called this method in the delegate.

elementName
> This is the name of the element. For instance, the `<address>...</address>` element's name is "address."

namespaceURI
> This is the namespace of the element, if any.

qName
> This is the qualified name of the element in its own namespace.

attributeDict
> This is a dictionary of attributes of the element. It is `null` if there are no attributes.

We will mainly focus on the element name and the dictionary of attributes in order to read the contents of the XML. If you are parsing XML files that rely on namespaces, you must use the other parameters passed to this method to be able to parse your XML file correctly, but for the sake of simplicity, we are not using namespaces in this recipe.

The code we have written in this method will essentially print out the element name, the number of attributes it has, and the actual contents of the attributes dictionary if there are any attributes. Figure 6-5 shows the result we will get in the console window.

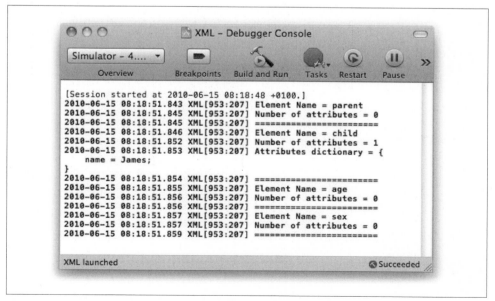

Figure 6-5. Printing out the contents of an XML file while parsing it

Go ahead and have a look at the other methods declared in the `NSXMLParserDelegate` protocol to get a better idea of how you can use an instance of the `NSXMLParser` class to parse XML files.

See Also

Recipe 6.2

6.2 Parsing an XML File into Objects

Problem

You would like to parse the contents of an XML file into Objective-C objects so that you can access them easily.

Solution

Create your object model and then use an instance of the `NSXMLParser` class to parse the contents of the XML. Once parsed, plug the values into your custom object model.

Discussion

An XML element can have these basic properties:

- A reference to the parent element
- Children (an array of element objects)
- An element name
- Element text
- Attributes (a dictionary of attribute objects)

To be able to parse an XML document into an object model (a series of objects representing the contents of an XML file), we must create our objects first. We will do that using two objects: an XML document object and an XML element object. Here is a simple XML element object's header file:

```
#import <Foundation/Foundation.h>

@interface XMLElement : NSObject {
@public
  NSString            *name;
  NSString            *text;
  XMLElement          *parent;
  NSMutableArray      *children;
  NSMutableDictionary *attributes;
}
```

```
@property (nonatomic, retain)  NSString            *name;
@property (nonatomic, retain)  NSString            *text;
@property (nonatomic, assign)  XMLElement          *parent;
@property (nonatomic, copy)  NSMutableArray      *children;
@property (nonatomic, copy)  NSMutableDictionary *attributes;

@end
```

We will now implement our XMLElement class:

```
#import "XMLElement.h"

@implementation XMLElement

@synthesize name;
@synthesize text;
@synthesize parent;
@synthesize children;
@synthesize attributes;

- (id) init {
  self = [super init];

  if (self != nil){
    NSMutableArray *childrenArray =
    [[NSMutableArray alloc] init];

    children = [childrenArray mutableCopy];
    [childrenArray release];

    NSMutableDictionary *newAttributes =
    [[NSMutableDictionary alloc] init];
    attributes = [newAttributes mutableCopy];
    [newAttributes release];
  }

  return(self);
}

- (void) dealloc {

  NSLog(@"Deallocated Element");

  [name release];
  [text release];
  [children release];
  [attributes release];

  [super dealloc];
}

@end
```

Fantastic! Now we have declared and implemented our XML element object. What's next? We need to create an object that defines an XML document rather than just an

XML element. An XML document contains a root XML element and in our example will also retain an `NSString` property containing the path to the XML file that it has to open and parse.

In addition to that, we will also have our XML document object parse the XML files. We will simply pass the full path of a local file to the XML document object and expect it to parse the XML file and tell us when the parsing operation is done. So, here is the declaration of our XML document object (the *.h* file):

```objc
#import <Foundation/Foundation.h>

@class XMLElement;

@interface XMLDocument : NSObject <NSXMLParserDelegate> {
@public
  NSString        *documentPath;
  XMLElement      *rootElement;
@private
  NSXMLParser     *xmlParser;
  XMLElement      *currentElement;
}

@property (nonatomic, retain) NSString        *documentPath;
@property (nonatomic, retain) XMLElement       *rootElement;

/* Private properties */
@property (nonatomic, retain) NSXMLParser      *xmlParser;
@property (nonatomic, retain) XMLElement       *currentElement;

- (BOOL) parseLocalXMLWithPath:(NSString *)paramLocalXMLPath;

@end
```

We want this object to be responsible for performing a single task: parsing a local XML file. That's why we declared the `parseLocalXMLWithPath:` method in the header file. We will implement this method later in the *.m* file.

We are also retaining the path of the XML file that will be given to us through the `parseLocalXMLWithPath:` method in the `documentPath` property (just in case we need it later).

We will also create and hold a retained reference to the root element of the XML file. Every XML document has a root element. Only through that root element can we access any other elements in the XML document. For this reason, we will hold the reference to the root element to allow the programmer to access the contents of the parsed XML file easily. Furthermore, if we use this class in another object, that object can get access to the XML file by referring to this class's `rootElement` property. In other words, if we create an instance named `document` of a class that loads the XML file into its `rootElement` property, the caller that created the `document` object can get access to the XML elements through a reference to `document.rootElement`.

It is very important to bear in mind that the `parse` method of an instance of the `NSXMLParser` class parses the XML file synchronously on the thread that calls it. This means it parses all the XML before returning, and then returns `YES` if the parse was successful and `NO` if there were any errors in the file or other problems were encountered.

Now we can go ahead and implement the `XMLDocument` object (the *.m* file):

```
#import "XMLDocument.h"
#import "XMLElement.h"

@implementation XMLDocument

@synthesize documentPath;
@synthesize rootElement;
@synthesize xmlParser;
@synthesize currentElement;

- (void)         parser:(NSXMLParser *)parser
        didStartElement:(NSString *)elementName
           namespaceURI:(NSString *)namespaceURI
          qualifiedName:(NSString *)qName
             attributes:(NSDictionary *)attributeDict{

  /* Have we already created a root element? */
  if (self.rootElement == nil){

    /* We are here because the element we have found is the
       root element. So let's create the root element and set the
       current element to the root element for later use */

    XMLElement *root = [[XMLElement alloc] init];
    self.rootElement = root;
    self.currentElement = self.rootElement;
    [root release];

  } else {

    /* If we are here, it means that we have already
       created  a root element. So let's create a new element */
    XMLElement *newElement = [[XMLElement alloc] init];

    newElement.parent = self.currentElement;

    /* Also add the new element that we just created as
       the child element of the current element */
    [self.currentElement.children addObject:newElement];

    /* Now the current element becomes the new
       element that we just created */
    self.currentElement = newElement;

    /* This is very important. Since the new element
```

```
      is added to the children array of its parent
      element, we make sure we release it as
      the array has now retained this element */
    [newElement release];
  }

  self.currentElement.name = elementName;

  if ([attributeDict count] > 0){
    /* Also if we have attributes, let's make sure we add
     them to the attributes dictionary of the current element */
    [self.currentElement.attributes
     addEntriesFromDictionary:attributeDict];
  }

}

- (void)          parser:(NSXMLParser *)parser
       foundCharacters:(NSString *)string{

  if (self.currentElement != nil){

    /* Here we have found some inner text for
     the current element. This doesn't happen unless
     the current element does indeed have some
     text inside it. For instance:

     <person name="foo"/>
     <person name="bar">a friend of mine</person>

     The first person element does not have inner text
     but the second one does (inner text of the
     second element is [a friend of mine])
     */

    if (self.currentElement.text == nil){
      self.currentElement.text = string;
    } else {
      self.currentElement.text =
        [self.currentElement.text
         stringByAppendingString:string];
    }

  }

}

- (void)           parser:(NSXMLParser *)parser
        didEndElement:(NSString *)elementName
          namespaceURI:(NSString *)namespaceURI
          qualifiedName:(NSString *)qName{

  if (self.currentElement != nil){
    /* For the description of this method, refer to the
     comments written for the parser:didStartElement:namespaceURI:
```

```
       qualifiedName:attributes: method */
     self.currentElement = self.currentElement.parent;
  }

}

- (BOOL) parseLocalXMLWithPath:(NSString*)paramLocalXMLPath{

  BOOL result = NO;

  /* Let's make sure that the path is not empty */
  if ([paramLocalXMLPath length] == 0){
    NSLog(@"The given XML path is nil or empty.");
    return(NO);
  }

  /* Does this file exist? */
  NSFileManager *fileManager = [[NSFileManager alloc] init];

  /* Make sure the file exists */
  if ([fileManager fileExistsAtPath:paramLocalXMLPath] == NO){
    NSLog(@"The given local path does not exist.");
    [fileManager release];
    return(NO);
  }

  [fileManager release];

  /* Form an NSData of the contents of the file */
  NSData *localXMLData =
  [NSData dataWithContentsOfFile:paramLocalXMLPath];

  /* Could we get any data */
  if ([localXMLData length] == 0){
    NSLog(@"The local XML file could not be loaded.");
    return(NO);
  }

  if (self.xmlParser != nil){
    self.xmlParser = nil;
  }

  if (self.rootElement != nil){
    self.rootElement = nil;
  }

  /* Now start parsing */
  NSXMLParser *parser =
  [[NSXMLParser alloc] initWithData:localXMLData];

  self.xmlParser = parser;

  [parser release];

  [self.xmlParser setDelegate:self];
```

```
    result = [self.xmlParser parse];

    return(result);

}

- (void) dealloc {
    NSLog(@"Deallocated document");

    [xmlParser release];
    [currentElement release];
    [rootElement release];
    [documentPath release];

    [super dealloc];
}

@end
```

Now, the `parser:didStartElement:namespaceURI:qualifiedName:attributes:` delegate method of `NSXMLParser` is particularly interesting. Since we always keep the reference to the element that was created the last time we were in this method, we know the parent of the new element we are parsing is the element we parsed the last time we were in this method. The trick here is that in the `parser:didEndElement:namespace URI:qualifiedName:` method, we set the current element's reference as the parent element of the current element.

Think of it this way:

```
Root
   + Child1
     + Child1.1
     + Child1.2
   + Child2
```

When we parse `Child1`, the parent of `Child1` is `Root`. Here, `Child1` has *not* been fully parsed (because it has children elements of its own), so we do *not* yet get to the `parser:didEndElement:namespaceURI:qualifiedName:` method for `Child1`. Now we start parsing `Child1.1`; the new element's parent element will become the current element (`Child1`) and we add `Child1.1` as the child of `Child1`. When `Child1.1` finishes getting parsed, we get to the `parser:didEndElement:namespaceURI:qualifiedName:` method. Here we set the reference of the current element (`Child1.1`) to the parent of the current element (`Child1`). So, now we are back at `Child1`. Then we start parsing `Child1.2` and the same story happens again. You can see how useful the `currentElement` reference is to us.

It's important to understand how information from the XML file is presented to your program. You get to know when an element in an XML file starts and ends. The previous code creates a linked list that will help us find the parent and children of any element. So, if we have the root that is the topmost element in an XML file, we can access the other elements by traversing this list through the items' `children` property. And from each child, we can move back up toward the root element through the `parent` property.

In terms of memory management, we will allocate but not release the root element. All the other elements, which are direct or indirect children of the root element, will be allocated, added to an array, and then released. This means that when we get around to deallocating the root element, this will cause a chain reaction that deallocates all the other elements.

Just imagine this XML document:

- Root
  —Child 1
  —Child 2

Now let's have a look at the normal lifetime of the XML document:

1. Allocation and initialization:
   a. The root element will have the retain count of 1.
   b. The subelements (children) of the root element will get allocated and initialized. At this point, they have the retain count of 1. Then they get added to the `children` array property of their parent element, at which point their retain count will be 2. Then we directly release the children, at which point they will have the retain count of 1 again.

2. Getting accessed:
   a. No retain count is modified here.

3. Release and deallocation:
   a. We release the root element of the instance of the `XMLDocument` class, at which point the root element will deallocate its children. Once that is done, the array will get deallocated and the retain count of all the children will become 0, at which point they get deallocated. If the children have children of their own, they will get deallocated, too. This is how we will prevent a memory leak.

 It's good to bear in mind that you can manage the memory in whatever way you want when writing your iOS applications. All you have to watch for is not to cause a memory leak. Static Analysis in Xcode can help you a lot in terms of checking for potential memory leaks, but don't rely on it too much. It can never be as intelligent as a human being.

The `currentElement` property of our `XMLDocument` class is just a pointer to the element that we have to be manipulating when a new element is parsed (and when the parsing of the new element finishes). Think of `currentElement` as a pointer to an element in the rather complex structure of an XML file. We move this pointer up and down the XML model as we parse the file so that we always have the reference to the current element that is being parsed.

Now for the moment of truth, we go ahead and attempt to parse our local XML file:

```
- (void)viewDidLoad {
  [super viewDidLoad];

  NSBundle *mainBundle = [NSBundle mainBundle];

  NSString *xmlFilePath = [mainBundle pathForResource:@"MyXML"
                                               ofType:@"xml"];

  XMLDocument    *document = [[XMLDocument alloc] init];

  if ([document parseLocalXMLWithPath:xmlFilePath] == YES){
    [self recursivelyLogXMLContents:document.rootElement];
  } else {
    NSLog(@"Could not parse the XML file.");
  }

  [document release];

}
```

Very simple, isn't it? The most important procedure we are calling here is **recursively LogXMLContents:**, which is a recursive method, as its name implies. This method will take the root element of an XML document and search through all the root elements' child elements, the children of those children, and so on. It will then dump the data to the console window. We have implemented this procedure in this way:

```
- (void) recursivelyLogXMLContents:(XMLElement*)paramElement{

  NSLog(@"Element Name = %@", paramElement.name);

  NSLog(@"Element Text = %@", paramElement.text);

  NSLog(@"Number of Attributes = %lu",
        (unsigned long)[paramElement.attributes count]);

  if (paramElement.parent != nil){
    NSLog(@"Parent Element Name = %@", paramElement.parent.name);
  }

  if ([paramElement.attributes count]){
    NSLog(@"Attributes Dictionary = %@", paramElement.attributes);
  }

  NSLog(@"Number of children elements = %lu",
        (unsigned long)[paramElement.children count]);

  for (XMLElement *child in paramElement.children){
    [self recursivelyLogXMLContents:child];
  }
```

}

The final block in this procedure loops through all the children elements of the current element (tail recursion). The rest of the short method prints out an element's name, attributes, and so forth. When reading the contents of the XML file that we created before, we get the content shown in Figure 6-6, written in the console window.

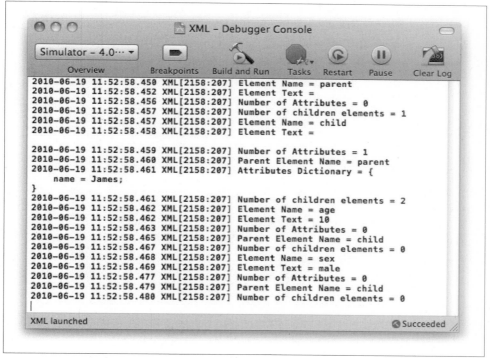

Figure 6-6. The parsed contents of an XML file

See Also

Recipe 6.5

6.3 Downloading Files Synchronously

Problem

You want to download a file in a single step.

Solution

Invoke the sendSynchronousRequest:returningResponse:error: class method of the NSURLConnection class:

```
- (void)viewDidLoad {
  [super viewDidLoad];

  /* This is the URL which we want to download the contents of */
  NSString *urlAsString = @"http://www.OReilly.com";

  NSURL     *url = [NSURL URLWithString:urlAsString];

  /* Create the request to pass to NSURLConnection */
  NSURLRequest *request = [NSURLRequest requestWithURL:url];

  NSError *error = nil;

  /* Block the current thread with the synchronous connection */
  NSData    *data = [NSURLConnection sendSynchronousRequest:request
                                          returningResponse:nil
                                                      error:&error];

  /* Did we get the data? */
  if (data != nil){
    /* Let's make an NSString out of the data */
    NSString *dataAsString =
    [[NSString alloc] initWithData:data
                          encoding:NSUTF8StringEncoding];

    NSLog(@"%@", dataAsString);

    [dataAsString release];

  } else {

    /* We could not get the data */
    NSLog(@"%@", error);

  } /* if (Data != nil){ */

}
```

Discussion

NSURLConnection can be used to download contents from a URL synchronously or asynchronously. The sendSynchronousRequest:returningResponse:error: class method of NSURLConnection, as its name implies, deals with synchronous calls. A synchronous call blocks its calling thread until the method returns, whereas an asynchronous call starts accessing and downloading the contents of a URL without blocking the calling thread.

When called on the main thread, a synchronous call blocks this thread until the synchronous request succeeds or fails. Blocking the main thread is highly discouraged because it is the UI thread. By blocking it, you keep the user from interacting with your application.

The return value of this method is either the NSData object representing the downloaded contents, or nil if a connection error occurs. We pass the reference to an NSError variable to the error parameter of this method just in case something goes wrong. We can then access this error for further information and to understand what went wrong.

See Also

Recipe 6.4; Recipe 7.2; Recipe 7.6

6.4 Downloading Files Asynchronously

Problem

You want to be able to download multiple files at the same time, or download a file while keeping your application's thread responsive.

Solution

Use the initWithRequest:delegate:startImmediately: method of the NSURLConnection class:

```
- (void)viewDidLoad {
  [super viewDidLoad];

  NSString *urlAsString = @"http://www.OReilly.com";

  NSURL    *url = [NSURL URLWithString:urlAsString];

  NSURLRequest *request = [NSURLRequest requestWithURL:url];

  /* The object where we will store the data that
   is being downloaded */
  NSMutableData *data = [[NSMutableData alloc] init];
  self.connectionData = data;
  [data release];

  NSURLConnection *newConnection = [[NSURLConnection alloc]
                                    initWithRequest:request
                                    delegate:self
                                    startImmediately:YES];
  self.connection = newConnection;
  [newConnection release];

  if (self.connection != nil){
```

```
      /* Successfully created the connection */
    } else {
      /* Could not create the connection */
    }

  }

- (void) viewDidUnload{
  [super viewDidUnload];

  [self.connection cancel];
  self.connection = nil;
  self.connectionData = nil;
}
```

The parsed XML is then handled by methods that the system calls in the object passed as the delegate parameter of this method.

Discussion

The `initWithRequest:delegate:startImmediately:` instance method of the `NSURLConnection` class asynchronously downloads the contents of a URL by creating a new thread. The return value of this method is either the connection that has been successfully created to download the contents or `nil` if an error has occurred.

The thread required to download the contents of the URL will be created by the framework itself, not by you. What you will get is a series of delegate messages that will keep you in the loop about the state of the connection: whether it has started, when it has downloaded some data, if it has failed, and so on.

This Discussion shows a subset of the useful messages available to the delegate object. For more information about the other delegate messages that are provided by `NSURLConnection`, read "NSURLConnection Class Reference" in the iOS SDK documentation.

Before we go any further, let's have a look at the two properties we declared in the current object, which create instances of the `NSURLConnection` and `NSMutableData` classes:

```
#import <UIKit/UIKit.h>

@interface RootViewController : UIViewController{
@public
  NSURLConnection      *connection;
  NSMutableData        *connectionData;
}

@property (nonatomic, retain) NSURLConnection      *connection;
@property (nonatomic, retain) NSMutableData        *connectionData;

@end
```

The NSURLConnection object will be instantiated in the viewDidLoad method of our view controller, as explained earlier. We will refer to the object through the connection property, while the connectionData property will hold the data that our connection will download. As you can see, the connection data is a mutable data object, so we can add to it as we download the XML.

Now that the connection has been created, we implement the four most important delegate messages of the NSURLConnection class for this particular example:

```
- (void)           connection:(NSURLConnection *)connection
            didFailWithError:(NSError *)error{

    NSLog(@"An error happened");
    NSLog(@"%@", error);

}

- (void)           connection:(NSURLConnection *)connection
            didReceiveData:(NSData *)data{

    NSLog(@"Received data");
    [self.connectionData appendData:data];

}

- (void)        connectionDidFinishLoading
            :(NSURLConnection *)connection{

    NSLog(@"Successfully downloaded the contents of the URL.");

    /* do something with the data here */

}

- (void)           connection:(NSURLConnection *)connection
        didReceiveResponse:(NSURLResponse *)response{

    [self.connectionData setLength:0];

}
```

The connection:didReceiveResponse: method, in some rare cases, gets called more than once for one request (e.g., in cases of redirection), so we have to dispose of the previously allocated data (if any) by setting the length of our connectionData property to zero.

After this, the connection:didReceiveData: delegate method will get called continuously as data is downloaded. In this method, all we do is append the returned data to the mutable data that we have. If an error occurs anytime during this process, the connection:didFailWithError: delegate method will get called.

After the contents of the URL are downloaded, our delegate object will receive the connectionDidFinishLoading: delegate method marking the end of the downloading process.

See Also

"NSURLConnection Class Reference," *http://developer.apple.com/library/mac/#docu mentation/Cocoa/Reference/Foundation/Classes/NSURLConnection_Class/Reference/ Reference.html*

6.5 Reading and Parsing Remote XML Files

Problem

You want to be able to parse an XML file that is available on the Internet.

Solution

Download the contents of the XML file with `NSURLConnection` and parse the data using `NSXMLParser`.

Discussion

We created the `XMLDocument` and `XMLElement` objects in Recipe 6.2. We implemented our `XMLDocument` object so that it could accept the path of a local XML file and parse the file. Now we need to take this implementation to the next level and allow our `XMLDocument` object to download files asynchronously and parse them when downloads are complete. Each file will be downloaded asynchronously on a separate thread (provided by `NSURLConnection`), and we will get delegate messages from `NSURLConnection` about the status of each thread. After a download is finished, we will instantiate an object of type `NSXMLParser` and parse the downloaded data.

Before we jump into the code, let's see what we really need from our `XMLDocument` object:

- The ability to parse local XML files (as previously implemented)
- The ability to parse remote XML files (which will be implemented in this recipe) by:
  - Downloading the files asynchronously
  - Parsing the downloaded data

We need to assign a delegate to our `XMLDocument` class so that it can stay in touch with whoever made a parsing request when the parsing process is done. Just like a table view with a delegate, we need to define our own protocol for the delegate of an instance of the `XMLDocument` class—the iOS SDK does not provide us with such a protocol. So, we'll define our own protocol with the arbitrary name of `XMLDocumentDelegate`, through which a delegate object will receive messages from the `XMLDocument` object:

```
@protocol XMLDocumentDelegate <NSObject>

@required

  - (void) xmlDocumentDelegateParsingFinished
```

```
    :(XMLDocument *)paramSender;

 - (void) xmlDocumentDelegateParsingFailed
          :(XMLDocument *)paramSender
  withError:(NSError *)paramError;

@end
```

Our `XMLDocumentDelegate` protocol declares two methods, one called when the parsing process finishes successfully and the other called when the parsing process fails. The important thing to bear in mind is that these methods should get defined in the delegate of our `XMLDocument` instance whether the file to be parsed is a local or a remote file.

Now that we have created our custom protocol through which our `XMLDocument` will stay in touch with its delegate, we can go ahead and declare the rest of the required items in the header file of the `XMLDocument` class. What we need from this class is:

- An XML parser.
- An instance of the `NSURLConnection` for downloading remote XML files asynchronously.
- An `XMLElement` instance to keep a reference to the parsed XML's root element (as stated in Recipe 6.2).
- An `XMLElement` instance that always points to the current element we are parsing (as stated in Recipe 6.2).
- A reference to an object conforming to the `XMLDocumentDelegate` protocol that we have created to which to report the success or failure of the downloading/parsing processes.
- An instance of the `NSMutableData` class to hold the data of a remotely downloaded XML file.
- A `BOOL` property that will hold the value `YES` if a parsing error has occurred and the value `NO` if not. We will use this to prevent sending both success and failure messages to our delegate in case a parsing error has occurred. If this value is `YES`, it means we have encountered a parsing error and we will not call `xmlDocument DelegateParsingFinished:` in our delegate. We are going to set this property to `YES` and `NO` manually in our `XMLDocument` object, as you will see soon.

```
#import <Foundation/Foundation.h>
#import "XMLElement.h"

@class XMLDocument;

@protocol XMLDocumentDelegate <NSObject>

@required
```

```objc
- (void) xmlDocumentDelegateParsingFinished
          :(XMLDocument *)paramSender;

- (void) xmlDocumentDelegateParsingFailed
          :(XMLDocument *)paramSender
  withError:(NSError *)paramError;

@end

@interface XMLDocument : NSObject <NSXMLParserDelegate> {
@public
  /* Keep the document path just in case we want to refer to it */
  NSString                  *documentPath;
  /* We will allocate and initialize the root element */
  XMLElement                *rootElement;
  /* Our delegate which will get called if parsing
     finishes successfully or fails */
  id<XMLDocumentDelegate>     delegate;
@private
  /* Our private XML parser used for local and remote files */
  NSXMLParser               *xmlParser;
  /* The pointer to the current element being parsed */
  XMLElement                *currentElement;
  /* The URL connection we will use to download remote XML files */
  NSURLConnection           *connection;
  /* The mutable data that will hold the XML which
     is being downloaded */
  NSMutableData             *connectionData;
  /* We will set this value to YES and NO manually
     to prevent calling the wrong delegate messages */
  BOOL                       parsingErrorHasHappened;
}

@property (nonatomic, retain) NSString                 *documentPath;
@property (nonatomic, retain) XMLElement               *rootElement;
@property (nonatomic, assign) id<XMLDocumentDelegate> delegate;

/* Private properties */
@property (nonatomic, retain) NSXMLParser       *xmlParser;
@property (nonatomic, retain) XMLElement         *currentElement;
@property (nonatomic, retain) NSURLConnection    *connection;
@property (nonatomic, retain) NSMutableData      *connectionData;
@property (nonatomic, assign) BOOL               parsingErrorHasHappened;

/* Designated Initializer */
- (id)   initWithDelegate:(id<XMLDocumentDelegate>)paramDelegate;

- (BOOL) parseLocalXMLWithPath:(NSString *)paramLocalXMLPath;
- (BOOL) parseRemoteXMLWithURL:(NSString *)paramRemoteXMLURL;

@end
```

The code is commented heavily, so please read the comments to get an even better idea of how we are planning to use the properties we defined for our custom object.

 We're using @class before declaring the XMLDocumentDelegate protocol because one of the parameters in the delegate messages is of type XML Document, and at the time of compiling this header file, XMLDocument is not yet defined. This directive is called a forward class declaration and will "promise" the compiler that at some point, the XMLDocument class will be declared.

One of the new methods we have is initWithDelegate:, which will accept our delegate object as its parameter. We will implement this method in its simplest form without any validation on the given delegate object. In production code, you should make sure the given object does in fact conform to the XMLDocumentDelegate protocol by using the conformsToProtocol: instance method of the given delegate. We must do this because calling a nonexistent selector on an object will result in a runtime error:

```
- (id) init {
  /* Call the designated initializer */
  return([self initWithDelegate:nil]);
}

- (id) initWithDelegate:(id<XMLDocumentDelegate>)paramDelegate{

  self = [super init];

  if (self != nil){
    delegate = paramDelegate;
  }

  return(self);

}
```

We have declared two important methods in the header file. The parseLocalXMLWith Path: method is the same as it is in Recipe 6.2. The new method that we have to implement is parseRemoteXMLWithURL:, which is responsible for performing the following steps:

1. Accepting a string for the URL pointing to an XML file.
2. Escaping this URL with percent signs so that we can successfully download it even if there are space characters, quotation marks, or other special characters that must be encoded in URLs (e.g., spaces tend to be turned into the string %20). The iOS library provides a useful instance method called stringByAddingPercentEscapes UsingEncoding: in NSString to escape all invalid URL characters so that they can be used properly.

3. Creating an instance of the NSURLConnection and starting the download of the contents of the given URL.

```
- (BOOL) parseRemoteXMLWithURL:(NSString *)paramRemoteXMLURL{

    BOOL result = NO;

    if ([paramRemoteXMLURL length] == 0){
      NSLog(@"The remote URL cannot be nil or empty.");
      return(NO);
    }

    /* escape the URL with percent signs */
    paramRemoteXMLURL =
      [paramRemoteXMLURL
      stringByAddingPercentEscapesUsingEncoding:NSUTF8StringEncoding];

    /* Make sure our connection hasn't been created before */
    self.connection = nil;

    NSURL        *url = [NSURL URLWithString:paramRemoteXMLURL];
    NSURLRequest *request = [NSURLRequest requestWithURL:url];

    /* Get rid of the previous download data (if any) */
    self.connectionData = nil;

    /* If we have already parsed another XML, then we have to
     get rid of its root element (all other child elements will then
     be deallocated automatically) */

    /* This is where we will store all our data */
    NSMutableData *newData = [[NSMutableData alloc] init];
    self.connectionData = newData;
    [newData release];

    /* Start the download process */
    NSURLConnection *newConnection = [[NSURLConnection alloc]
                                      initWithRequest:request
                                      delegate:self
                                      startImmediately:YES];
    self.connection = newConnection;
    [newConnection release];

    return(result);

}
```

Since we have chosen the current class to be the delegate of our NSURLConnection instance, we must implement some of the most important and useful delegate methods of the NSURLConnection class. We implement these NSURLConnection delegate methods in our XMLDocument class:

`connection:didReceiveResponse:`

This method, in rare cases such as redirection, might get invoked more than once on a single request, so it must dispose of any previously allocated/downloaded data by calling the `setLength:` method of our `NSMutableData` instance to set the length of this data to zero.

`connection:didReceiveData:`

This method gets called whenever we have downloaded a part of the requested URL. We will append the downloaded data, which is given to us through the `didReceiveData` parameter, to our mutable data declared as the `connectionData` property in the `XMLDocument` class.

`connection:didFailWithError:`

If an error occurs while downloading the data—for instance, disconnection from the server—this method gets called in our `XMLDocument` object, which is set as the delegate of the instance of `NSURLConnection` in the `parseRemoteXMLWithURL:` method.

`connectionDidFinishLoading:`

This method gets called whenever the download process has finished successfully. At this point, we can start to parse the downloaded data by creating an instance of the `NSXMLParser` class.

```
- (void)  connection:(NSURLConnection *)connection
  didReceiveResponse:(NSURLResponse *)response{

    [self.connectionData setLength:0];

}

- (void)  connection:(NSURLConnection *)connection
      didReceiveData:(NSData *)data{

    [self.connectionData appendData:data];

}

- (void)  connection:(NSURLConnection *)connection
    didFailWithError:(NSError *)error{

    NSLog(@"A connection error has occurred.");

    [self.delegate xmlDocumentDelegateParsingFailed:self
                                          withError:error];

}

- (void)  connectionDidFinishLoading:(NSURLConnection *)connection{

    /* Now that we have finished downloading, let's start parsing
     the downloaded data */

    if (self.connectionData != nil){
```

```
NSXMLParser *newParser = [[NSXMLParser alloc]
                          initWithData:self.connectionData];

self.xmlParser = newParser;
[newParser release];

[self.xmlParser setShouldProcessNamespaces:NO];
[self.xmlParser setShouldReportNamespacePrefixes:NO];
[self.xmlParser setShouldResolveExternalEntities:NO];
[self.xmlParser setDelegate:self];
if ([self.xmlParser parse] == YES){
  NSLog(@"Successfully parsed the remote file.");
} else{
  NSLog(@"Failed to parse the remote file.");
}

}

}
```

Now that we have downloaded the XML file, we need to parse it. The parsing will be done in the same way as in Recipe 6.2. However, we need to implement a few more NSXMLParserDelegate methods in order to be able to keep our delegate up to date with the state of the parsing process. These extra methods that we will implement are:

parser:parseErrorOccurred:
:   This method gets called whenever a parsing error occurs. For instance, if we forget to enclose an attribute with quotation marks, we will get a parsing error. We will set the value of the parsingErrorHasHappened property to YES in this case, too. Furthermore, we will report the error to the delegate of our XMLDocument object and attempt to abort the parsing process.

parser:validationErrorOccurred:
:   This method will get called whenever a validation error occurs while parsing the XML file. Validation is usually performed on XML documents with a DTD or an XML schema. In case of a parsing failure, we will set the value of the parsingErrorHasHappened property to YES. We will also immediately report the failure to our delegate (conforming to the XMLDocumentDelegate protocol).

parserDidStartDocument:
:   This method will get called whenever the parsing of the XML document starts. We will use this method to set the parsingErrorHasHappened property to NO because at this stage of the parsing process, no errors have occurred.

parserDidEndDocument:
:   Last but not least, this method will get called whenever the parsing has finished. We will invoke the xmlDocumentDelegateParsingFinished: method in our XML Document delegate whenever this happens.

The `parserDidEndDocument:` method gets called even if parsing and/or validation errors occur while parsing an XML document. For this reason, we will set the value of the `parsingErrorHasHappened` property to YES whenever a validation or parsing error occurs, and we will check this property's value in the `parserDidEndDocument:` method before calling the `xmlDocumentDelegateParsingFinished:` delegate method on the `XMLDocument` delegate object. In the case of a parsing or validation error, we will call the `xmlDocumentDelegateParsingFailed:withError:` method of our delegate. It would be poor programming to tell our delegate that an error has occurred but that we have also been able to parse the XML successfully! That would probably confuse the delegate.

```objc
- (void)          parser:(NSXMLParser *)parser
     parseErrorOccurred:(NSError *)parseError{

    NSLog(@"Parsing error has occurred.");

    self.parsingErrorHasHappened = YES;

    /* Abort the parsing straight away */
    [parser abortParsing];

    [self.delegate xmlDocumentDelegateParsingFailed:self
                                      withError:parseError];

}

- (void)          parser:(NSXMLParser *)parser
  validationErrorOccurred:(NSError *)validationError{

    NSLog(@"Validation error has occurred.");

    self.parsingErrorHasHappened = YES;

    [self.delegate xmlDocumentDelegateParsingFailed:self
                                      withError:validationError];

}

- (void)parserDidStartDocument:(NSXMLParser *)parser{

    self.parsingErrorHasHappened = NO;

}

- (void)parserDidEndDocument:(NSXMLParser *)parser{

    if (self.parsingErrorHasHappened == NO){
      [self.delegate xmlDocumentDelegateParsingFinished:self];
    }

}
```

```objc
- (void)           parser:(NSXMLParser *)parser
      didStartElement:(NSString *)elementName
         namespaceURI:(NSString *)namespaceURI
        qualifiedName:(NSString *)qName
           attributes:(NSDictionary *)attributeDict{

  if (self.rootElement == nil){
    XMLElement *newElement = [[XMLElement alloc] init];
    self.rootElement = newElement;
    self.currentElement = self.rootElement;
    [newElement release];
  } else {
    XMLElement *newElement = [[XMLElement alloc] init];
    newElement.parent = self.currentElement;
    [self.currentElement.children addObject:newElement];
    self.currentElement = newElement;
    [newElement release];
  }

  self.currentElement.name = elementName;

  if ([attributeDict count] > 0){
    [self.currentElement.attributes
     addEntriesFromDictionary:attributeDict];
  }

}

- (void)        parser:(NSXMLParser *)parser
      foundCharacters:(NSString *)string{

  if (self.currentElement != nil){

    if (self.currentElement.text == nil){
      self.currentElement.text = string;
    } else {
      self.currentElement.text =
      [self.currentElement.text
       stringByAppendingString:string];
    }

  }

}

- (void)           parser:(NSXMLParser *)parser
        didEndElement:(NSString *)elementName
         namespaceURI:(NSString *)namespaceURI
        qualifiedName:(NSString *)qName{

  if (self.currentElement != nil){
    self.currentElement = self.currentElement.parent;
  }
```

```
    }
```

Of course, we will also equip the `dealloc` method of our `XMLDocument` object with a few lines of code to deallocate the objects that we might have allocated during the lifetime of our object:

```
- (void) dealloc {

  if (connection != nil){
    [connection cancel];
  }
  [connection release];

  [connectionData release];
  [xmlParser release];
  [rootElement release];
  [currentElement release];
  [documentPath release];

  [super dealloc];
}
```

We have not fully implemented the `XMLDocument` object. Let's use it in a view controller. You don't necessarily have to use this class in a view controller, though.

```
#import <UIKit/UIKit.h>
#import "XMLDocument.h"

@interface RootViewController : UIViewController
                          <XMLDocumentDelegate> {
@public
  XMLDocument      *xmlDocument;

}

@property (nonatomic, retain) XMLDocument      *xmlDocument;

@end
```

And finally, we will implement the view controller and instantiate our `XMLDocument` class:

```
#import "RootViewController.h"
#import "XMLElement.h"

@implementation RootViewController

@synthesize xmlDocument;

- (void) xmlDocumentDelegateParsingFinished
        :(XMLDocument *)paramSender{

  NSLog(@"Finished downloading and parsing the remote XML");

}
```

```
- (void) xmlDocumentDelegateParsingFailed
        :(XMLDocument *)paramSender
 withError:(NSError *)paramError{

  NSLog(@"Failed to download/parse the remote XML.");

}

- (void)viewDidLoad {
  [super viewDidLoad];

  NSString *xmlPath = @"THE URL OF AN XML FILE ON THE INTERNET";

  XMLDocument *newDocument = [[XMLDocument alloc] initWithDelegate:self];
  self.xmlDocument = newDocument;
  [newDocument release];

  [self.xmlDocument parseRemoteXMLWithURL:xmlPath];

}

- (void) viewDidUnload{
  [super viewDidUnload];

  self.xmlDocument = nil;

}

- (BOOL)shouldAutorotateToInterfaceOrientation:
  (UIInterfaceOrientation)interfaceOrientation {
  return (YES);
}

- (void)dealloc {
  [xmlDocument release];
  [super dealloc];
}

@end
```

See Also

Recipe 6.2; Recipe 6.4

6.6 Caching Files in Memory

Problem

You would like to make your application more responsive by caching downloaded data so that the program can avoid downloading data over and over again from the same URL.

Solution

iOS offers servers ways to cache remote data. You can use the help of NSURLRequest and NSURLCache or implement your own caching mechanism.

Discussion

Caching is not a magical mechanism. Maybe if you knew the purpose of caching data it would be easier for you to choose an existing method or even implement one for yourself. The purpose of caching is to avoid downloading remote data when a valid and recent copy of the data is present on the local storage accessible to the application. There are various ways to implement caching, some more intelligent and more difficult to implement and some straightforward and easier to implement.

The two methods provided automatically on the iPhone and iPad to cache a URL are the local cache and the protocol-based cache. Caching on the iOS memory can be achieved by using the NSURLCache class, as we will see soon. For protocol-specific caching—for instance, caching in HTTP/1.1—refer to each protocol's respective documentation. You can find information on HTTP/1.1's caching mechanism, as described in RFC 2616, at *http://www.w3.org/Protocols/rfc2616/rfc2616-sec13.html*.

One of the most straightforward ways to implement caching when dealing with remote data is to use the NSURLRequest caching mechanism combined with NSURLCache. The requestWithURL:cachePolicy:timeoutInterval: class method of the NSURLRequest class accepts a caching parameter to specify how caching must be done on this URL. The values you can pass to the cachePolicy parameter of this method are:

NSURLRequestUseProtocolCachePolicy
> This flag will use the underlying protocol's caching mechanism if the protocol supports it.

NSURLRequestReloadIgnoringLocalCacheData
> This flag specifies that the local cached copy of the resource that is about to be downloaded must be disregarded and the remote cache policy must be effective. If there is a local copy of the resource, managed by the framework itself, it will be ignored.

NSURLRequestReturnCacheDataElseLoad
> This flag specifies that the cached data must be used before attempting to load the data from the original source. The cached data could be protocol-based cached or locally cached. If there is no cached data, the data will be downloaded from the original source.

NSURLRequestReturnCacheDataDontLoad
> This flag specifies that only the local cached data must be used. If the data has not been cached, the request will fail. This is a great flag to use whenever your application wants to perform operations in offline mode (such as the Offline Mode in web browsers).

NSURLRequestReloadIgnoringLocalAndRemoteCacheData

This flag disregards any type of caching involved in the process, local and remote, and always attempts to download the data from the original source.

NSURLRequestReloadRevalidatingCacheData

This flag specifies that the original source of data must validate the local cache (if any) before an attempt is made to download the data from the original source. If there is a copy of the original source cached locally and the remote source specifies that the cached data is valid, the data won't be downloaded again. In any other case, the data will be downloaded from the original source.

The NSURLCache class also offers a variety of methods to implement effective caching in our applications.

 Without NSURLCache, none of the caching mechanisms implemented in NSURLRequest and NSURLConnection will work. NSURLCache provides the mechanism through which the two aforementioned classes can store and retrieve data to and from the cache.

Let's go ahead and implement a very simple method that will allow our data to be cached locally for this session only:

```
- (void) downloadURL:(NSString *)paramURLAsString{

    if ([paramURLAsString length] == 0){
      NSLog(@"Nil or empty URL is given");
      return;
    }

    /* Get the shared URL Cache object. No need to create a new one */
    NSURLCache *urlCache = [NSURLCache sharedURLCache];

    /* We will store up to 1 Megabyte of data into the cache */
    [urlCache setMemoryCapacity:1*1024*1024];

    /* For our request, we need an instance of NSURL so let's retrieve
     that from the string that we created before */
    NSURL *url = [NSURL URLWithString:paramURLAsString];

    /* And this is our request */
    NSMutableURLRequest *request =
    [NSMutableURLRequest
     requestWithURL:url
     cachePolicy:NSURLRequestUseProtocolCachePolicy
     timeoutInterval:60.0f];

    /* Try to get a cached response to our request.
     This might come back as nil */
    NSCachedURLResponse *response =
    [urlCache cachedResponseForRequest:request];
```

```
  /* Do we have a cached response? */
  if (response != nil){
    NSLog(@"Cached response exists. Loading data from cache...");
    [request setCachePolicy:NSURLRequestReturnCacheDataDontLoad];
  }

  self.connection = nil;

  /* Start the connection with the request */
  NSURLConnection *newConnection =
  [[NSURLConnection alloc] initWithRequest:request
                                  delegate:self
                           startImmediately:YES];

  self.connection = newConnection;
  [newConnection release];

}
```

The NSURLCache class can be used as a singleton or a shared instance. In other words, you do not have to allocate an instance of this class to be able to make use of the facilities that it provides, just like the sharedApplication class method of the UIApplication class.

The setMemoryCapacity: instance method of NSURLCache allows you to set the maximum number of bytes that the local cache can hold in memory. If your application has received a low-memory warning from iOS, reducing the size of any cache that you have for your application is one way to reduce overall memory usage.

In iOS, NSURLCache supports caching data only in memory and not on disk.

After setting the maximum amount of data we would like to hold in memory for the cached data, we can start our request and connection to work in conjunction in order to download our content. But before doing that, we have to decide which caching policy we would like to use.

In this example, we want to reuse any file the application has previously downloaded during this run, and download any file from the original source only if that file's cached version doesn't exist in memory. For this reason, we will initially set the cache policy to NSURLRequestUseProtocolCachePolicy in order to use the underlying protocol's caching policy (if any).

However, after constructing our request, we will use the cachedResponseForRequest: instance method of NSURLCache to find out whether a cached version of this request already exists in memory. If it does, we get an instance of the NSCachedURLResponse class. If the response can be found, we can be sure that a cached version of our request exists in memory. For this reason, to avoid downloading the original version from the source, we switch our cache policy to NSURLRequestReturnCacheDataDontLoad as explained before.

 We are using an instance of NSMutableURLRequest instead of NSURL Request because we can change the properties of our mutable request after it has been created, unlike in NSURLRequest.

To get a better understanding of the course of events, we must implement some delegate methods of NSURLConnection and log messages to the console window:

```objc
- (void)  connection:(NSURLConnection *)connection
  didReceiveResponse:(NSURLResponse *)response{

  NSLog(@"Did Receive Response");

}

- (NSURLRequest *)connection:(NSURLConnection *)connection
            willSendRequest:(NSURLRequest *)request
          redirectResponse:(NSURLResponse *)redirectResponse{

  NSLog(@"Will Send Request");


  return(request);

}

- (void)connection:(NSURLConnection *)connection
    didReceiveData:(NSData *)data{

  NSLog(@"Did Receive Data");

  NSLog(@"Data Length = %lu", (unsigned long)[data length]);

}

- (NSCachedURLResponse *)connection:(NSURLConnection *)connection
    willCacheResponse:(NSCachedURLResponse *)cachedResponse{

  NSLog(@"Will Cache Response");

  return(cachedResponse);

}

- (void)connectionDidFinishLoading:(NSURLConnection *)connection{

  NSLog(@"Did Finish Loading.");

}
```

```
- (void)connection:(NSURLConnection *)connection
  didFailWithError:(NSError *)error{

    NSLog(@"Did Fail With Error");

}
```

By executing the block of code that downloads files in the action method of a button and viewing the console, we can get a very good understanding of how caching will work. It doesn't necessarily have to be in a button's action, but it should at least be in a place where we could execute the code more than once to be able to see that the caching mechanisms in place will avoid downloading the data over and over again. The first time we run our application, we will see the following log messages (depending on the URL you are trying to fetch, you will see more or less downloaded data):

```
Did Receive Response
Did Receive Data
Data Length = 2648
Did Receive Data
Data Length = 1318
Will Cache Response
Did Finish Loading.
```

Now if we attempt to download the same URL without terminating the application, we will see the following messages:

```
Cached response exists. Loading data from cache...
Did Receive Response
Did Receive Data
Data Length = 3966
Did Finish Loading.
```

As you can see, the length of the downloaded data (which isn't really downloaded, but retrieved from the cache) is equal to the sum of all data packets downloaded by the previous connection.

6.7 Caching Files on Disk

Problem

You would like to be able to cache downloaded files on disk and retrieve them from disk whenever required.

Solution

Create a custom cache management mechanism. The cache we create in this example will save the downloaded files to disk, along with metadata pertaining to when the files should expire from the cache. This means the cache will persist between runs.

Discussion

The simplest way to implement a cache manager is to take advantage of the `connection:willSendRequest:redirectResponse:` delegate method of `NSURLConnection`. This method accepts an instance of `NSURLRequest` and will use this request to fetch the data. To make it simpler to understand, let's go through a scenario:

1. We ask our cache manager to download the file *foo.bar*.
2. The cache manager passes this URL to an instance of `NSURLConnection` and the download begins.
3. Before `NSURLConnection` downloads any data, our main class, named `CachedDown loadManager` in this example, attempts to search in the local dictionary that it keeps (a dictionary of all the files that have been downloaded fully, with their metadata) to find out whether the requested file has already been downloaded.
4. If this file has been downloaded, our custom cache manager will also determine whether the file has expired (we always keep this flag in the dictionary that we save to and load from the disk, as explained in step 3). If the file has expired in the cache, the cache manager will attempt to download it again.

To make our disk caching mechanism more intelligent, we will create a dictionary of dictionaries. You might find this phrase intimidating, but let me explain why I think it is a good method for storing metadata of cached URLs. We will create a dictionary, save it to the disk, and load it from the disk whenever the caching mechanism is loaded and terminated. For every URL that we download, we will create a new dictionary and add it to the root dictionary. The key of each child dictionary will be the remote URL of the file. The subdictionary itself will hold the local URL of the cached data, the date the download was initiated, and the date the download was finished. Perhaps I can illustrate the data cached by performing an `NSLog` on such a root dictionary:

```
{
  "http://www.cnn.com" =       {
    DownloadEndDate = "2010-06-28 07:51:57 +0100";
    DownloadStartDate = "2010-06-28 07:51:55 +0100";
    ExpiresInSeconds = 20;
    ExpiryDate = "2010-06-28 07:52:17 +0100";
    LocalURL = "/var/mobile/Applications/ApplicationID/Documents/
                httpwww.cnn.com.cache";
  };
  "http://www.oreilly.com" =       {
    DownloadEndDate = "2010-06-28 07:51:49 +0100";
    DownloadStartDate = "2010-06-28 07:51:44 +0100";
    ExpiresInSeconds = 20;
    ExpiryDate = "2010-06-28 07:52:09 +0100";
    LocalURL = "/var/mobile/Applications/ApplicationID/Documents/
                httpwww.oreilly.com.cache";
  };
}
```

By having such a dictionary, we can see when the contents expire, where they were downloaded from, and where they are downloaded to. The expiration date of each URL will be specified by the programmer (that's us!), not the remote server.

Let's start implementing our cache. These are the requirements that we have from our cache manager:

- Ability to cache data to the disk.
- Ability to indicate that a resource was already downloaded if the same resource is asked to be downloaded over and over again.
- Ability to specify when we want each resource to expire. To make it simple, we accept the number of seconds that have to be added to the date when the resource is fully downloaded. We add this number of seconds to the final download date to construct an expiration date of type NSDate.
- Ability to return the cached data from disk if the data has not expired yet.
- Ability to "sanitize" the metadata of downloaded items and the items that failed to be downloaded after the cache manager has been deallocated. The next time the cache manager is initialized, it must remove those items that were not fully downloaded from the list of cached items and allow them to be downloaded again successfully. This will leave some room for us to extend the functionality of our cache manager to enable resuming incomplete downloads.
- Ability to download more than one item at a time.

Let's start with the declaration (*.h* file) of the cache manager:

```
#import <Foundation/Foundation.h>
#import "CacheItem.h"

@class CachedDownloadManager;

@protocol CachedDownloadManagerDelegate <NSObject>
@required

- (void) cachedDownloadManagerSucceeded
        :(CachedDownloadManager *)paramSender
        remoteURL:(NSURL *)paramRemoteURL
        localURL:(NSURL *)paramLocalURL
        aboutToBeReleasedData:(NSData *)paramAboutToBeReleasedData
        isCachedData:(BOOL)paramIsCachedData;

- (void) cachedDownloadManagerFailed
        :(CachedDownloadManager *)paramSender
        remoteURL:(NSURL *)paramRemoteURL
        localURL:(NSURL *)paramLocalURL
        withError:(NSError *)paramError;

@end

@interface CachedDownloadManager : NSObject
                            <CacheItemDelegate> {
```

```
@public
  id<CachedDownloadManagerDelegate>  delegate;
@private
  NSMutableDictionary                *cacheDictionary;
  NSString                           *cacheDictionaryPath;
}

@property (nonatomic, assign)
id<CachedDownloadManagerDelegate> delegate;

@property (nonatomic, copy)
NSMutableDictionary *cacheDictionary;

@property (nonatomic, retain)
NSString *cacheDictionaryPath;

/* Private methods */

- (BOOL) saveCacheDictionary;

/* Public methods */

- (BOOL)            download:(NSString *)paramURLAsString
    urlMustExpireInSeconds:(NSTimeInterval)paramURLMustExpireInSeconds
updateExpiryDateIfInCache:(BOOL)paramUpdateExpiryDateIfInCache;

@end
```

Let's go through the properties of our class quickly and see what each one is responsible for:

delegate

As its name implies, this is the object to which the cache manager will report its events. Whenever a file is retrieved from the cache, is downloaded, or fails to get downloaded, the delegate must be kept informed.

cacheDictionary

This is the main dictionary that we will read from and write to. We will save this to and load it from the disk. This dictionary will contain a list of other dictionaries. The key of each subdictionary is the remote URL of the file to be downloaded, and the dictionary associated with that key is the dictionary that holds the metadata of that remote URL (e.g., its expiration date). Using the remote URL as the key of each metadata dictionary allows us to retrieve that URL's metadata from this dictionary with ease, simply by passing the URL to the dictionary as a key and retrieving the value (the subdictionary in this case) associated with the key.

cacheDictionaryPath

This is the local path where we load and save our cache dictionary. We will retain this property throughout the lifetime of the cache manager.

The `download:urlMustExpireInSeconds:updateExpiryDateIfInCache:` instance method of our cache manager is the most important method that I believe needs explanation. From this method, we need to find out whether:

- An item has already been downloaded. If it was, we don't attempt to download it again unless it has expired.

- An item has not yet been downloaded and needs to be downloaded for the first time.

- An item is not cached properly and needs to be downloaded again.

- An item has been cached and has not expired. In this case, if the `updateExpiry` `DateIfInCache` parameter is `YES`, we will update the expiry date to the current date plus the number of seconds specified by the `urlMustExpireInSeconds` parameter. This will allow us to keep a frequently accessed item in the cache in the expectation that it will be requested again soon.

We save all our caching metadata into the `cacheDictionary` property as explained earlier, so let's see how we are loading and saving it. We will use the `init` instance method to initialize the dictionary and read its contents from the disk, if it has already been saved to the disk.

As mentioned earlier, each subdictionary stores one file and its associated metadata. The last thing we do for each file is set the expiration date in `DownloadEndDate`. This means that, if the dictionary contains a subdictionary with a `DownloadEndDate` value of `nil`, the download started but did not complete. As you can see in the following method, which runs when our applications starts, we use this fact to determine whether a file is incomplete and remove its subdictionary:

```
- (void) removeCorruptedCachedItems{

  /* Now let's clean the dictionary by removing any items
   that have not been downloaded fully */
  NSMutableArray *keysToRemove = [NSMutableArray array];

  for (NSMutableDictionary *itemKey in self.cacheDictionary){

    NSMutableDictionary *itemDictionary =
    [self.cacheDictionary objectForKey:itemKey];

    if ([itemDictionary
        objectForKey:CachedKeyDownloadEndDate] == nil){
      NSLog(@"This file didn't get downloaded fully. Removing.");
      [keysToRemove addObject:itemKey];
    }
  }

  [self.cacheDictionary removeObjectsForKeys:keysToRemove];

}
```

```objc
- (id) init{

  self = [super init];

  if (self != nil){

    NSString *documentsDirectory =
    [self documentsDirectoryWithTrailingSlash:YES];

    cacheDictionaryPath =
    [[documentsDirectory
      stringByAppendingString:@"CachedDownloads.dic"] retain];

    NSFileManager *fileManager = [[NSFileManager alloc] init];

    if ([fileManager
         fileExistsAtPath:self.cacheDictionaryPath] == YES){

      NSMutableDictionary *dictionary =
      [[NSMutableDictionary alloc]
       initWithContentsOfFile:self.cacheDictionaryPath];

      cacheDictionary = [dictionary mutableCopy];

      [dictionary release];

      [self removeCorruptedCachedItems];

    } else {

      NSMutableDictionary *dictionary =
      [[NSMutableDictionary alloc] init];

      cacheDictionary = [dictionary mutableCopy];

      [dictionary release];

    }

    [fileManager release];

    if ([cacheDictionaryPath length] > 0 &&
        cacheDictionary != nil){
      NSLog(@"Successfully initialized the cached dictionary.");
    } else {
      NSLog(@"Failed to initialize the cached dictionary.");
    }

  }

  return(self);

}
```

This method will attempt to load the dictionary from the application's *directory* folder. The *documents* directory is retrieved using the `documentsDirectoryWithTrailing Slash:` method, which we have implemented in this way:

```objc
- (NSString *) documentsDirectoryWithTrailingSlash
                :(BOOL)paramWithTrailingSlash{

  NSString *result = nil;

  NSArray *documents =
  NSSearchPathForDirectoriesInDomains(NSDocumentDirectory,
                                      NSUserDomainMask,
                                      YES);

  if ([documents count] > 0){
    result = [documents objectAtIndex:0];

    if (paramWithTrailingSlash == YES){
      result = [result stringByAppendingString:@"/"];
    }

  }

  return(result);
}
```

After we have the *documents* directory, we append a filename to it and keep it stored in the `cacheDictionaryPath` property. We will later use this property to save the contents of the dictionary to the same path as we modify those contents. We use the **saveCache Dictionary** method to save the contents of our metadata dictionary:

```objc
- (BOOL) saveCacheDictionary{

  BOOL result = NO;

  if ([self.cacheDictionaryPath length] == 0 ||
      self.cacheDictionary == nil){
    return(NO);
  }

  result = [self.cacheDictionary
            writeToFile:self.cacheDictionaryPath
            atomically:YES];

  return(result);

}
```

We must store various keys in each dictionary in order to be able to store the metadata of each file:

```objc
const NSString *CachedKeyLocalURL =
              @"LocalURL";

const NSString *CachedKeyDownloadStartDate =
```

```
                @"DownloadStartDate";

    const NSString *CachedKeyDownloadEndDate =
                @"DownloadEndDate";

    const NSString *CachedKeyExpiresInSeconds =
                @"ExpiresInSeconds";

    const NSString *CachedKeyExpiryDate =
                @"ExpiryDate";
```

We will be using these fields to determine the current state of the cached item:

LocalURL

Refers to the local address of the cached file after it has been downloaded

DownloadStartDate

Specifies when the download process of that particular URL will start

DownloadEndDate

Specifies when the download process of that specific URL has successfully finished

ExpiresInSeconds

Stores the number of seconds that have to be added to DownloadEndDate to specify the date when the cached contents will expire

ExpiryDate

The date that is calculated by adding the ExpiresInSeconds field to DownloadEndDate

Now that we can load and save our metadata dictionary, we will go ahead and implement the method that will take care of arranging the files to be downloaded. We will do this using the download:urlMustExpireInSeconds:updateExpiryDateIfInCache: instance method. By recording the state of the file while a download is in progress, we allow multiple instances of our application to run without duplicating the work of downloading an individual file.

```
- (BOOL)          download:(NSString *)paramURLAsString
    urlMustExpireInSeconds:(NSTimeInterval)paramURLMustExpireInSeconds
updateExpiryDateIfInCache:(BOOL)paramUpdateExpiryDateIfInCache{

  BOOL result = NO;

  if (self.cacheDictionary == nil ||
      [paramURLAsString length] == 0){
    return(NO);
  }

  paramURLAsString = [paramURLAsString lowercaseString];

  NSMutableDictionary *itemDictionary =
  [self.cacheDictionary objectForKey:paramURLAsString];

  /* Use of Boolean values to make it easier for us
   to understand the logic behind caching */
  BOOL    fileHasBeenCached = NO;
```

```
BOOL    cachedFileHasExpired = NO;
BOOL    cachedFileExists = NO;
BOOL    cachedFileDataCanBeLoaded = NO;
NSData  *cachedFileData = nil;
BOOL    cachedFileIsFullyDownloaded = NO;
BOOL    cachedFileIsBeingDownloaded = NO;

NSDate    *expiryDate = nil;
NSDate    *downloadEndDate = nil;
NSDate    *downloadStartDate = nil;
NSString  *localURL = nil;
NSNumber  *expiresInSeconds = nil;
NSDate    *now = [NSDate date];

if (itemDictionary != nil){
  fileHasBeenCached = YES;
}

if (fileHasBeenCached == YES){

  expiryDate = [itemDictionary
              objectForKey:CachedKeyExpiryDate];

  downloadEndDate = [itemDictionary
                  objectForKey:CachedKeyDownloadEndDate];

  downloadStartDate = [itemDictionary
                    objectForKey:CachedKeyDownloadStartDate];

  localURL = [itemDictionary
            objectForKey:CachedKeyLocalURL];

  expiresInSeconds = [itemDictionary
                    objectForKey:CachedKeyExpiresInSeconds];

  if (downloadEndDate != nil &&
      downloadStartDate != nil){
    cachedFileIsFullyDownloaded = YES;
  }

  /* If we know the number of seconds, after the download
   process is finished, when the content expires and we
   still do not have the time when the download has finished,
   that means we have added this content to the list but have not
   yet downloaded it fully */
  if (expiresInSeconds != nil &&
      downloadEndDate == nil){
    cachedFileIsBeingDownloaded = YES;
  }

  /* If the expiry date is less than today's date then this
   means that the content has expired */
  if (expiryDate != nil &&
      [now timeIntervalSinceDate:expiryDate] > 0.0){
    cachedFileHasExpired = YES;
```

```objc
    }

    if (cachedFileHasExpired == NO){
      /* The Cached File Has Not Expired. Try to load it and also
       update the expiry date */
      NSFileManager *fileManager = [[NSFileManager alloc] init];

      if ([fileManager fileExistsAtPath:localURL] == YES){
        cachedFileExists = YES;
        cachedFileData = [NSData dataWithContentsOfFile:localURL];
        if (cachedFileData != nil){
          cachedFileDataCanBeLoaded = YES;
        } /* if (cachedFileData != nil){ */
      } /* if ([fileManager fileExistsAtPath:localURL] == YES){ */

      [fileManager release];

      /* Update the expiry date. If the file has been
       downloaded and has not expired yet but a new expiry
       number of seconds is proposed and we are asked to
       update the expiry date,
       then we must update the expiry date */

      if (paramUpdateExpiryDateIfInCache == YES){

        NSDate *newExpiryDate =
        [NSDate dateWithTimeIntervalSinceNow:
         paramURLMustExpireInSeconds];

        NSLog(@"Updating the expiry date from %@ to %@.",
              expiryDate,
              newExpiryDate);

        [itemDictionary setObject:newExpiryDate
                           forKey:CachedKeyExpiryDate];

        NSNumber *expires =
        [NSNumber numberWithFloat:paramURLMustExpireInSeconds];

        [itemDictionary setObject:expires
                           forKey:CachedKeyExpiresInSeconds];
      }

    } /* if (cachedFileHasExpired == NO){ */

  }

  if (cachedFileIsBeingDownloaded == YES){
    NSLog(@"This file is already being downloaded...");
    return(YES);
  }

  if (fileHasBeenCached == YES){

    if (cachedFileHasExpired == NO &&
```

```objc
        cachedFileExists == YES &&
        cachedFileDataCanBeLoaded == YES &&
        [cachedFileData length] > 0 &&
        cachedFileIsFullyDownloaded == YES){

    /* The item has been cached and its data has been read.
     Also, it has not yet been expired so let's
     just return this data */

    NSLog(@"This file has been cached and not expired yet.");

    [self.delegate
     cachedDownloadManagerSucceeded:self
     remoteURL:[NSURL URLWithString:paramURLAsString]
     localURL:[NSURL URLWithString:localURL]
     aboutToBeReleasedData:cachedFileData
     isCachedData:YES];

    return(YES);

  } else {
    /* The file was not cached or not cached properly */
    NSLog(@"This file is not properly cached.");
    [self.cacheDictionary removeObjectForKey:paramURLAsString];
    [self saveCacheDictionary];
  } /* if (cachedFileHasExpired == NO && */

} /* if (fileHasBeenCached == YES){ */

/* Now let's try to download the file */


NSNumber *expires =
[NSNumber numberWithFloat:paramURLMustExpireInSeconds];

NSMutableDictionary *newDictionary =
[[[NSMutableDictionary alloc] init] autorelease];

[newDictionary setObject:expires
                forKey:CachedKeyExpiresInSeconds];


localURL = [paramURLAsString
            stringByAddingPercentEscapesUsingEncoding:
            NSUTF8StringEncoding];

localURL = [localURL stringByReplacingOccurrencesOfString:@"://"
                                               withString:@""];

localURL = [localURL stringByReplacingOccurrencesOfString:@"/"
                                               withString:@"$"];

localURL = [localURL stringByAppendingPathExtension:@"cache"];

NSString *documentsDirectory =
```

```
    [self documentsDirectoryWithTrailingSlash:NO];

    localURL = [documentsDirectory
                    stringByAppendingPathComponent:localURL];

    [newDictionary setObject:localURL
                        forKey:CachedKeyLocalURL];

    [newDictionary setObject:now
                        forKey:CachedKeyDownloadStartDate];

    [self.cacheDictionary setObject:newDictionary
                                forKey:paramURLAsString];

    [self saveCacheDictionary];

    CacheItem *item = [[[CacheItem alloc] init] autorelease];
    [item setDelegate:self];
    [item startDownloadingURL:paramURLAsString];

    return(result);

}
```

We are changing the remote URL names to acceptable local URLs. The rules we apply to any remote URL are simple:

1. We remove all occurrences of :// and replace them with an empty string.

   We are simply cutting them out of the URL because we are going to use this URL as the filename on the disk. The filename must conform to Mac OS X conventions, so the operating system may alter certain characters to make it conform. For instance, if you try to save a file on your Mac OS X computer with the name *foo:bar.txt* (note the colon), the operating system will change the colon character to a dash (-) immediately as you type.

2. We replace all occurrences of the forward slash character (/) with a dollar sign.

   Aside from the fact that we cannot have consecutive forward slashes in a URL path on iOS, in order to avoid confusion when we create the final local URLs, it is better that we have forward slashes just as a way to distinguish folders, not files. For instance, if the remote resource "*http://www.OReilly.com/authors*" is downloaded, after removing the "://" part, as explained before, we will end up with "httpwww.OReilly.com/authors". Now imagine storing a file under this name in the *documents* folder of your application whose path might look like this: "/var/Mobile/Applications/<ApplicationID>/Documents/httpwww.OReilly.com/authors". This will later confuse us by making us think there is a folder under the *documents* folder, named *authors*. To avoid this confusion, we change all occurrences of the forward slash character in the URL to a dollar sign so that the resultant remote URL looks like this: "/var/Mobile/Applications/<ApplicationID>/Documents/httpwww.OReilly.com$authors".

3. We append the ".cache" extension to every local URL that we construct to make it easier for us to later spot the files in the *documents* directory of our application should we need to perform a file-search operation.

So, extending the "*http://www.OReilly.com/authors*" example in the preceding step, our final URL will be similar to this: "/var/Mobile/Applications/<ApplicationID>/Documents/httpwww.OReilly.com$authors.cache".

This code might look big, but in reality, all it is doing is checking the metadata dictionary of the given item to see whether it has already been downloaded. Concisely explained, this method does the following:

1. It checks whether the given remote URL is the key of any metadata dictionary in the `cacheDictionary` property. If not, it will attempt to download it.

2. If the file's metadata dictionary has already been added to the `cacheDictionary` property, it attempts to see whether a download end date value has been set for its `DownloadEndDate` key. If not, this means the file caching has begun but the downloading process has not yet finished. In this case, the code will return without downloading the resource because it believes the resource is being downloaded at the moment.

3. If the file has all the required metadata including the download end date, the corresponding cached local version of the remote file exists on disk, and the file has not expired yet, the code will read the contents of the local copy of the file and avoid downloading the file again.

4. If the file seems to have already been downloaded and cached but has expired, it will remove the file from the cache dictionary and will attempt to download it again.

Once you have this road map in mind, it will be much easier for you to understand how this code works.

As you may have noticed, at the end of the `download:urlMustExpireInSeconds:update` `ExpiryDateIfInCache:` method, we use `autorelease` instances of the `CacheItem` class to download the contents of the requested remote file. This class is just responsible for downloading files and returning their data to us. It doesn't really do any caching. The header file of the `CacheItem` class is:

```
#import <Foundation/Foundation.h>

@class CacheItem;

@protocol CacheItemDelegate <NSObject>

- (void) cacheItemDelegateSucceeded
  :(CacheItem *)paramSender
  withRemoteURL:(NSURL *)paramRemoteURL
  withAboutToBeReleasedData:(NSData *)paramAboutToBeReleasedData;


- (void) cacheItemDelegateFailed
```

```
   :(CacheItem *)paramSender
   remoteURL:(NSURL *)paramRemoteURL
   withError:(NSError *)paramError;

@end

@interface CacheItem : NSObject {
@public
  id<CacheItemDelegate> delegate;
  NSString              *remoteURL;
@private
  BOOL                  isDownloading;
  NSMutableData         *connectionData;
  NSURLConnection       *connection;
}

@property (nonatomic, retain) id<CacheItemDelegate> delegate;
@property (nonatomic, retain) NSString  *remoteURL;
@property (nonatomic, assign) BOOL         isDownloading;
@property (nonatomic, retain) NSMutableData *connectionData;
@property (nonatomic, retain) NSURLConnection *connection;

- (BOOL) startDownloadingURL:(NSString *)paramRemoteURL;

@end
```

 The delegate property is of type retain because the CacheDownloadMan ager class creates instances of the CacheItem class as autorelease objects and does not retain them anywhere. For this reason, if the CacheDown loadManager, which will become the delegate of every instance of the CacheItem class, gets deallocated, the CacheItem instances will crash our application by sending delegate messages to a deallocated instance of the CacheDownloadManager class. Therefore, we will retain our delegate (CacheDownloadManager) here and release it once we are done.

Now we will implement the CacheItem class in this way:

```
#import "CacheItem.h"

@implementation CacheItem

@synthesize delegate;
@synthesize remoteURL;
@synthesize isDownloading;
@synthesize connectionData;
@synthesize connection;

- (void)connection:(NSURLConnection *)connection
didReceiveResponse:(NSURLResponse *)response{

  [self.connectionData setLength:0];
}
```

```objc
- (void)connection:(NSURLConnection *)connection
  didReceiveData:(NSData *)data{

  [self.connectionData appendData:data];

}

- (void)connectionDidFinishLoading:(NSURLConnection *)connection{

  self.isDownloading = NO;

  if (self.delegate != nil){
    [self.delegate
     cacheItemDelegateSucceeded:self
     withRemoteURL:[NSURL URLWithString:self.remoteURL]
     withAboutToBeReleasedData:self.connectionData];
  }

  self.connectionData = nil;
}
- (void)connection:(NSURLConnection *)connection
  didFailWithError:(NSError *)error{

  if (self.delegate != nil){
    [self.delegate
     cacheItemDelegateFailed:self
     remoteURL:[NSURL URLWithString:self.remoteURL]
     withError:error];
  }

  self.connectionData = nil;

  self.isDownloading = NO;

}

- (BOOL) startDownloadingURL:(NSString *)paramRemoteURL{

  /* we must have a delegate */
  BOOL result = NO;

  if (self.isDownloading == YES ||
      [paramRemoteURL length] == 0){
    return(NO);
  }

  self.isDownloading = YES;

  self.remoteURL = paramRemoteURL;

  self.connectionData = nil;

  NSMutableData *data = [[NSMutableData alloc] init];
```

```
    self.connectionData = data;
    [data release];

    NSURL *url = [NSURL URLWithString:paramRemoteURL];

    NSURLRequest *request = [NSURLRequest requestWithURL:url];

    NSURLConnection *newConnection = [[NSURLConnection alloc]
                                      initWithRequest:request
                                      delegate:self
                                      startImmediately:YES];
    self.connection = newConnection;
    [newConnection release];

    if (self.connection != nil){
      result = YES;
    }

    return(result);

}

- (void) dealloc{
  if (connection != nil){
    [connection cancel];
  }
  [connection release];
  [connectionData release];
  [delegate release];
  [remoteURL release];
  [super dealloc];
}

@end
```

As you can see, we are not doing anything complex in the CacheItem class. We are just downloading items and sending messages back to our delegate. Nevertheless, the way CacheDownloadManager handles the delegate messages of every CacheItem is important:

```
- (void) cacheItemDelegateSucceeded:(CacheItem *)paramSender
         withRemoteURL:(NSURL *)paramRemoteURL
         withAboutToBeReleasedData:(NSData *)paramAboutToBeReleasedData{

  /* Now let's see which item we just downloaded */

  NSMutableDictionary *dictionary =
  [self.cacheDictionary objectForKey:[paramRemoteURL absoluteString]];

  NSDate *now = [NSDate date];

  NSNumber *expiresInSeconds = [dictionary
                                objectForKey:CachedKeyExpiresInSeconds];

  NSTimeInterval expirySeconds = [expiresInSeconds floatValue];
```

```objectivec
[dictionary setObject:[NSDate date]
                forKey:CachedKeyDownloadEndDate];

[dictionary setObject:[now dateByAddingTimeInterval:expirySeconds]
                forKey:CachedKeyExpiryDate];

[self saveCacheDictionary];

NSString *localURL = [dictionary objectForKey:CachedKeyLocalURL];

/* Now let's save the contents of the file to a disk */
if ([paramAboutToBeReleasedData writeToFile:localURL
                                  atomically:YES] == YES){
  NSLog(@"Successfully cached the file on disk.");
} else{
  NSLog(@"Failed to cache the file on disk.");
}

[self.delegate
 cachedDownloadManagerSucceeded:self
 remoteURL:paramRemoteURL
 localURL:[NSURL URLWithString:localURL]
 aboutToBeReleasedData:paramAboutToBeReleasedData
 isCachedData:NO];


}

- (void) cacheItemDelegateFailed:(CacheItem *)paramSender
                       remoteURL:(NSURL *)paramRemoteURL
                       withError:(NSError *)paramError{

  /* We have to remove the current item from the list and send
   a delegate message */

  if (self.delegate != nil){

    NSMutableDictionary *dictionary =
    [self.cacheDictionary
     objectForKey:[paramRemoteURL absoluteString]];

    NSString *localURL = [dictionary
                            objectForKey:CachedKeyLocalURL];

    [self.delegate
     cachedDownloadManagerFailed:self
     remoteURL:paramRemoteURL
     localURL:[NSURL URLWithString:localURL]
     withError:paramError];
  }

  [self.cacheDictionary
   removeObjectForKey:[paramRemoteURL absoluteString]];

}
```

Upon completion of the download of any of the files, we will complete the information that we store as the files' metadata in our `cacheDictionary` property. This information includes their download end date and the expiration time of the resource. We will also remove this item from the metadata dictionary if the download process fails. In both cases, we are calling the cache download manager's delegate about the actual event that has occurred.

Last but not least, and before we forget, we need to implement the deallocation method of our cache download manager object:

```
- (void) dealloc {

  if (cacheDictionary != nil){
    [self saveCacheDictionary];
  }
  [cacheDictionary release];
  [cacheDictionaryPath release];

  [super dealloc];

}
```

Now go ahead and give the `CacheDownloadManager` class a try. Try to download many files with it at the same time and experiment with different expiration dates for contents.

 Although caching files on disk could be implemented in an even simpler way, keeping the metadata of files the way we did is generally a good idea, as your application could understand and fulfill the request for a file to be downloaded more intelligently.

One simple implementation of a caching mechanism that utilizes disk storage would be to download files to disk and save them in the *documents* folder. The next time the same resource is requested, search in the *documents* folder for that file, and if it exists, do not attempt to download it again. However, as you can see, no mechanism here is mentioned for dismissing files that might have been downloaded long before the new resource is requested, and with this method, we are running the risk of not always having an up-to-date copy of our remote resources. But, as mentioned, depending on how you would like your remote resources to be downloaded, you can implement different types of caching mechanisms.

CHAPTER 7

Operations, Threads, and Timers

7.0 Introduction

This chapter shows how to run multiple tasks on iOS. The basic operating system mechanism for spinning off tasks is threading, but most iOS programmers prefer higher-level abstractions such as *operations*. Timers are another useful feature for handling separate tasks.

Operations can be configured to run a block of code synchronously or asynchronously. You can manage operations manually or place them on *operation queues*, which facilitate concurrency so that you do not need to think about the underlying thread management. In this chapter, you will see how to use operations and operation queues, as well as basic threads and timers, to synchronously and asynchronously execute tasks in applications.

Cocoa provides three different types of operations:

Block operations
> These facilitate the execution of one or more block objects.

Invocation operations
> These allow you to invoke a method in another, currently existing object.

Plain operations
> These are plain operation classes that need to be subclassed. The code to be executed will be written inside the `main` method of the operation object.

Operations, as mentioned before, can be managed with operation queues, which have the data type `NSOperationQueue`. After instantiating any of the aforementioned operation types (block, invocation, or plain operation), you can add them to an operation queue and have the queue manage the operation.

An operation object can have dependencies on other operation objects and be instructed to wait for the completion of one or more operations before executing the task associated with it. Unless you add a dependency, you have no control over the order in which operations run. For instance, adding them to a queue in a certain order does not guarantee that they will execute in that order, despite the use of the term *queue*.

There are a few important things to bear in mind while working with operation queues and operations:

- Operations, by default, run on the thread that starts them, using their `start` instance method. If you want the operations to work asynchronously, you will have to use either an operation queue or a subclass `NSOperation` and detach a new thread on the `main` instance method of the operation.

- An operation can wait for the execution of another operation to finish before it starts its execution. Be careful not to create interdependent operations, a common mistake known as a *race condition*. In other words, do not tell operation A to depend on operation B if B already depends on A; this will cause both to wait forever, taking up memory and possibly hanging your application.

- Operations can be cancelled. So, if you have subclassed `NSOperation` to create custom operation objects, you have to make sure to use the `isCancelled` instance method to check whether the operation has cancelled before executing the task associated with the operation. For instance, if your operation's task is to check for the availability of an Internet connection every 20 seconds, it must call the `isCancelled` instance method at the beginning of each run to make sure it has not been cancelled before attempting to check for an Internet connection again. If the operation takes more than a few seconds (such as when you download a file), you should also check `isCancelled` periodically while running the task.

- Operation objects are key-value observing (KVO) compliant on various key paths such as `isFinished`, `isReady`, and `isExecuting`. Discussion of KVO is outside the scope of this book. Consult Apple's "Introduction to Key-Value Observing Programming Guide" document to get a better understanding of how KVO works.

- If you plan to subclass `NSOperation` and provide a custom implementation for the operation, you must create your own autorelease pool in the `main` method of the operation, which gets called from the `start` method. We will discuss this in detail later in this chapter.

- Always keep a reference to the operation objects you create. The concurrent nature of operation queues might make it impossible for you to retrieve a reference to an operation after it has been added to the queue.

Threads and timers are objects, subclassing `NSObject`. Threads are more low-level than timers. When an application runs under iOS, the operating system creates at least one thread for that application, called the main thread. Every thread and timer must be added to a run loop. A run loop, as its name implies, is a loop during which different events can occur, such as a timer firing or a thread running. Discussion of run loops is beyond the scope of this chapter, but we will refer to them here and there in recipes.

Think of a run loop as a kind of loop that has a starting point, a condition for finishing, and a series of events to process during its lifetime. A thread or timer is attached to a run loop and in fact requires a run loop to function.

The main thread of an application is the thread that handles the UI events. If you perform a long-running task on the main thread, you will notice that the UI of your application will become unresponsive or slow to respond. To avoid this, you can create separate threads and/or timers, each of which performs its own task (even if it is a long-running task) but will not block the main thread.

7.1 Running Tasks Synchronously

Problem

You want to run a series of tasks synchronously.

Solution

Create operations and start them manually:

```
#import <UIKit/UIKit.h>

@interface OperationsAppDelegate : NSObject
          <UIApplicationDelegate> {

  UIWindow *window;
  NSInvocationOperation   *simpleOperation;

}

@property (nonatomic, retain)
IBOutlet UIWindow *window;

@property (nonatomic, retain)
NSInvocationOperation   *simpleOperation;

@end
```

The implementation of the application delegate is as follows:

```
- (void) simpleOperationEntry:(id)paramObject{

  NSLog(@"Parameter Object = %@", paramObject);
  NSLog(@"Main Thread = %@", [NSThread mainThread]);
  NSLog(@"Current Thread = %@", [NSThread currentThread]);

}

- (BOOL)            application:(UIApplication *)application
  didFinishLaunchingWithOptions:(NSDictionary *)launchOptions {

  NSNumber *simpleObject = [NSNumber numberWithInteger:123];
```

```
NSInvocationOperation *newOperation =
[[NSInvocationOperation alloc]
 initWithTarget:self
 selector:@selector(simpleOperationEntry:)
 object:simpleObject];

self.simpleOperation = newOperation;
[newOperation release];

[self.simpleOperation start];

[window makeKeyAndVisible];

return YES;
}

- (void)dealloc {
[simpleOperation release];
[window release];
[super dealloc];
}
```

The output of this program (in the console window) will be similar to this:

```
Parameter Object = 123
Main Thread = <NSThread: 0x5f0d720>{name = (null), num = 1}
Current Thread = <NSThread: 0x5f0d720>{name = (null), num = 1}
```

As the name of this class implies (`NSInvocationOperation`), the main responsibility of an object of this type is to invoke a method in an object. This is the most straightforward way to invoke a method inside an object using operations.

Discussion

An invocation operation, as described in this chapter's Introduction, is able to invoke a method inside an object. "What is so special about this?" you might ask. The invocation operation's power can be demonstrated when it is added to an operation queue. With an operation queue, an invocation operation can invoke a method in a target object asynchronously and in parallel to the thread that started the operation. If you have a look at the output printed to the console (in this recipe's Solution), you will notice that the current thread inside the method invoked by the invocation operation is the same as the main thread since the main thread in the `application:didFinishLaun chingWithOptions:` method started the operation using its `start` method. In Recipe 7.2, we will learn how to take advantage of operation queues to run tasks asynchronously.

In addition to invocation operations, you can use block or plain operations to perform tasks synchronously. Here is an example using a block operation to count numbers from zero to 999 (inside the *.h* file of our application delegate):

```objc
#import <UIKit/UIKit.h>

@interface OperationsAppDelegate : NSObject
            <UIApplicationDelegate> {

  UIWindow *window;
  NSBlockOperation    *simpleOperation;

}

@property (nonatomic, retain)
IBOutlet UIWindow *window;

@property (nonatomic, retain)
NSBlockOperation    *simpleOperation;

@end
```

Here is the implementation of our application delegate (*.m* file):

```objc
#import "OperationsAppDelegate.h"

@implementation OperationsAppDelegate

@synthesize window;
@synthesize simpleOperation;

- (BOOL)              application:(UIApplication *)application
  didFinishLaunchingWithOptions:(NSDictionary *)launchOptions {

  /* Here is our block */
  NSBlockOperation *newBlockOperation =
  [NSBlockOperation blockOperationWithBlock:^{
    NSLog(@"Main Thread = %@", [NSThread mainThread]);
    NSLog(@"Current Thread = %@", [NSThread currentThread]);
    NSUInteger counter = 0;
    for (counter = 0;
         counter < 1000;
         counter++){
      NSLog(@"Count = %lu", (unsigned long)counter);
    }
  }];

  /* Make sure we keep the reference somewhere */
  self.simpleOperation = newBlockOperation;

  /* Start the operation */
  [self.simpleOperation start];

  /* Print something out just to test if we have to wait
   for the block to execute its code or not */
  NSLog(@"Main thread is here");

  [window makeKeyAndVisible];

  return YES;
```

```
}

- (void)dealloc {
  [simpleOperation release];
  [window release];
  [super dealloc];
}

@end
```

If we run our application, we will see the values 0 to 999 printed out to the screen
followed by the "Main thread is here" message, like this:

```
Main Thread = <NSThread: 0x5f0d720>{name = (null), num = 1}
Current Thread = <NSThread: 0x5f0d720>{name = (null), num = 1}
...
Count = 986
Count = 987
Count = 988
Count = 989
Count = 990
Count = 991
Count = 992
Count = 993
Count = 994
Count = 995
Count = 996
Count = 997
Count = 998
Count = 999
Main thread is here
```

This proves that since the block operation was started in the `application:didFinish`
`LaunchingWithOptions:` method, which itself runs on the main thread, the code inside
the block was also running on the main thread. The main point to take from the log
messages is that our operation blocked the main thread and the main thread's code
continued to be executed after the work for the block operation was done. This is a
very bad programming practice. In fact, iOS developers must perform any trick and use
any technique that they know of to keep the main thread responsive so that it can do
the key job of processing users' input. Here is what Apple has to say about this:

> You should be careful what work you perform from the main thread of your application.
> The main thread is where your application handles touch events and other user input.
> To ensure that your application is always responsive to the user, you should never use
> the main thread to perform long-running tasks or to perform tasks with a potentially
> unbounded end, such as tasks that access the network. Instead, you should always move
> those tasks onto background threads. The preferred way to do so is to wrap each task in
> an operation object and add it to an operation queue, but you can also create explicit
> threads yourself.

To read more about this subject, browse through the "Tuning for Performance and
Responsiveness" document in the iOS Reference Library, available at this URL:

---

*http://developer.apple.com/iphone/library/documentation/iphone/conceptual/ipho
neosprogrammingguide/Performance/Performance.html*

In addition to invocation and block operations, you can also subclass `NSOperation` and perform your task in that class. Before getting started, you must keep a few things in mind while subclassing `NSOperation`:

- If you are not planning on using an operation queue, you have to detach a new thread of your own in the `start` method of the operation. If you do not want to use an operation queue and you do not want your operation to run asynchronously with other operations that you start manually, you can simply call the `main` method of your operation inside the `start` method.

- Two important methods in an instance of `NSOperation` must be overridden by your own implementation of the operation: `isExecuting` and `isFinished`. These can be called from any other object. In these methods, you must return a thread-safe value that you can manipulate from inside the operation. As soon as your operation starts, you must, through KVO, inform any listeners that you are changing the values that these two methods return. We will see how this works in the example code.

- You must provide your own autorelease pool inside the `main` method of the operation in case your operation will be added to an operation queue at some point in the future. You must make sure your operations work in both ways: whether you start them manually or they get started by an operation queue.

- You must have an initialization method for your operations. There must be only one designated initializer method per operation. All other initializer methods, including the default `init` method of an operation, must call the designated initializer that has the most number of parameters. Other initializer methods must make sure they pass appropriate parameters (if any) to the designated initializer.

Here is the declaration of our operation object (*.h* file):

```
#import <Foundation/Foundation.h>

@interface CountingOperation : NSOperation {
@protected
    NSUInteger    startingCount;
    NSUInteger    endingCount;
    BOOL          finished;
    BOOL          executing;
}

/* Designated Initializer */
- (id) initWithStartingCount:(NSUInteger)paramStartingCount
              endingCount:(NSUInteger)paramEndingCount;

@end
```

The implementation (.*m* file) of the operation might be a bit long, but hopefully it's easy to understand:

```objc
#import "CountingOperation.h"

@implementation CountingOperation

- (id) init {
  return([self initWithStartingCount:0
                        endingCount:1000]);
}

- (id) initWithStartingCount:(NSUInteger)paramStartingCount
                 endingCount:(NSUInteger)paramEndingCount{

  self = [super init];

  if (self != nil){

    if (paramEndingCount <= paramStartingCount){
      /* The given data is not valid so fail
       the initialization */
      [self release];
      return(nil);
    }

    /* Keep these values for the main method */
    startingCount = paramStartingCount;
    endingCount = paramEndingCount;

  }

  return(self);

}

- (void) start {

  /* If we are cancelled before starting, then
   we have to return immediately and generate the
   required KVO notifications */
  if ([self isCancelled] == YES){
    /* If this operation *is* cancelled */
    /* KVO compliance */
    [self willChangeValueForKey:@"isFinished"];
    finished = YES;
    [self didChangeValueForKey:@"isFinished"];
    return;

  } else {
    /* If this operation is *not* cancelled */
    /* KVO compliance */
    [self willChangeValueForKey:@"isExecuting"];
    executing = YES;
    /* Call the main method from inside the start method */
```

```
    [self main];
    [self didChangeValueForKey:@"isExecuting"];

  }

}

- (void) main {

  @try {

    /* Here is our autorelease pool */
    NSAutoreleasePool *pool = [[NSAutoreleasePool alloc] init];

    /* Keep a local variable here that must get set to YES
     whenever we are done with the task */
    BOOL taskIsFinished = NO;

    /* Create a while loop here that only exists
     if the taskIsFinished variable is set to YES or
     the operation has been cancelled */
    while (taskIsFinished == NO &&
           [self isCancelled] == NO){

      /* Perform the task here */
      NSLog(@"Main Thread = %@", [NSThread mainThread]);
      NSLog(@"Current Thread = %@", [NSThread currentThread]);
      NSUInteger counter = startingCount;
      for (counter = startingCount;
           counter < endingCount;
           counter++){
        NSLog(@"Count = %lu", (unsigned long)counter);
      }
      /* Very important. This way we can get out of the
       loop while complying with the cancellation
       rules of operations */
      taskIsFinished = YES;

    }

    /* KVO compliance. Generate the
     required KVO notifications */
    [self willChangeValueForKey:@"isFinished"];
    [self willChangeValueForKey:@"isExecuting"];
    finished = YES;
    executing = NO;
    [self didChangeValueForKey:@"isFinished"];
    [self didChangeValueForKey:@"isExecuting"];

    /* Make sure we are releasing the autorelease pool */
    [pool release];

  }
  @catch (NSException * e) {
    NSLog(@"Exception %@", e);
```

```
    }

  }

  - (BOOL) isFinished{
    /* Simply return the value */
    return(finished);
  }

  - (BOOL) isExecuting{
    /* Simply return the value */
    return(executing);
  }

  @end
```

We can start this operation like so:

```
#import "OperationsAppDelegate.h"
#import "CountingOperation.h"

@implementation OperationsAppDelegate

@synthesize window;
@synthesize simpleOperation;

- (BOOL)                    application:(UIApplication *)application
  didFinishLaunchingWithOptions:(NSDictionary *)launchOptions {

  CountingOperation *newOperation = [[CountingOperation alloc]
                                      initWithStartingCount:0
                                      endingCount:1000];
  self.simpleOperation = newOperation;
  [newOperation release];

  [self.simpleOperation start];

  /* Print something out just to test if we have to wait
   for the block to execute its code or not */
  NSLog(@"Main thread is here");

  [window makeKeyAndVisible];

  return YES;
}

- (void)dealloc {
  [simpleOperation release];
  [window release];
  [super dealloc];
}

@end
```

If we run our code, we will see the following results in the console window, just as we did when we used a block operation:

```
Main Thread = <NSThread: 0x5f0d720>{name = (null), num = 1}
Current Thread = <NSThread: 0x5f0d720>{name = (null), num = 1}
...
Count = 986
Count = 987
Count = 988
Count = 989
Count = 990
Count = 991
Count = 992
Count = 993
Count = 994
Count = 995
Count = 996
Count = 997
Count = 998
Count = 999
Main thread is here
```

See Also

Recipe 7.2; Recipe 7.6

7.2 Running Tasks Asynchronously

Problem

You want to execute operations concurrently.

Solution

Use operation queues. Alternatively, subclass NSOperation and detach a new thread on the main method.

Discussion

As mentioned in Recipe 7.1, operations, by default, run on the thread that calls the start method. Usually we start operations on the main thread, but at the same time we expect our operations to run on their own threads and not take the main thread's time slice. The best solution for us would be to use operation queues. However, if you want to manage your operations manually, which I do not recommend, you can subclass NSOperation and detach a new thread on the main method. Please refer to Recipe 7.6 for more information about detached threads.

Let's go ahead and use an operation queue and add two simple invocation operations to it. (For more information about invocation operations, please refer to this chapter's Introduction. For additional example code on invocation operations, please refer to

Recipe 7.1.) Here is the declaration (.*h* file) of our application delegate that utilizes an operation queue and two invocation operations:

```
#import <UIKit/UIKit.h>

@interface OperationsAppDelegate : NSObject <UIApplicationDelegate> {
  UIWindow *window;
  NSOperationQueue        *operationQueue;
  NSInvocationOperation   *firstOperation;
  NSInvocationOperation   *secondOperation;

}

@property (nonatomic, retain) IBOutlet UIWindow *window;
@property (nonatomic, retain) NSOperationQueue        *operationQueue;
@property (nonatomic, retain) NSInvocationOperation *firstOperation;
@property (nonatomic, retain) NSInvocationOperation *secondOperation;

@end
```

The implementation (.*m* file) of the application delegate is as follows:

```
#import "OperationsAppDelegate.h"

@implementation OperationsAppDelegate

@synthesize window;
@synthesize firstOperation;
@synthesize secondOperation;
@synthesize operationQueue;

- (void) firstOperationEntry:(id)paramObject{

  NSLog(@"%s", __FUNCTION__);
  NSLog(@"Parameter Object = %@", paramObject);
  NSLog(@"Main Thread = %@", [NSThread mainThread]);
  NSLog(@"Current Thread = %@", [NSThread currentThread]);

}

- (void) secondOperationEntry:(id)paramObject{

  NSLog(@"%s", __FUNCTION__);
  NSLog(@"Parameter Object = %@", paramObject);
  NSLog(@"Main Thread = %@", [NSThread mainThread]);
  NSLog(@"Current Thread = %@", [NSThread currentThread]);

}

- (BOOL)              application:(UIApplication *)application
  didFinishLaunchingWithOptions:(NSDictionary *)launchOptions {

  NSNumber *firstNumber = [NSNumber numberWithInteger:111];
  NSNumber *secondNumber = [NSNumber numberWithInteger:222];

  /* Instantiate the first invocation operation */
```

```
    NSInvocationOperation *firstInvocation =
    [[NSInvocationOperation alloc]
     initWithTarget:self
     selector:@selector(firstOperationEntry:)
     object:firstNumber];
    self.firstOperation = firstInvocation;
    [firstInvocation release];

    /* Instantiate the second invocation operation */
    NSInvocationOperation *secondInvocation =
    [[NSInvocationOperation alloc]
     initWithTarget:self
     selector:@selector(secondOperationEntry:)
     object:secondNumber];
    self.secondOperation = secondInvocation;
    [secondInvocation release];

    /* Instantiate the operation queue */
    NSOperationQueue  *newOperationQueue =
    [[NSOperationQueue alloc] init];
    self.operationQueue = newOperationQueue;
    [newOperationQueue release];

    /* Add the operations to the queue */
    [self.operationQueue addOperation:self.firstOperation];
    [self.operationQueue addOperation:self.secondOperation];

    NSLog(@"Main thread is here");

    [window makeKeyAndVisible];

    return YES;
}

- (void)dealloc {
    [firstOperation release];
    [secondOperation release];
    [operationQueue release];
    [window release];
    [super dealloc];
}
```

Here is what is happening in the implementation of our code:

- We have two methods: firstOperationEntry: and secondOperationEntry:. Each method accepts an object as a parameter and prints out the current thread, the main thread, and the parameter to the console window. These are the entry methods of our invocation operations that will be added to an operation queue.

- We initialize two objects of type NSInvocationOperation and set the target selector to each operation entry point described previously.

- We then initialize an object of type NSOperationQueue. (It could also be created before the entry methods.) The queue object will be responsible for managing the concurrency in the operation objects.

- We invoke the addOperation: instance method of NSOperationQueue to add each invocation operation to the operation queue. At this point, the operation queue may or may not immediately start the invocation operations through their start methods. However, it is very important to bear in mind that after adding operations to an operation queue, you must not start the operations manually. You must leave this to the operation queue.

Now let's run the example code once and see the results in the console window:

```
Main thread is here
-[OperationsAppDelegate firstOperationEntry:]
-[OperationsAppDelegate secondOperationEntry:]
Parameter Object = 111
Parameter Object = 222
Main Thread = <NSThread: 0x5f0d720>{name = (null), num = 1}
Main Thread = <NSThread: 0x5f0d720>{name = (null), num = 1}
Current Thread = <NSThread: 0x5f2d9c0>{name = (null), num = 3}
Current Thread = <NSThread: 0x5f47110>{name = (null), num = 4}
```

Brilliant! This proves that our invocation operations are running on their own threads in parallel to the main thread without blocking the main thread at all. Now let's run the same code a couple more times and observe the output in the console window. If you do this, chances are that you will get a result that is completely different, such as this:

```
Main thread is here
-[OperationsAppDelegate firstOperationEntry:]
Parameter Object = 111
Main Thread = <NSThread: 0x5f0d720>{name = (null), num = 1}
Current Thread = <NSThread: 0x5f18e40>{name = (null), num = 3}
-[OperationsAppDelegate secondOperationEntry:]
Parameter Object = 222
Main Thread = <NSThread: 0x5f0d720>{name = (null), num = 1}
Current Thread = <NSThread: 0x6a3b5f0>{name = (null), num = 4}
```

You can clearly observe that the main thread is not blocked and that both invocation operations are running in parallel with the main thread. This just proves the concurrency in the operation queue when two nonconcurrent operations are added to it. The operation queue manages the threads required to run the operations.

If we were to subclass NSOperation and add the instances of the new class to an operation queue, we would do things slightly differently. Keep a few things in mind:

- Plain operations that subclass NSOperation, when added to an operation queue, will run asynchronously. For this reason, you must override the isConcurrent instance method of NSOperation and return the value YES.
- You must prepare your operation for cancellation by checking the value of the isCancelled method periodically while performing the main task of the operation and in the start method before you even run the operation. The start method will get called by the operation queue in this case after the operation is added to the queue. In this method, check whether the operation is cancelled using the

`isCancelled` method. If the operation is cancelled, simply return from the `start` method. If not, call the `main` method from inside the `start` method.

- Override the `main` method with your own implementation of the main task that is to be carried out by the operation. Make sure to allocate and initialize your own autorelease pool in this method and to release the pool just before returning.

- Override the `isFinished` and `isExecuting` methods of your operation and return appropriate `BOOL` values to reveal whether the operation is finished or is executing at the time.

Here is the declaration (.*h* file) of our operation:

```
#import <Foundation/Foundation.h>

@interface SimpleOperation : NSOperation {
@protected
  NSObject        *givenObject;
  BOOL            finished;
  BOOL            executing;
}

/* Designated Initializer */
- (id) initWithObject:(NSObject *)paramObject;

@end
```

The implementation of the operation is as follows:

```
#import "SimpleOperation.h"

@implementation SimpleOperation

- (id) init {
  NSNumber *dummyObject = [NSNumber numberWithInteger:123];
  return([self initWithObject:dummyObject]);
}

- (id) initWithObject:(NSObject *)paramObject{

  self = [super init];

  if (self != nil){

    if (paramObject == nil){
      /* The given data is not valid so fail
       the initialization */
      [self release];
      return(nil);
    }

    /* Keep these values for the main method */
    givenObject = [paramObject retain];

  }
```

```objc
    return(self);

}

- (void) start {

  /* If we are cancelled before starting, then
   we have to return immediately and generate the
   required KVO notifications */
  if ([self isCancelled] == YES){
    /* If this operation *is* cancelled */
    /* KVO compliance */
    [self willChangeValueForKey:@"isFinished"];
    finished = YES;
    [self didChangeValueForKey:@"isFinished"];
    return;

  } else {
    /* If this operation is *not* cancelled */
    /* KVO compliance */
    [self willChangeValueForKey:@"isExecuting"];
    executing = YES;
    /* Call the main method from inside the start method */
    [self main];
    [self didChangeValueForKey:@"isExecuting"];

  }

}

- (void) main {

  @try {

    /* Here is our autorelease pool */
    NSAutoreleasePool *pool = [[NSAutoreleasePool alloc] init];

    /* Keep a local variable here that must get set to YES
     whenever we are done with the task */
    BOOL taskIsFinished = NO;

    /* Create a while loop here that only exists
     if the taskIsFinished variable is set to YES or
     the operation has been cancelled */
    while (taskIsFinished == NO &&
           [self isCancelled] == NO){

      /* Perform the task here */
      NSLog(@"%s", __FUNCTION__);
      NSLog(@"Parameter Object = %@", givenObject);
      NSLog(@"Main Thread = %@", [NSThread mainThread]);
      NSLog(@"Current Thread = %@", [NSThread currentThread]);

      /* Very important. This way we can get out of the
       loop and we are still complying with the cancellation
```

```
          rules of operations */
      taskIsFinished = YES;


    }

      /* KVO compliance. Generate the
       required KVO notifications */
      [self willChangeValueForKey:@"isFinished"];
      [self willChangeValueForKey:@"isExecuting"];
      finished = YES;
      executing = NO;
      [self didChangeValueForKey:@"isFinished"];
      [self didChangeValueForKey:@"isExecuting"];

      /* Make sure we are releasing the autorelease pool */
      [pool release];

    }
    @catch (NSException * e) {
      NSLog(@"Exception %@", e);
    }

}

- (BOOL)  isConcurrent{
   return(YES);
}

- (BOOL)  isFinished{
   /* Simply return the value */
   return(finished);
}

- (BOOL)  isExecuting{
   /* Simply return the value */
   return(executing);
}

@end
```

You can now use this operation class in any other class, such as your application delegate. Here is the declaration of our application delegate to utilize this new operation class and add it in an operation queue:

```
#import <UIKit/UIKit.h>

@class SimpleOperation;

@interface OperationsAppDelegate : NSObject <UIApplicationDelegate> {
   UIWindow *window;
   NSOperationQueue   *operationQueue;
   SimpleOperation    *firstOperation;
   SimpleOperation    *secondOperation;

}
```

```objc
@property (nonatomic, retain) IBOutlet UIWindow *window;
@property (nonatomic, retain) NSOperationQueue *operationQueue;
@property (nonatomic, retain) SimpleOperation *firstOperation;
@property (nonatomic, retain) SimpleOperation *secondOperation;

@end
```

The implementation of our application delegate is as follows:

```objc
#import "OperationsAppDelegate.h"
#import "SimpleOperation.h"

@implementation OperationsAppDelegate

@synthesize window;
@synthesize firstOperation;
@synthesize secondOperation;
@synthesize operationQueue;

- (BOOL)              application:(UIApplication *)application
  didFinishLaunchingWithOptions:(NSDictionary *)launchOptions {

  NSNumber *firstNumber = [NSNumber numberWithInteger:111];
  NSNumber *secondNumber = [NSNumber numberWithInteger:222];

  SimpleOperation *operationOne =
  [[SimpleOperation alloc]
   initWithObject:firstNumber];
  self.firstOperation = operationOne;
  [operationOne release];

  SimpleOperation *operationTwo =
  [[SimpleOperation alloc]
   initWithObject:secondNumber];
  self.secondOperation = operationTwo;
  [operationTwo release];

  /* Instantiate the operation queue */
  NSOperationQueue *newOperationQueue =
  [[NSOperationQueue alloc] init];
  self.operationQueue = newOperationQueue;
  [newOperationQueue release];

  /* Add the operations to the queue */
  [self.operationQueue addOperation:self.firstOperation];
  [self.operationQueue addOperation:self.secondOperation];

  NSLog(@"Main thread is here");

  [window makeKeyAndVisible];

  return YES;
}

- (void)dealloc {
```

```
    [firstOperation release];
    [secondOperation release];
    [operationQueue release];
    [window release];
    [super dealloc];
}
```

The results printed to the console window will be similar to what we saw earlier when
we used concurrent invocation operations:

```
Main thread is here
-[SimpleOperation main]
-[SimpleOperation main]
Parameter Object = 111
Parameter Object = 222
Main Thread = <NSThread: 0x5f0d730>{name = (null), num = 1}
Main Thread = <NSThread: 0x5f0d730>{name = (null), num = 1}
Current Thread = <NSThread: 0x5f34df0>{name = (null), num = 3}
Current Thread = <NSThread: 0x5f20390>{name = (null), num = 4}
```

7.3 Creating a Dependency Between Tasks

Problem

You want to start a certain task only after another task has finished executing.

Solution

If operation B has to wait for operation A before it can run the task associated with it,
operation B has to add operation A as its dependency using the addDependency: instance
method of NSOperation, as shown here:

```
    [self.firstOperation addDependency:self.secondOperation];
```

Both the firstOperation and the secondOperation properties are of type NSInvocation
Operation, as we will see in this recipe's Discussion. In this example code, the first
operation will not be executed by the operation queue until after the second operation's
task is finished.

Discussion

An operation will not start executing until all the operations on which it depends have
successfully finished executing the tasks associated with them. By default, an operation,
after initialization, has no dependency on other operations.

If we want to introduce dependencies to the example code described in Recipe 7.2, we can slightly modify our application delegate's implementation and use the **addDependency:** instance method to have the first operation wait for the second operation:

```objc
#import "OperationsAppDelegate.h"

@implementation OperationsAppDelegate

@synthesize window;
@synthesize firstOperation;
@synthesize secondOperation;
@synthesize operationQueue;

- (void) firstOperationEntry:(id)paramObject{

  NSLog(@"First Operation - Parameter Object = %@",
        paramObject);

  NSLog(@"First Operation - Main Thread = %@",
        [NSThread mainThread]);

  NSLog(@"First Operation - Current Thread = %@",
        [NSThread currentThread]);

}

- (void) secondOperationEntry:(id)paramObject{

  NSLog(@"Second Operation - Parameter Object = %@",
        paramObject);

  NSLog(@"Second Operation - Main Thread = %@",
        [NSThread mainThread]);

  NSLog(@"Second Operation - Current Thread = %@",
        [NSThread currentThread]);

}

- (BOOL)              application:(UIApplication *)application
  didFinishLaunchingWithOptions:(NSDictionary *)launchOptions {

  NSNumber *firstNumber = [NSNumber numberWithInteger:111];
  NSNumber *secondNumber = [NSNumber numberWithInteger:222];

  /* Instantiate the first invocation operation */
  NSInvocationOperation *firstInvocation =
  [[NSInvocationOperation alloc]
   initWithTarget:self
   selector:@selector(firstOperationEntry:)
   object:firstNumber];
  self.firstOperation = firstInvocation;
  [firstInvocation release];
```

```
/* Instantiate the second invocation operation */
NSInvocationOperation *secondInvocation =
[[NSInvocationOperation alloc]
 initWithTarget:self
 selector:@selector(secondOperationEntry:)
 object:secondNumber];
self.secondOperation = secondInvocation;
[secondInvocation release];

[self.firstOperation addDependency:self.secondOperation];

/* Instantiate the operation queue */
NSOperationQueue  *newOperationQueue =
[[NSOperationQueue alloc] init];
self.operationQueue = newOperationQueue;
[newOperationQueue release];

/* Add the operations to the queue */
[self.operationQueue addOperation:self.firstOperation];
[self.operationQueue addOperation:self.secondOperation];

NSLog(@"Main thread is here");

[window makeKeyAndVisible];

return YES;
}

- (void)dealloc {
[firstOperation release];
[secondOperation release];
[operationQueue release];
[window release];
[super dealloc];
}
```

Now if you execute the program, you will see a result similar to this in the console window:

```
Main thread is here
Second Operation - Parameter Object = 222
Second Operation - Main Thread = <NSThread: 0x5f0d720>
                                {name = (null), num = 1}
Second Operation - Current Thread = <NSThread: 0x5f1efd0>
                                   {name = (null), num = 3}
First Operation - Parameter Object = 111
First Operation - Main Thread = <NSThread: 0x5f0d720>
                               {name = (null), num = 1}
First Operation - Current Thread = <NSThread: 0x5f1efd0>
                                  {name = (null), num = 3}
```

It's quite obvious that although the operation queue attempted to run both operations in parallel, the first operation had a dependency on the second operation, and therefore the second operation had to finish before the first operation could be run.

If, at any time, you want to break the dependency between two operations, you can use the removeDependency: instance method of an operation object.

See Also

Recipe 7.1; Recipe 7.2

7.4 Performing a Task After a Delay

Problem

You have a method in one of your objects that needs to be called after a certain number of seconds have passed.

Solution

Use the performSelector:withObject:afterDelay: instance method of an NSObject.

Discussion

Suppose you have attempted to download a file and the network connection has failed to access the remote URL. Now you have two choices: abort the whole operation or attempt to download the file again. If you choose to reattempt the download, you have two options again. Either you initialize another connection immediately after the network failure or wait a couple of seconds before you try one more time.

Performing a task immediately is not a problem, but what if you want to initialize another connection after, for instance, three seconds? You can use the performSelector:withObject:afterDelay: instance method of any NSObject to perform a selector on that object after a given number of seconds:

```
- (void) connectionHasFailedWithError:(NSError *)paramError
                        onRemoteURL:(NSURL *)paramRemoteURL{

  /* We failed to download the file. Attempt to download it again
   after 3 seconds */

  [self performSelector:@selector(attemptToDownloadRemoteURL:)
          withObject:paramRemoteURL
          afterDelay:3.0f];

}

- (void) attemptToDownloadRemoteURL:(NSURL *)paramRemoteURL{

  /* Attempt to download the remote file again here by initializing
   a new connection ... */

}
```

As you can see, we are passing an object to a method that has to be invoked after three seconds.

The receiver object and the object passed with the `performSelector:with Object:afterDelay:` method will be retained after the method issuing the call terminates, and will be released after the target method is called. If you decide, after performing a selector with delay, that you want to cancel the invocation of this selector, you can call either the `cancelPre viousPerformRequestsWithTarget:` or `cancelPreviousPerformRequests WithTarget:selector:object:` class method of `NSObject`. This is especially useful if the object that has fired an action with delay to call a method no longer needs that method to be called. For instance, if you want to check the availability of an Internet connection two seconds after your application starts, and after only a second your application is sent to the background, you can cancel the delayed action you initialized when the application was first run.

The `performSelector:withObject:afterDelay:` method accepts only one object that needs to be passed to the target selector. What if we need to pass more than one object? We can have our target selector accept a dictionary of values instead of a single object. This way, we can pack all our parameters into a dictionary and pass them all using one object (the dictionary itself, obviously). If your target method doesn't accept arguments, pass `nil` as the `withObject` parameter.

The target selector passed to the `performSelector:withObject:after Delay:` method should not have a return value.

7.5 Performing Periodic Tasks

Problem

You would like to perform a specific task repeatedly with a certain delay. For instance, you want to update a view on your screen every second for as long as your application is running.

Solution

Use a timer:

```
- (void) startPainting{

    NSTimer *newTimer = [NSTimer
                        scheduledTimerWithTimeInterval:1.0
                        target:self
```

```
                    selector:@selector(paint:)
                    userInfo:nil
                    repeats:YES];

  self.paintingTimer = newTimer;

}

- (void) paint:(NSTimer *)paramTimer{

  /* Do something here */
  NSLog(@"Painting");

}

- (void) stopPainting{

  if (self.paintingTimer != nil){
    [self.paintingTimer invalidate];
  }

}

- (void)applicationWillResignActive:(UIApplication *)application {

  [self stopPainting];

}

- (void)applicationDidBecomeActive:(UIApplication *)application {

  [self startPainting];

}
```

The invalidate method of the timer will also release the timer so that we don't have to do that manually. As you can see, we have defined a property called paintingTimer that is declared in this way in our header file (.h file):

```
#import <UIKit/UIKit.h>

@interface ThreadsAppDelegate : NSObject <UIApplicationDelegate> {
  UIWindow            *window;
  NSTimer             *paintingTimer;
}

@property (nonatomic, retain) NSTimer           *paintingTimer;
@property (nonatomic, retain) IBOutlet UIWindow *window;

@end
```

Discussion

A timer is an object that fires an event at specified intervals. A timer must be scheduled in a run loop. Defining an NSTimer object creates a nonscheduled timer that does nothing but is available to the program when you want to schedule it. Once you issue a call such as `scheduledTimerWithTimeInterval:target:selector:userInfo:repeats:`, the time becomes a scheduled timer and will fire the event you request. A scheduled timer is a timer that is added to a run loop. To get any timer to fire its target event, we must schedule that timer on a run loop. This is demonstrated in a later example where we create a nonscheduled timer and then schedule it on the main run loop of the application manually.

Once a timer is created and added to a run loop either explicitly or implicitly, the timer will start calling a method in its target object, specified by the programmer, every n seconds, where n is specified by the programmer as well. Because n is floating-point, you can specify a fraction of a second.

There are various ways to create, initialize, and schedule timers. One of the easiest ways is through the `scheduledTimerWithTimeInterval:target:selector:userInfo:repeats:` class method of `NSTimer`. Here are the different parameters of this method:

`scheduledTimerWithTimeInterval`

This is the number of seconds the timer has to wait before it fires an event. For example, if you want the timer to call a method in its target object twice per second, you have to set this parameter to 0.5 (1 second divided by 2); if you want the target method to be called four times per second, this parameter should be set to 0.25 (1 second divided by 4).

`target`

This is the object that will receive the event.

`selector`

This is the method signature in the target object that will receive the event.

`userInfo`

This is the object that will be retained in the timer for later reference (in the target method of the target object).

`repeats`

This specifies whether the timer must call its target method repeatedly (in which case this parameter has to be set to `YES`), or just once and then stop (in which case this parameter has to be set to `NO`).

Once a timer is created and added to a run loop, you can stop and release that timer using the `invalidate` instance method of the `NSTimer` class. This not only will release the timer, but also will release the object, if any, that was passed for the timer to retain during its lifetime (e.g., the object passed to the `userInfo` parameter of the `scheduledTimer WithTimeInterval:target:selector:userInfo:repeats:` class method of `NSTimer`). If you pass `NO` to the `repeats` parameter, the timer will invalidate itself after the first pass and subsequently will release the object it had retained (if any).

There are other methods you can use to create a scheduled timer. One of them is the `scheduledTimerWithTimeInterval:invocation:repeats:` class method of `NSTimer`:

```
- (void) startPainting{

  /* Here is the selector that we want to call */
  SEL selectorToCall = @selector(paint:);

  /* Here we compose a method signature out of the selector. We
   know that the selector is in the current class so it is easy
   to construct the method signature */
  NSMethodSignature *methodSignature =
    [[self class] instanceMethodSignatureForSelector:selectorToCall];

  /* Now base our invocation on the method signature. We need this
   invocation to schedule a timer */
  NSInvocation *invocation =
    [NSInvocation invocationWithMethodSignature:methodSignature];
  [invocation setTarget:self];
  [invocation setSelector:selectorToCall];

  /* Start a scheduled timer now */
  NSTimer *newTimer =
  [NSTimer scheduledTimerWithTimeInterval:1.0
                                 invocation:invocation
                                    repeats:YES];

  self.paintingTimer = newTimer;

}
```

Scheduling a timer can be compared to starting a car's engine. A scheduled timer is a running car engine. A nonscheduled timer is a car engine that is ready to be started but is not running yet. We can schedule and unschedule timers whenever we want in our application, just like we might need the engine of a car to be on or off depending on what situation we are in. If you want to schedule a timer manually at a certain time in your application, you can use the `timerWithTimeInterval:target:selector:userInfo:repeats:` class method of `NSTimer`, and when you are ready you can add the timer to your run loop of choice:

```
- (void) startPainting{

  NSTimer *newTimer = [NSTimer timerWithTimeInterval:1.0
                                              target:self
                                            selector:@selector(paint:)
                                            userInfo:nil
                                             repeats:YES];
  self.paintingTimer = newTimer;

  /* Do your processing here and whenever you are ready,
   use the addTimer:forMode instance method of the NSRunLoop class
   in order to schedule the timer on that run loop */
```

```
[[NSRunLoop currentRunLoop] addTimer:self.paintingTimer
                          forMode:NSDefaultRunLoopMode];

}
```

 The `currentRunLoop` and `mainRunLoop` class methods of `NSRunLoop` return the current and main run loops of the application, respectively, as their names imply.

Just like you can use the `scheduledTimerWithTimeInterval:invocation:repeats:` variant of creating scheduled timers using invocations, you can also use the `timerWithTimeInterval:invocation:repeats:` class method of `NSTimer` to create an unscheduled timer using an invocation:

```
- (void) startPainting{

    /* Here is the selector that we want to call */
    SEL selectorToCall = @selector(paint:);

    /* Here we compose a method signature out of the selector. We
     know that the selector is in the current class so it is easy
     to construct the method signature */
    NSMethodSignature *methodSignature =
      [[self class] instanceMethodSignatureForSelector:selectorToCall];

    /* Now base our invocation on the method signature. We need this
     invocation to schedule a timer */
    NSInvocation *invocation =
      [NSInvocation invocationWithMethodSignature:methodSignature];

    [invocation setTarget:self];
    [invocation setSelector:selectorToCall];

    NSTimer *newTimer = [NSTimer timerWithTimeInterval:1.0
                                            invocation:invocation
                                               repeats:YES];
    self.paintingTimer = newTimer;

    /* Do your processing here and whenever you are ready,
     use the addTimer:forMode instance method of the NSRunLoop class
     in order to schedule the timer on that run loop */

    [[NSRunLoop currentRunLoop] addTimer:self.paintingTimer
                              forMode:NSDefaultRunLoopMode];

}
```

The target method of a timer receives the instance of the timer that calls it as its parameter. For instance, the `paint:` method introduced initially in this recipe demonstrates how the timer gets passed to its target method, by default, as the target method's one and only parameter:

```
- (void) paint:(NSTimer *)paramTimer{

    /* Do something here */
    NSLog(@"Painting");

}
```

This parameter provides you with a reference to the timer that is firing this method. You can, for instance, prevent the timer from running again using the `invalidate` method, if needed. You can also invoke the `userInfo` method of the `NSTimer` instance in order to retrieve the object being retained by the timer (if any). This object is just an object passed to the initialization methods of `NSTimer`, and it gets directly passed to the timer for future reference.

See Also

Recipe 7.6

7.6 Performing Periodic Tasks Efficiently

Problem

You would like to have maximum control over how separate tasks run in your application. For instance, you would like to run a long calculation requested by the user while freeing the main UI thread to interact with the user and do other things.

Solution

Utilize threads in your application:

```
- (void) downloadNewFile:(id)paramObject{

    NSAutoreleasePool *pool = [[NSAutoreleasePool alloc] init];

    NSString *fileURL = (NSString *)paramObject;

    NSURL    *url = [NSURL URLWithString:fileURL];

    NSURLRequest *request = [NSURLRequest requestWithURL:url];

    NSURLResponse *response = nil;
    NSError       *error = nil;

    NSData *downloadedData =
    [NSURLConnection sendSynchronousRequest:request
```

```
                  returningResponse:&response
                         error:&error];

  if ([downloadedData length] > 0){
    /* Fully downloaded */
  } else {
    /* Nothing was downloaded. Check the Error value */
  }

  [pool release];

}

- (void)viewDidLoad {
  [super viewDidLoad];

  NSString *fileToDownload = @"http://www.OReilly.com";

  [NSThread detachNewThreadSelector:@selector(downloadNewFile:)
                           toTarget:self
                         withObject:fileToDownload];

}
```

Discussion

Any iOS application is made out of one or more threads. In iOS 4, a normal application with one view controller could initially have up to four or five threads created by the system libraries to which the application is linked. At least one thread will be created for your application whether you use multiple threads or not. It is called the "main UI thread" attached to the main run loop.

To understand how useful threads are, let's do an experiment. Suppose we have three loops:

```
- (void) firstCounter{

  NSUInteger counter = 0;
  for (counter = 0;
       counter < 1000;
       counter++){
    NSLog(@"First Counter = %lu", (unsigned long)counter);
  }

}

- (void) secondCounter{

  NSUInteger counter = 0;
  for (counter = 0;
       counter < 1000;
       counter++){
    NSLog(@"Second Counter = %lu", (unsigned long)counter);
  }
```

```
}

- (void) thirdCounter{

  NSUInteger counter = 0;
  for (counter = 0;
       counter < 1000;
       counter++){
    NSLog(@"Third Counter = %lu", (unsigned long)counter);
  }

}
```

Very simple, aren't they? All they do is go from zero to 1,000, printing their counter numbers. Now suppose you want to run these counters as we would normally do:

```
- (void)viewDidLoad {

  [super viewDidLoad];

  [self firstCounter];
  [self secondCounter];
  [self thirdCounter];

}
```

This code does not necessarily have to be in a view controller's viewDid
Load method.

Now open the console window and run this application. You will see the first counter's complete run, followed by the second counter and then the third counter. This means these loops are being run on the same thread. Each one blocks the rest of the thread's code from being executed until it finishes its loop.

What if we wanted all these counters to run at the same time? Of course, we would have to create separate threads for each one. But wait a minute! We already learned that our application creates threads for us when it loads and that whatever code we have been writing so far in our application, wherever it was, was being executed in a thread. So, we just have to create two threads for the first and second counters and leave the third counter to do its job in the main thread:

```
- (void) firstCounter{

  NSAutoreleasePool *pool = [[NSAutoreleasePool alloc] init];

  NSUInteger counter = 0;
  for (counter = 0;
       counter < 1000;
       counter++){
```

```objc
    NSLog(@"First Counter = %lu", (unsigned long)counter);
  }

  [pool release];

}

- (void) secondCounter{

  NSAutoreleasePool *pool = [[NSAutoreleasePool alloc] init];

  NSUInteger counter = 0;
  for (counter = 0;
       counter < 1000;
       counter++){
    NSLog(@"Second Counter = %lu", (unsigned long)counter);
  }

  [pool release];

}

- (void) thirdCounter{

  NSUInteger counter = 0;
  for (counter = 0;
       counter < 1000;
       counter++){
    NSLog(@"Third Counter = %lu", (unsigned long)counter);
  }

}

- (void)viewDidLoad {

  [super viewDidLoad];

  [NSThread detachNewThreadSelector:@selector(firstCounter)
                           toTarget:self
                         withObject:nil];


  [NSThread detachNewThreadSelector:@selector(secondCounter)
                           toTarget:self
                         withObject:nil];

  /* Run this on the main thread */
  [self thirdCounter];

}
```

The `thirdCounter` method does not have an autorelease pool since it is not run in a new detached thread. This method will be run in the application's main thread, which has an autorelease pool created for it automatically at the startup of every Cocoa Touch application. Also bear in mind that since we are detaching our threads in the `viewDidLoad` method of a view controller, we must handle the deallocation of these threads, if necessary, in the `viewDidUnload` method of the same view controller. The view runs `viewDidUnload` whenever the system sends it a low-memory warning, and failing to deallocate detached threads in that method will lead to a memory leak if those threads were created in the `viewDidLoad` method. For this you will need a reference to the `NSThread` object you have allocated and initialized. Please use the `init WithTarget:selector:object:` instance method of `NSThread` to initialize an object of type `NSThread`. This way, you will get a reference to this thread and you can start it manually using its `start` instance method.

The calls to `detachNewThreadSelector` near the end of the code run the first and second counters as separate threads. Now if you run the application, you will notice output such as the following, in the console window:

```
Second Counter = 921
Third Counter = 301
Second Counter = 922
Second Counter = 923
Second Counter = 924
First Counter = 956
Second Counter = 925
Counter = 957
Second Counter = 926
First Counter = 958
Third Counter = 302
Second Counter = 927
Third Counter = 303
Second Counter = 928
```

In other words, all three counters run at once, and interleave their output randomly.

Every thread must create an autorelease pool. An autorelease pool internally keeps a reference to objects that are being autoreleased before the pool itself is released. This is a very important mechanism in a reference-counted memory management environment such as Cocoa Touch where objects can be autoreleased. Whenever we allocate instances of objects, the retain count of the objects gets set to 1. If we mark the objects as autorelease, the retain count remains at 1, but when the autorelease pool in which the object was created is released, the autorelease object is also sent a `release` message. If its retain count is still 1 at that point, the object gets deallocated.

Every thread requires an autorelease pool to be created for it as the first object that is allocated in that thread. If you don't do this, any object that you allocate in your thread will leak when the thread exists. To understand this better, let's have a look at the following code:

```
- (void) autoreleaseThread:(id)paramSender{

    NSBundle *mainBundle = [NSBundle mainBundle];
    NSString *filePath = [mainBundle pathForResource:@"Default"
                                              ofType:@"png"];

    UIImage *image = [[[UIImage alloc]
                         initWithContentsOfFile:filePath]
                         autorelease];

    /* Do something with the image */
    NSLog(@"Image = %@", image);

}

- (void)viewDidLoad {

    [super viewDidLoad];

    [NSThread detachNewThreadSelector:@selector(autoreleaseThread:)
                             toTarget:self
                           withObject:self];

}
```

If you run this code and keep an eye on the console window, you will receive a message similar to this:

```
*** __NSAutoreleaseNoPool(): Object 0x5b2c990 of
class NSCFString autoreleased with no pool in place - just leaking
*** __NSAutoreleaseNoPool(): Object 0x5b2ca30 of
class NSPathStore2 autoreleased with no pool in place - just leaking
*** __NSAutoreleaseNoPool(): Object 0x5b205c0 of
class NSPathStore2 autoreleased with no pool in place - just leaking
*** __NSAutoreleaseNoPool(): Object 0x5b2d650 of
class UIImage autoreleased with no pool in place - just leaking
```

This shows that the autorelease UIImage instance we created is creating a memory leak—and in addition, so is the NSString instance called FilePath and other objects that would normally "magically" get deallocated. This is because in our thread, we forgot to allocate and initialize an autorelease pool as the first thing we did. The following is the correct code, which you can test for yourself to make sure it doesn't leak:

```
- (void) autoreleaseThread:(id)paramSender{

    NSAutoreleasePool *pool = [[NSAutoreleasePool alloc] init];

    NSBundle *mainBundle = [NSBundle mainBundle];
    NSString *filePath = [mainBundle pathForResource:@"Default"
                                              ofType:@"png"];

    UIImage *image = [[[UIImage alloc]
```

```
                                initWithContentsOfFile:filePath]
                       autorelease];

    /* Do something with the image */
    NSLog(@"Image = %@", image);

    [pool release];

}
```

7.7 Initializing Threads Implicitly

Problem

You want to know an easy way to create threads, without having to deal with threads directly.

Solution

Use the performSelectorInBackground:withObject: instance method of NSObject:

```
- (void)viewDidLoad {

    [super viewDidLoad];

    [self performSelectorInBackground:@selector(firstCounter)
                           withObject:nil];

    [self performSelectorInBackground:@selector(secondCounter)
                           withObject:nil];

    [self performSelectorInBackground:@selector(thirdCounter)
                           withObject:nil];

}
```

The counter methods are implemented in this way:

```
- (void) firstCounter{

    NSAutoreleasePool *pool = [[NSAutoreleasePool alloc] init];

    NSUInteger counter = 0;
    for (counter = 0;
         counter < 1000;
         counter++){
      NSLog(@"First Counter = %lu", (unsigned long)counter);
    }

    [pool release];

}

- (void) secondCounter{
```

```
NSAutoreleasePool *pool = [[NSAutoreleasePool alloc] init];

NSUInteger counter = 0;
for (counter = 0;
     counter < 1000;
     counter++){
  NSLog(@"Second Counter = %lu", (unsigned long)counter);
}

[pool release];

}

- (void) thirdCounter{

NSAutoreleasePool *pool = [[NSAutoreleasePool alloc] init];

NSUInteger counter = 0;
for (counter = 0;
     counter < 1000;
     counter++){
  NSLog(@"Third Counter = %lu", (unsigned long)counter);
}

[pool release];

}
```

Discussion

The performSelectorInBackground:withObject: method creates a new thread in the background for us. This is equivalent to our creating a new thread for our selectors. The most important thing that we have to keep in mind is that since this method creates a thread on the given selector, the selector must have an autorelease pool just like any other thread in a reference-counted memory environment.

7.8 Exiting Threads and Timers

Problem

You would like to stop a thread or a timer, or prevent one from firing again.

Solution

For timers, use the invalidate instance method of NSTimer. For threads, use the cancel method. Avoid using the exit method of threads, as it does not give the thread a chance to clean up after itself and your application will end up leaking resources:

```
NSThread *thread = /* Get the reference to your thread here */;
[thread cancel];
```

```
NSTimer *timer = /* Get the reference to your timer here */;
[timer invalidate];
```

Discussion

Exiting a timer is quite straightforward; you can simply call the timer's **invalidate** instance method. After you call that method, the timer will not fire any more events to its target object.

However, threads are a bit more complicated to exit. When a thread is sleeping and its **cancel** method is called, the thread's loop will still perform its task fully before exiting. Let me demonstrate this for you:

```
- (void) threadEntryPoint{

  NSAutoreleasePool *pool = [[NSAutoreleasePool alloc] init];

  NSLog(@"Thread Entry Point");

  while ([[NSThread currentThread] isCancelled] == NO){

    [NSThread sleepForTimeInterval:0.2];
    NSLog(@"Thread Loop");

  }

  NSLog(@"Thread Finished");

  [pool release];

}

- (void) stopThread{
  NSLog(@"Cancelling the Thread");
  [self.myThread cancel];
  NSLog(@"Releasing the thread");
  self.myThread = nil;
}

- (void)viewDidLoad {

  NSThread *newThread = [[NSThread alloc]
                          initWithTarget:self
                          selector:@selector(threadEntryPoint)
                          object:nil];
  self.myThread = newThread;
  [newThread release];

  [self performSelector:@selector(stopThread)
              withObject:nil
              afterDelay:3.0f];

  [self.myThread start];
```

```
    }

    - (void) viewDidUnload{
      [super viewDidUnload];

      self.myThread = nil;
    }
```

This code creates an instance of NSThread and starts the thread immediately. Our thread sleeps for four seconds in every loop before performing its task. However, before the thread is started, we are calling the stopThread method of our view controller (which we have written) with a three-second delay. This method calls the cancel method of our thread in an attempt to make the thread exit its loop. Now let's run the application and see what gets printed to the console screen:

```
Thread Entry Point
Cancelling the Thread
Releasing the thread
Thread Loop
Thread Finished
```

You can clearly see that our thread finished its current loop before exiting, even though the request to cancel it was fired in the middle of the loop. This is a very common pitfall that can be avoided simply by checking whether the thread is cancelled before attempting to perform a task with external side effects inside the thread's loop. We can rewrite our example as follows so that the operation with an external effect (writing to the log) checks first to make sure the thread hasn't been cancelled.

```
    - (void) threadEntryPoint{

      NSAutoreleasePool *pool = [[NSAutoreleasePool alloc] init];

      NSLog(@"Thread Entry Point");

      while ([[NSThread currentThread] isCancelled] == NO){

        [NSThread sleepForTimeInterval:0.2];

        if ([[NSThread currentThread] isCancelled] == NO){
          NSLog(@"Thread Loop");
        }

      }

      NSLog(@"Thread Finished");

      [pool release];

    }
```

7.9 Avoiding Memory Leaks in Threads

Problem

You are experiencing memory leaks in your threads.

Solution

Make sure your thread has an autorelease pool allocated for it before its run loop starts, and release the autorelease pool after the run loop finishes. It is always best to have an autorelease pool allocated and initialized for a thread as the first line of code you write in that thread. Also make sure a thread with a long lifetime does not allocate many autorelease objects. You should avoid this because these objects will not be deallocated until after the autorelease pool of the thread that created them is released.

Discussion

With threads, we must be very careful as to how we manage the memory in a non-garbage-collected environment such as Cocoa Touch on iOS. Any thread calling a selector on an object is responsible for cleaning up the autorelease objects created in that method. For instance, if you create a new detached thread with an entry point of myThread: in your code and you have this thread call a method inside your code, but not on the main thread, myThread:'s autorelease pool will be responsible for cleaning up after the autorelease objects allocated (if any) in the method that was called. There are a couple of simple things you have to bear in mind while working with threads:

- Always allocate and initialize an object of type NSAutoreleasePool before any other code in a thread's entry method.

- Always release the autorelease pool for a thread as the last line of code you have in the method that serves as the thread's entry point.

- Make sure you do not create autorelease objects inside a thread whose lifetime could be very long. In this case, make sure you allocate and initialize and eventually release the objects yourself. Do not let them be managed by the autorelease pool of the thread, since this pool might not be released anytime soon. If you do go ahead and autorelease objects in the run loop of a thread, I advise that you create (allocate and initialize) an autorelease pool inside the thread's run loop in addition to the thread's main autorelease pool. Release the internal autorelease pool in every iteration of the thread's run loop (as demonstrated later in this recipe) to make sure the autorelease objects in the run loop of the thread are managed by the secondary autorelease pool that does get released in every iteration.

Most of the calls for starting a thread, such as performSelector:, use the calling thread's autorelease pool. It is your responsibility to create and release the pool, as shown in Recipe 7.6.

If your thread makes a lot of calls on objects, without making the main thread responsible for the autorelease objects, none of the autorelease objects will be deallocated until after the thread has finished its work and releases its autorelease pool. In other words, if you have a thread that has to run throughout the lifetime of your application and should never quit, it is beneficial to avoid making calls to methods in other objects because the autorelease objects in those methods will not be released and deallocated until after your application has terminated and your thread has finished its execution, increasing the memory footprint of your application in every iteration of the thread's loop. To solve this, you can call the methods using the main thread as demonstrated before, designate another thread to become the caller of the method, or create an autorelease pool inside the run loop of your thread. Let me demonstrate the last strategy with an example. Create a new Objective-C class and name it `MyObject`. Let's define this object in this way (.*h* file):

```
#import <Foundation/Foundation.h>

@interface MyObject : NSObject {

}

- (void) doSomething;

@end
```

Here is the implementation of the new object (.*m* file):

```
#import "MyObject.h"

@implementation MyObject

- (void) doSomething{
  /* Empty implementation */
}
- (id) init {
  self = [super init];

  if (self != nil){
    NSLog(@"Initialized");
  }

  return(self);
}
- (void) dealloc {

  NSLog(@"Deallocated");
  [super dealloc];
}

@end
```

Now let's go ahead and create a detached thread and autorelease objects of type MyObject in the run loop of the thread:

```objc
- (void) myThread:(id)paramObject{

  NSAutoreleasePool *pool = [[NSAutoreleasePool alloc] init];

  while ([[NSThread currentThread] isCancelled] == NO){

    [NSThread sleepForTimeInterval:0.20f];

    if ([[NSThread currentThread] isCancelled] == NO){
      MyObject *Object = [[[MyObject alloc] init] autorelease];
      [Object doSomething];
    }

  }

  [pool release];

}
```

After running this application, have a look at the console window. Messages similar to these will be printed out:

```
Initialized
Initialized
Initialized
Initialized
Initialized
Initialized
Initialized
Initialized
Initialized
Initialized
Initialized
```

What you don't see is the deallocation of instances of MyObject. The reason, as explained before, is that the autorelease objects here are getting added to the single autorelease pool we created in our thread's entry point. The autorelease objects that are created in the run loop of this thread will not get deallocated until after the main autorelease pool of the thread's run loop gets released. To fix this issue, we can have an autorelease pool inside the run loop of a thread, too:

```objc
- (void) myThread:(id)paramObject{

  NSAutoreleasePool *pool = [[NSAutoreleasePool alloc] init];

  while ([[NSThread currentThread] isCancelled] == NO){

    NSAutoreleasePool *insidePool = [[NSAutoreleasePool alloc] init];

    [NSThread sleepForTimeInterval:0.20f];

    if ([[NSThread currentThread] isCancelled] == NO){
```

```
        MyObject *Object = [[[MyObject alloc] init] autorelease];
        [Object doSomething];
    }

    [insidePool release];

}

[pool release];

}
```

Running this code now will result in messages similar to these printed out to the console window:

```
Initialized
Deallocated
Initialized
Deallocated
Initialized
Deallocated
Initialized
Deallocated
Initialized
Deallocated
```

The `performSelectorOnMainThread:withObject:waitUntilDone:` instance method of NSObject makes the main thread's autorelease pool responsible for cleaning up the autorelease objects in the invoked method. This may lead to a waste of memory because your new thread may end long before the main thread ends, and the new thread's autorelease objects will hang around in memory until the main thread ends, which is the end of the entire application. However, using this method relieves you of the need to create and release another autorelease pool.

To see the difference between these ways of invoking a thread, take a look at the following code:

```
- (void) newThreadEntryPoint{

    /* A thread without an autorelease pool to test the following code */

    //NSAutoreleasePool *pool = [[NSAutoreleasePool alloc] init];

    /* This WILL cause a memory leak */
    [self performSelector:@selector(allocateSomething)];

    /* This will NOT cause a memory leak */
    [self performSelectorOnMainThread:@selector(allocateSomething)
                           withObject:nil
                        waitUntilDone:YES];

    //[pool release];

}
```

```
- (void) allocateSomething{

    NSBundle *mainBundle = [NSBundle mainBundle];

    NSString *imagePath = [mainBundle pathForResource:@"MyImage"
                                               ofType:@"png"];

    NSData    *imageData = [NSData dataWithContentsOfFile:imagePath];

    UIImage   *myImage = [[[UIImage alloc] initWithData:imageData] autorelease];

    /* Do something with the image here */

}


- (void)viewDidLoad {

    [NSThread detachNewThreadSelector:@selector(newThreadEntryPoint)
                            toTarget:self
                          withObject:nil];

}
```

Here is a breakdown of what is happening in this code:

1. The `viewDidLoad` method of our view controller creates a new detached thread on the `newThreadEntryPoint` selector.
2. The created thread's implementation does not have an autorelease pool. This is a very common mistake programmers make when working with threads and the biggest cause of memory leaks in multithreaded applications.
3. The thread calls the `allocateSomething` method using the `performSelector:` instance method of `NSObject`. The `allocateSomething` method creates an autorelease instance of `UIImage`, which is typical application behavior.
4. The thread then calls the same method using the `performSelectorOnMain Thread:withObject:waitUntilDone:` instance method of `NSObject`.

Our application will leak in step 3, but not step 4. This is because step 3 uses `perform Selector:` to invoke the method in the new thread using the calling thread's autorelease pool. Step 4 uses `performSelectorOnMainThread:withObject:waitUntilDone:`, which registers the autorelease objects of the new thread in the main thread's autorelease pool, and therefore does not cause a leak.

CHAPTER 8

Audio and Video

8.0 Introduction

The AV Foundation framework on the iOS SDK allows developers to play and/or record audio and video with ease. In addition, the Media Player framework allows developers to play audio and video files.

Before you can run the code in this chapter, you must add the AVFoundation.framework and MediaPlayer.framework frameworks to your Xcode project. You can do this by following these steps:

1. In Xcode, find your target.
2. Right-click on the target and choose Add→Existing Frameworks.
3. Hold down the Command key and choose AVFoundation.framework and Media-Player.framework from the list.
4. Select Add.

8.1 Playing Audio Files

Problem

You want to be able to play an audio file in your application.

Solution

Use the AV Foundation (Audio and Video Foundation) framework's `AVAudioPlayer` class.

Discussion

The `AVAudioPlayer` class in the AV Foundation framework can play back all audio formats supported by iOS. The `delegate` property of an instance of `AVAudioPlayer` allows you to get notified by events, such as whenever the audio playback is interrupted or if an error occurs as a result of playing an audio file. Let's have a look at a simple example that demonstrates how we can play an audio file that is in our application's bundle:

```
- (void) startPlayingAudio{

    NSAutoreleasePool *pool = [[NSAutoreleasePool alloc] init];

    NSBundle *mainBundle = [NSBundle mainBundle];

    NSString *filePath = [mainBundle pathForResource:@"MySong"
                                              ofType:@"mp3"];

    NSData   *fileData = [NSData dataWithContentsOfFile:filePath];

    NSError  *error = nil;

    /* Start the audio player */

    AVAudioPlayer *newPlayer =
    [[AVAudioPlayer alloc] initWithData:fileData
                                  error:&error];
    self.audioPlayer = newPlayer;
    [newPlayer release];

    /* Did we get an instance of AVAudioPlayer? */
    if (self.audioPlayer != nil){
      /* Set the delegate and start playing */
      self.audioPlayer.delegate = self;
      if ([self.audioPlayer prepareToPlay] == YES &&
          [self.audioPlayer play] == YES){
        /* Successfully started playing */
      } else {
        /* Failed to play */
      }
    } else {
      /* Failed to instantiate AVAudioPlayer */
    }

    [pool release];

}
```

As you can see, the file's data is loaded into an instance of NSData and then passed on to AVAudioPlayer's initWithData:error: method. Because we need the actual, absolute path of the MP3 file to extract the data from that file, we invoke the mainBundle class method of NSBundle to retrieve the information from our application's configuration. The pathForResource:ofType: instance method of NSBundle can then be used to retrieve the absolute path to a resource of a specific type, as demonstrated in the example code.

The startPlayingAudio method in this code has an NSAutoreleasePool instance to allow us to turn this method into a thread's entry point. As an example of why this is useful, say we load a big audio file into an instance of NSData and then pass it to the `initWithData:error:` instance method of the `AVAudioPlayer` class. The thread making these calls will obviously be blocked until after the data is returned and `AVAudio Player` is initialized with the given data. If we put these operations into a separate thread, our main thread will not be blocked and can continue to interact with the user.

Here is how we will branch the code out into another thread for loading and playing the audio file:

```objc
#import "AudioAndVideoViewController.h"

@implementation AudioAndVideoViewController

@synthesize audioPlayer;

- (void)audioPlayerDidFinishPlaying:(AVAudioPlayer *)player
                        successfully:(BOOL)flag{

  NSLog(@"Finished playing the song");

  /* The [flag] parameter tells us if the playback was successfully
   finished or not */

  if ([player isEqual:self.audioPlayer] == YES){
    self.audioPlayer = nil;
  } else {
    [player release];
  }

}

- (void) startPlayingAudio{

  NSAutoreleasePool *pool = [[NSAutoreleasePool alloc] init];

  NSBundle *mainBundle = [NSBundle mainBundle];

  NSString *filePath = [mainBundle pathForResource:@"MySong"
                                            ofType:@"mp3"];

  NSData    *fileData = [NSData dataWithContentsOfFile:filePath];

  NSError   *error = nil;

  /* Start the audio player */

  AVAudioPlayer *newPlayer =
  [[AVAudioPlayer alloc] initWithData:fileData
                               error:&error];
  self.audioPlayer = newPlayer;
  [newPlayer release];

  /* Did we get an instance of AVAudioPlayer? */
  if (self.audioPlayer != nil){
    /* Set the delegate and start playing */
    self.audioPlayer.delegate = self;
```

```
      if ([self.audioPlayer prepareToPlay] == YES &&
          [self.audioPlayer play] == YES){
        /* Successfully started playing */
      } else {
        /* Failed to play */
      }
    } else {
      /* Failed to instantiate AVAudioPlayer */
    }

    [pool release];

}

- (void)viewDidLoad {
  [super viewDidLoad];

  [NSThread detachNewThreadSelector:@selector(startPlayingAudio)
                           toTarget:self
                         withObject:nil];

}

- (void) viewDidUnload{
  [super viewDidUnload];

  if(self.audioPlayer != nil){
    if([self.audioPlayer isPlaying] == YES){
      [self.audioPlayer stop];
    }
    self.audioPlayer = nil;
  }

}

- (BOOL)shouldAutorotateToInterfaceOrientation:
  (UIInterfaceOrientation)interfaceOrientation {
  return(YES);
}

- (void)dealloc {
  [audioPlayer release];
  [super dealloc];
}

@end
```

Since we are assigning the instance of AVAudioPlayer to a property named audio Player, we must also see how this property is defined:

```
#import <UIKit/UIKit.h>
#import <AVFoundation/AVFoundation.h>

@interface AudioAndVideoViewController : UIViewController
                                        <AVAudioPlayerDelegate> {

@public
```

```
AVAudioPlayer          *audioPlayer;
}

@property (nonatomic, retain) AVAudioPlayer        *audioPlayer;

@end
```

As you can see, we have made our view controller the delegate of the audio player. This way, we can receive messages from the system whenever our audio player, for instance, is interrupted or has finished playing the song. With this information in hand, we can make appropriate decisions in our application, such as starting to play another audio file.

See Also

Recipe 8.2; Recipe 8.5

8.2 Handling Interruptions While Playing Audio Files

Problem

You want your AVAudioPlayer instance to resume playing after an interruption on an iOS device, such as an incoming call.

Solution

Implement the audioPlayerBeginInterruption: and audioPlayerEndInterruption:with Flags: methods of the AVAudioPlayerDelegate protocol in the delegate object of your AVAudioPlayer instance:

```
- (void)audioPlayerBeginInterruption:(AVAudioPlayer *)player{

    /* Audio Session is interrupted. The player will be paused here */

}

- (void)audioPlayerEndInterruption:(AVAudioPlayer *)player{
    /* For iOS lower than iOS 4 */
    [player play];
}

- (void)audioPlayerEndInterruption:(AVAudioPlayer *)player
                          withFlags:(NSUInteger)flags{

    /* For iOS 4 and newer */
    /* Check the flags, if we can resume the audio, then we should
     do it here */

    if (flags == AVAudioSessionInterruptionFlags_ShouldResume &&
        player != nil){
      [player play];
```

```
    }

}
```

Discussion

On an iOS device, such as an iPhone, a phone call could interrupt the execution of the foreground application. The audio session(s) associated with the application will be deactivated in that case and audio files will not be played until after the interruption is ended. In the beginning and the end of an interruption, we receive delegate messages from the AVAudioPlayer informing us of the different states the audio session is passing through. After the end of an interruption, we can simply resume the playback of audio.

 Incoming phone calls cannot be simulated with iPhone Simulator. You must always test your applications on a real device.

When an interruption occurs, the audioPlayerBeginInterruption: delegate method of an AVAudioPlayer instance will be called. Here, your audio session has been deactivated. In case of a phone call, the user can just hear her ring tone. When the interruption ends (the phone call is finished or the user rejects the call), the audioPlayerEndInterrup tion:withFlags: delegate method of your AVAudioPlayer will be invoked. If the with Flags parameter contains the value AVAudioSessionInterruptionFlags_ShouldResume, you can immediately resume the playback of your audio player using the play instance method of AVAudioPlayer.

 The playback of audio files using AVAudioPlayer might show memory leaks in Instruments when the application is being run on iPhone Simulator. Testing the same application on an iOS device proves that the memory leaks are unique to the simulator, not the device. I strongly suggest that you run, test, debug, and optimize your applications on real devices before releasing them on the App Store.

8.3 Recording Audio Files

Problem

You want to be able to record audio files on an iOS device.

Solution

Make sure you have added the CoreAudio.framework framework to your target file, and use the AVAudioRecorder class in the AV Foundation framework:

```
NSError    *error = nil;

NSString *pathAsString = [self audioRecordingPath];

NSURL *audioRecordingURL = [NSURL fileURLWithPath:pathAsString];

AVAudioRecorder *newRecorder =
[[AVAudioRecorder alloc] initWithURL:audioRecordingURL
                            settings:[self audioRecordingSettings]
                               error:&error];

self.audioRecorder = newRecorder;
[newRecorder release];
```

For information about the `audioRecordingSettings` and `audioRecordingPath` methods used in this example, refer to this recipe's Discussion.

Discussion

The `AVAudioRecorder` class in the AV Foundation framework facilitates audio recording in iOS applications compiled with iOS SDK 3.0 and later. To start a recording, you need to pass various pieces of information to the `initWithURL:settings:error:` instance method of `AVAudioRecorder`:

The URL of the file where the recording should be saved
> This is a local URL. The AV Foundation framework will decide which audio format should be used for the recording based on the file extension provided in this URL, so choose the extension carefully.

The settings that must be used before and while recording
> Examples include the sampling rate, channels, and other information that will help the audio recorder start the recording. This is a dictionary object.

The address of an instance of `NSError` where any initialization errors should be saved to
> The error information could be valuable later, and you can retrieve it from this instance method in case something goes wrong.

The `settings` parameter of the `initWithURL:settings:error:` instance method is particularly interesting. There are many keys that could be saved in the `settings` dictionary, but we will discuss only some of the most important ones in this recipe:

`AVFormatIDKey`
> The format of the recorded audio. Some of the values that can be specified for this key are:
> * `kAudioFormatLinearPCM`
> * `kAudioFormatAppleLossless`

`AVSampleRateKey`
> The sample rate that needs to be used for the recording.

AVNumberOfChannelsKey

The number of channels that must be used for the recording.

AVEncoderAudioQualityKey

The quality with which the recording must be made. Some of the values that can be specified for this key are:

- AVAudioQualityMin
- AVAudioQualityLow
- AVAudioQualityMedium
- AVAudioQualityHigh
- AVAudioQualityMax

With all this information in hand, we can go on and write an application that can record audio input into a file and then play it using AVAudioPlayer. What we want to do, specifically, is:

1. Start recording audio in Apple Lossless format.
2. Save the recording into a file named *Recording.m4a* in our application's *Documents* directory.
3. Five seconds after the recording starts, finish the recording process and immediately start playing the file into which we recorded the audio input.

We will start by declaring the required properties in the *.h* file of a simple view controller:

```
#import <UIKit/UIKit.h>
#import <AVFoundation/AVFoundation.h>
#import <CoreAudio/CoreAudioTypes.h>

@interface AudioAndVideoViewController : UIViewController
            <AVAudioRecorderDelegate, AVAudioPlayerDelegate> {
@public
  AVAudioRecorder        *audioRecorder;
  AVAudioPlayer          *audioPlayer;
}

@property (nonatomic, retain) AVAudioRecorder    *audioRecorder;
@property (nonatomic, retain) AVAudioPlayer      *audioPlayer;

- (NSString *)     audioRecordingPath;
- (NSDictionary *) audioRecordingSettings;

@end
```

When the view inside our view controller is loaded for the first time, we will attempt to start the recording process and then stop the process, if successfully started, after five seconds:

```
- (void)viewDidLoad {
  [super viewDidLoad];
```

```objc
    NSError   *error = nil;

    NSString *pathAsString = [self audioRecordingPath];

    NSURL *audioRecordingURL = [NSURL fileURLWithPath:pathAsString];

    AVAudioRecorder *newRecorder =
    [[AVAudioRecorder alloc] initWithURL:audioRecordingURL
                                settings:[self audioRecordingSettings]
                                   error:&error];

    self.audioRecorder = newRecorder;
    [newRecorder release];

    if (self.audioRecorder != nil){

      self.audioRecorder.delegate = self;
      /* Prepare the recorder and then start the recording */

      if ([self.audioRecorder prepareToRecord] == YES &&
          [self.audioRecorder record] == YES){
        NSLog(@"Successfully started to record.");

        /* After 5 seconds, let's stop the recording process */
        [self performSelector:@selector(stopRecordingOnAudioRecorder:)
                   withObject:self.audioRecorder
                   afterDelay:5.0f];

      } else {
        NSLog(@"Failed to record.");
        self.audioRecorder = nil;
      }

    } else {
      NSLog(@"Failed to create an instance of the audio recorder.");
    }

}

- (void) viewDidUnload{
    [super viewDidUnload];

    if (self.audioRecorder != nil){
      if ([self.audioRecorder isRecording] == YES){
        [self.audioRecorder stop];
      }
      self.audioRecorder = nil;
    }


    if (self.audioPlayer != nil){
      if ([self.audioPlayer isPlaying] == YES){
        [self.audioPlayer stop];
      }
```

```
        self.audioPlayer = nil;
    }

}
```

In the viewDidLoad method of our view controller, we attempt to instantiate an object of type AVAudioRecorder and assign it to the audioRecorder property that we declared in the .h file of the same view controller earlier.

We are using an instance method called audioRecordingPath to determine the NSString representation of the local URL where we want to store our recording. This method is implemented like so:

```
- (NSString *) audioRecordingPath{

  NSString *result = nil;

  NSArray *folders =
  NSSearchPathForDirectoriesInDomains(NSDocumentDirectory,
                                      NSUserDomainMask,
                                      YES);
  NSString *documentsFolder = [folders objectAtIndex:0];

  result = [documentsFolder
            stringByAppendingPathComponent:@"Recording.m4a"];

  return(result);

}
```

The return value of this function is the document path of your application with the name of the destination file appended to it. For instance, if the document path of your application is:

```
/var/mobile/Applications/ApplicationID/Documents/
```

the destination audio recording path will be:

```
/var/mobile/Applications/ApplicationID/Documents/Recording.m4a
```

When instantiating our AVAudioRecorder, we are using a dictionary for the settings parameter of the initialization method of the audio recorder, as explained before. This dictionary is constructed using our audioRecordingSettings instance method, implemented in this way:

```
- (NSDictionary *) audioRecordingSettings{

  NSDictionary *result = nil;

  /* Let's prepare the audio recorder options in the dictionary.
   Later we will use this dictionary to instantiate an audio
   recorder of type AVAudioRecorder */

  NSMutableDictionary *settings =
  [[[NSMutableDictionary alloc] init] autorelease];
```

```
[settings
 setValue:[NSNumber numberWithInteger:kAudioFormatAppleLossless]
 forKey:AVFormatIDKey];

[settings
 setValue:[NSNumber numberWithFloat:44100.0f]
 forKey:AVSampleRateKey];

[settings
 setValue:[NSNumber numberWithInteger:1]
 forKey:AVNumberOfChannelsKey];

[settings
 setValue:[NSNumber numberWithInteger:AVAudioQualityLow]
 forKey:AVEncoderAudioQualityKey];

result = [NSDictionary dictionaryWithDictionary:settings];

return(result);

}
```

You can see that five seconds after the recording starts successfully in the viewDid
Load method of the view controller, we are calling the stopRecordingOnAudioRecorder
method, implemented like so:

```
- (void) stopRecordingOnAudioRecorder
            :(AVAudioRecorder *)paramRecorder{

/* Just stop the audio recorder here */
[paramRecorder stop];

}
```

Now that we have asked our audio recorder to stop recording, we will wait for its
delegate messages to tell us when the recording has actually stopped. It's good not to
assume that the stop instance method of AVAudioRecorder instantly stops the recording.
Instead, I recommend that you wait for the audioRecorderDidFinishRecording:success
fully: delegate method (declared in the AVAudioRecorderDelegate protocol) before
proceeding.

When the audio recording has actually stopped, we will attempt to play what was
recorded:

```
- (void)audioRecorderDidFinishRecording:(AVAudioRecorder *)recorder
                       successfully:(BOOL)flag{

if (flag == YES){

    NSLog(@"Successfully stopped the audio recording process.");

    /* Let's try to retrieve the data for the recorded file */
    NSError *playbackError = nil;
```

```
NSError *readingError = nil;
NSData  *fileData =
[NSData dataWithContentsOfFile:[self audioRecordingPath]
                      options:NSDataReadingMapped
                        error:&readingError];

/* Form an audio player and make it play the recorded data */
AVAudioPlayer *newPlayer =
[[AVAudioPlayer alloc] initWithData:fileData
                             error:&playbackError];
self.audioPlayer = newPlayer;
[newPlayer release];

/* Could we instantiate the audio player? */
if (self.audioPlayer != nil){
  self.audioPlayer.delegate = self;

  /* Prepare to play and start playing */
  if ([self.audioPlayer prepareToPlay] == YES &&
      [self.audioPlayer play] == YES){
    NSLog(@"Started playing the recorded audio.");
  } else {
    NSLog(@"Could not play the audio.");
  }

} else {
  NSLog(@"Failed to create an audio player.");
}

} else {
  NSLog(@"Failed to stop the recording of audio.");
}

/* Here we don't need the audio recorder anymore */
self.audioRecorder = nil;

}
```

After the audio player is finished playing the song (if it does so successfully), the audio
PlayerDidFinishPlaying:successfully: delegate method will be called in the delegate
object of our audio player. We will implement this method like so (this method is
defined in the AVAudioPlayerDelegate protocol):

```
- (void)audioPlayerDidFinishPlaying:(AVAudioPlayer *)player
                       successfully:(BOOL)flag{

if (flag == YES){
  NSLog(@"Audio player stopped correctly.");
} else {
  NSLog(@"Audio player did not stop correctly.");
}

if ([player isEqual:self.audioPlayer] == YES){
  self.audioPlayer = nil;
} else {
```

```
    [player release];
  }

}
```

As explained in Recipe 8.2, when playing audio files using `AVAudioPlayer`, we also need to handle interruptions, such as incoming phone calls, when deploying our application on an iOS device and before releasing the application on the App Store:

```
- (void)audioPlayerBeginInterruption:(AVAudioPlayer *)player{

  /* The audio session has been deactivated here */

}

- (void)audioPlayerEndInterruption:(AVAudioPlayer *)player{

  /* For iOS versions earlier than 4 */
  [player play];

}

- (void)audioPlayerEndInterruption:(AVAudioPlayer *)player
                          withFlags:(NSUInteger)flags{

  /* For iOS 4 and later */
  if (flags == AVAudioSessionInterruptionFlags_ShouldResume){
    [player play];
  }

}
```

Instances of `AVAudioRecorder` must also handle interruptions, just like instances of `AVAudioPlayer`. These interruptions can be handled as explained in Recipe 8.4.

See Also

Recipe 8.2; Recipe 8.4

8.4 Handling Interruptions While Recording Audio Files

Problem

You want your `AVAudioRecorder` instance to be able to resume recording after an interruption, such as an incoming phone call.

Solution

Implement the `audioRecorderBeginInterruption:` and `audioRecorderEndInterruption :withFlags:` methods of the `AVAudioRecorderDelegate` protocol in the delegate object of your audio recorder, and resume the recording process by invoking the record instance method of your AVAudioRecorder when the interruption has ended:

```objectivec
- (void)audioRecorderBeginInterruption:(AVAudioRecorder *)recorder{

    NSLog(@"Recording process is interrupted");

}

- (void)audioRecorderEndInterruption:(AVAudioRecorder *)recorder{

    /* For iOS versions earlier than 4 */
    [recorder record];
}

- (void)audioRecorderEndInterruption:(AVAudioRecorder *)recorder
                            withFlags:(NSUInteger)flags{

    /* For iOS 4 and later */
    if (flags == AVAudioSessionInterruptionFlags_ShouldResume){
        NSLog(@"Resuming the recording...");
        [recorder record];
    }

}
```

Discussion

Just like audio players (instances of AVAudioPlayer), audio recorders of type AVAudio Recorder also receive delegate messages whenever the audio session associated with them is deactivated because of an interruption. The two methods mentioned in this recipe's Solution are the best places to handle such interruptions. In the case of an interruption to the audio recorder, you can invoke the record instance method of AVAudioRecorder after the interruption to continue the recording process. However, the recording will overwrite the previous recording and all data that was recorded before the interruption will be lost.

It is very important to bear in mind that when the delegate of your audio recorder receives the audioRecorderBeginInterruption: method, the audio session has already been deactivated, and invoking the resume instance method will not work on your audio recorder. After the interruption is ended, you must invoke the record instance method of your AVAudioRecorder to resume recording.

8.5 Playing Audio over Other Sounds That Are Playing

Problem

You either want to put other applications in silent mode while you play audio or play audio on top of other applications' audio playback (if any).

Solution

Use audio sessions to set the type of audio category your application uses.

Discussion

The `AVAudioSession` class was introduced in the AV Foundation framework starting with iOS SDK 3.0. Every iOS application has one audio session. This audio session can be accessed using the `sharedInstance` class method of the `AVAudioSession` class, like so:

```
AVAudioSession *audioSession = [AVAudioSession sharedInstance];
```

After retrieving an instance of the `AVAudioSession` class, you can invoke the `setCate gory:error:` instance method of the audio session object to choose among the different categories available to iOS applications. Different values that can be set as the audio session category of an application are listed here:

AVAudioSessionCategorySoloAmbient
> This category is exactly like the `AVAudioSessionCategoryAmbient` category, except that this category will stop the audio playback of all other applications, such as the iPod. When the device is put into silent mode, your audio playback will be paused. This also happens when the screen is locked. This is the default category that iOS chooses for an application.

AVAudioSessionCategoryRecord
> This stops other applications' audio (e.g., the iPod) and also will not allow your application to initiate an audio playback (e.g., using `AVAudioPlayer`). You can only record audio in this mode. Using this category, calling the `prepareToPlay` instance method of `AVAudioPlayer` will return `YES` and the `play` instance method will return `NO`. The main UI interface will function as usual. The recording of your application will continue even if the iOS device's screen is locked by the user.

AVAudioSessionCategoryPlayback
> This category will silence other applications' audio playback (such as the audio playback of iPod applications). You can then use the `prepareToPlay` and `play` instance methods of `AVAudioPlayer` to play a sound in your application. The main UI thread will function as normal. The audio playback will continue even if the screen is locked by the user and even if the device is in silent mode.

AVAudioSessionCategoryPlayAndRecord

This category allows audio to be played and recorded at the same time in your application. This will stop the audio playback of other applications when your audio recording or playback begins. The main UI thread of your application will function as normal. The playback and the recording will continue even if the screen is locked or the device is in silent mode.

AVAudioSessionCategoryAudioProcessing

This category can be used for applications that do audio processing, but not audio playback or recording. By setting this category, you cannot play or record any audio in your application. Calling the prepareToPlay and play instance methods of AVAudioPlayer will return NO. Audio playback of other applications, such as the iPod, will also stop if this category is set.

AVAudioSessionCategoryAmbient

This category will not stop the audio from other applications, but it will allow you to play audio over the audio being played by other applications, such as the iPod. The main UI thread of your application will function normally. The prepareTo Play and play instance methods of AVAudioPlayer will return with the value YES. The audio being played by your application will stop when the user locks the screen. The silent mode silences the audio playback of your application only if your application is the only application playing an audio file. If you start playing audio while the iPod is playing a song, putting the device in silent mode does not stop your audio playback.

To give you an example of using AVAudioSession, let's start an audio player that will play its audio file over other applications' audio playback. We will begin with the *.h* file of a view controller:

```
#import <UIKit/UIKit.h>
#import <AVFoundation/AVFoundation.h>

@interface AudioAndVideoViewController : UIViewController
              <AVAudioPlayerDelegate> {
@public
  AVAudioPlayer          *audioPlayer;
}

@property (nonatomic, retain) AVAudioPlayer     *audioPlayer;

@end
```

Here is the implementation (*.m* file) of our view controller:

```
#import "AudioAndVideoViewController.h"

@implementation AudioAndVideoViewController

@synthesize audioPlayer;

- (void)audioPlayerBeginInterruption:(AVAudioPlayer *)player{
```

```objc
    /* The audio session has been deactivated here */

}

- (void)audioPlayerEndInterruption:(AVAudioPlayer *)player{

  /* For iOS versions earlier than 4 */

}

- (void)audioPlayerEndInterruption:(AVAudioPlayer *)player
                         withFlags:(NSUInteger)flags{

  /* For iOS 4 and later */
  if (flags == AVAudioSessionInterruptionFlags_ShouldResume){
    [player play];
  }

}

- (void)audioPlayerDidFinishPlaying:(AVAudioPlayer *)player
                       successfully:(BOOL)flag{

  if (flag == YES){
    NSLog(@"Audio player stopped correctly.");
  } else {
    NSLog(@"Audio player did not stop correctly.");
  }

  if ([player isEqual:self.audioPlayer] == YES){
    self.audioPlayer = nil;
  } else {
    [player release];
  }

}

- (void)viewDidLoad {
  [super viewDidLoad];

  NSBundle *mainBundle = [NSBundle mainBundle];

  NSString *filePath = [mainBundle pathForResource:@"MySong"
                                            ofType:@"mp3"];

  NSError *audioSessionError = nil;
  AVAudioSession *audioSession = [AVAudioSession sharedInstance];
  if ([audioSession setCategory:AVAudioSessionCategoryAmbient
                          error:&audioSessionError] == YES){
    NSLog(@"Successfully set the audio session.");
  } else {
    NSLog(@"Could not set the audio session");
  }
```

```
            NSData *fileData = [NSData dataWithContentsOfFile:filePath];

            NSError *audioPlayerError = nil;

            AVAudioPlayer *newPlayer =
            [[AVAudioPlayer alloc] initWithData:fileData
                                          error:&audioPlayerError];
            self.audioPlayer = newPlayer;
            [newPlayer release];

            if (self.audioPlayer != nil){

              self.audioPlayer.delegate = self;

                if ([self.audioPlayer prepareToPlay] == YES &&
                  [self.audioPlayer play] == YES){
                NSLog(@"Successfully started playing.");

              } else {
                NSLog(@"Failed to play the audio file.");
                self.audioPlayer = nil;
              }

            } else {
              NSLog(@"Could not instantiate the audio player.");
            }

          }

          - (void) viewDidUnload{
            [super viewDidUnload];

            if (self.audioPlayer != nil){
              if ([self.audioPlayer isPlaying] == YES){
                [self.audioPlayer stop];
              }
              self.audioPlayer = nil;
            }

          }

          - (BOOL)shouldAutorotateToInterfaceOrientation:
            (UIInterfaceOrientation)interfaceOrientation {
            return(YES);
          }

          - (void)dealloc {

            if (audioPlayer != nil){
              if ([audioPlayer isPlaying] == YES){
                [audioPlayer stop];
              }
            }
            [audioPlayer release];
```

```
    [super dealloc];
}

@end
```

You can see that we are using the shared instance of the AVAudioSession class in the viewDidLoad instance method of our view controller to set the audio category of our application to AVAudioSessionCategoryAmbient in order to allow our application to play audio files over other applications' audio playback.

8.6 Playing Video Files

Problem

You would like to be able to play video files in your iOS application.

Solution

Use an instance of the MPMoviePlayerController class.

 If you simply want to display a full-screen movie player, you can use the MPMoviePlayerViewController class and push your movie player view controller into the stack of view controllers of a navigation controller (for instance), or simply present your movie player view controller as a modal controller on another view controller using the presentMovie PlayerViewControllerAnimated: instance method of UIViewController. In this recipe, we will use MPMoviePlayerController instead of MPMovie PlayerViewController in order to get full access to various settings that a movie player view controller does not offer, such as windowed-mode video playback (not full-screen).

Discussion

The Media Player framework in the iOS SDK allows programmers to play audio and video files, among other interesting things. To be able to play a video file, we will instantiate an object of type MPMoviePlayerController like so:

```
MPMoviePlayerController *newMoviePlayer =
[[MPMoviePlayerController alloc] initWithContentURL:url];

self.moviePlayer = newMoviePlayer;

[newMoviePlayer release];
```

In this code, moviePlayer is a property of type MPMoviePlayerController defined and synthesized for the current view controller. Prior to iOS 3.1, programmers had very little control over how movies were played using the Media Player framework. With the introduction of the iPad, the whole framework changed drastically to give more control to programmers and allow them to present their contents with more flexibility than before.

An instance of `MPMoviePlayerController` has a property called `view`. This view is of type `UIView` and is the view in which the media, such as video, will be played. As a programmer, you are responsible for inserting this view into your application's view hierarchy to present your users with the content being played. Since you get a reference to an object of type `UIView`, you can shape this view however you want. For instance, you can simply change the background color of this view to a custom color.

Many multimedia operations depend on the notification system. For instance, `MPMovie PlayerController` does not work with delegates; instead, it relies on notifications. This allows for a very flexible decoupling between the system libraries and the applications that iOS programmers write. For classes such as `MPMoviePlayerController`, we start listening for notifications that get sent by instances of that class. We use the default notification center and add ourselves as an observer for a notification.

To be able to test our recipe, we need a sample *.mov* file to play with the movie player. You can download an Apple-provided sample file from *http://support.apple.com/kb/ HT1425*. Make sure you download the H.264 file format. If this file is zipped, unzip it and rename it to *Sample.m4v*. Now drag and drop this file into your application bundle in Xcode.

After doing this, we can go ahead and write a simple program that attempts to play the video file for us. Here is our *.h* file:

```
#import <UIKit/UIKit.h>
#import <MediaPlayer/MediaPlayer.h>

@interface AudioAndVideoViewController : UIViewController{
@public
  MPMoviePlayerController *moviePlayer;
}

@property (nonatomic, retain) MPMoviePlayerController *moviePlayer;

- (IBAction) startPlayingVideo:(id)paramSender;
- (IBAction) stopPlayingVideo:(id)paramSender;

@end
```

Here is the implementation of the `startPlayingVideo:` method that we defined in our *.h* file:

```
- (IBAction) startPlayingVideo:(id)paramSender{

  /* First let's construct the URL of the file in our application bundle
   that needs to get played by the movie player */
  NSBundle *mainBundle = [NSBundle mainBundle];

  NSString *urlAsString = [mainBundle pathForResource:@"Sample"
                                               ofType:@"m4v"];
```

```
NSURL    *url = [NSURL fileURLWithPath:urlAsString];

/* If we have already created a movie player before,
 let's try to stop it */
if (self.moviePlayer != nil){
  [self stopPlayingVideo:nil];
}

/* Now create a new movie player using the URL */
MPMoviePlayerController *newMoviePlayer =
[[MPMoviePlayerController alloc] initWithContentURL:url];

self.moviePlayer = newMoviePlayer;

[newMoviePlayer release];

if (self.moviePlayer != nil){

  /* Listen for the notification that the movie player sends us
   whenever it finishes playing an audio file */
  [[NSNotificationCenter defaultCenter]
   addObserver:self
   selector:@selector(videoHasFinishedPlaying:)
   name:MPMoviePlayerPlaybackDidFinishNotification
   object:self.moviePlayer];

  NSLog(@"Successfully instantiated the movie player.");

  /* Scale the movie player to fit the aspect ratio */
  self.moviePlayer.scalingMode = MPMovieScalingModeAspectFit;

  /* Let's start playing the video in full screen mode */
  [self.moviePlayer play];

  [self.view addSubview:self.moviePlayer.view];

  [self.moviePlayer setFullscreen:YES
                         animated:YES];

} else {
  NSLog(@"Failed to instantiate the movie player.");
}

}
```

As you can see, we manage the movie player's view ourselves. If we add the view of the movie player to our view controller's view, we have to remove the view manually. This view will not get removed from our view controller's view even if we release the movie player. The following method stops the video and then removes the associated view:

```
- (IBAction) stopPlayingVideo:(id)paramSender {

  if (self.moviePlayer != nil){
```

```
[[NSNotificationCenter defaultCenter]
 removeObserver:self
 name:MPMoviePlayerPlaybackDidFinishNotification
 object:self.moviePlayer];

[self.moviePlayer stop];

if (self.moviePlayer.view != nil &&
    self.moviePlayer.view.superview != nil &&
    [self.moviePlayer.view.superview isEqual:self.view] == YES){
  [self.moviePlayer.view removeFromSuperview];
}

}

}
```

 Please be extra careful when using the **superview** property of your movie player's view. In cases such as when your view is unloaded because of low-memory warnings, you must not use the **view** property of your view controller in any way after it is set to **nil**. Doing so will cause the iOS SDK to throw out warnings in the console window.

Here is our implementation of the **viewDidUnload** method:

```
- (void) viewDidUnload{

  [self stopPlayingVideo:nil];
  self.moviePlayer = nil;
  [super viewDidUnload];

}
```

In the startPlayingVideo: instance method of our view controller we are listening for the MPMoviePlayerPlaybackDidFinishNotification notification that MKMoviePlayerView Controller will send to the default notification center. We listen to this notification on the videoHasFinishedPlaying: instance method of our view controller. Here we can get notified when the movie playback has finished and perhaps dispose of our movie player object:

```
- (void) videoHasFinishedPlaying:(NSNotification *)paramNotification{

  /* Find out what the reason was for the player to stop */
  NSNumber *reason =
  [paramNotification.userInfo
   valueForKey:MPMoviePlayerPlaybackDidFinishReasonUserInfoKey];

  if (reason != nil){
    NSInteger reasonAsInteger = [reason integerValue];

    switch (reasonAsInteger){
      case MPMovieFinishReasonPlaybackEnded:{
```

```
      /* The movie ended normally */
      break;
    }
    case MPMovieFinishReasonPlaybackError:{
      /* An error happened and the movie ended */
      break;
    }
    case MPMovieFinishReasonUserExited:{
      /* The user exited the player */
      break;
    }
  }

  NSLog(@"Finish Reason = %ld", (long)reasonAsInteger);
  [self stopPlayingVideo:nil];
} /* if (reason != nil){ */

}
```

You might have already noticed that we are invoking the `stopPlayingVideo:` instance method that we've implemented in the `videoHasFinishedPlaying:` notification handler. We do this because the `stopPlayingVideo:` instance method takes care of unregistering our object from the notifications received by the media player and removes the media player from the superview. In other words, when the video stops playing, it does not necessarily mean the resources we allocated for that player have been deallocated. We need to take care of that manually. Bear in mind that the `MPMoviePlayerController` class does not work in iPhone Simulator. You need to run this code on a real device and check the results for yourself.

See Also

Recipe 8.7

8.7 Capturing Thumbnails from a Video File Asynchronously

Problem

You are playing a video file using an instance of the `MPMoviePlayerController` class and would like to capture a screenshot from the movie at a certain time.

Solution

Use the `requestThumbnailImagesAtTimes:timeOption:` instance method of `MPMovie PlayerController` like so:

```
/* Capture the frame at the third second into the movie */
NSNumber *thirdSecondThumbnail = [NSNumber numberWithFloat:3.0f];

/* We can ask to capture as many frames as we want. But for now, we are just
   asking to capture one frame */
```

```
NSArray  *requestedThumbnails =
[NSArray arrayWithObject:thirdSecondThumbnail];

/* Ask the movie player to capture this frame for us */
[self.moviePlayer
 requestThumbnailImagesAtTimes:requestedThumbnails
 timeOption:MPMovieTimeOptionExact];
```

Discussion

An instance of `MPMoviePlayerController` is able to capture thumbnails from the currently playing movie, synchronously and asynchronously. In this recipe, we are going to focus on asynchronous image capture for this class.

We can use the `requestThumbnailImagesAtTimes:timeOption:` instance method of `MPMoviePlayerController` to asynchronously access thumbnails. When I say "asynchronously," I mean that during the time the thumbnail is being captured and reported to your designated object (as we will soon see), the movie player will continue its work and will not block the playback. We must observe the `MPMoviePlayerThumbnailImage RequestDidFinishNotification` notification message the movie player sends to the default notification center in order to find out when our thumbnails are available:

```
- (IBAction) startPlayingVideo:(id)paramSender{

  /* First let's construct the URL of the file in our application bundle
   that needs to get played by the movie player */
  NSBundle *mainBundle = [NSBundle mainBundle];

  NSString *urlAsString = [mainBundle pathForResource:@"Sample"
                                               ofType:@"m4v"];

  NSURL    *url = [NSURL fileURLWithPath:urlAsString];

  /* If we have already created a movie player before,
   let's try to stop it */
  self.moviePlayer = nil;

  /* Now create a new movie player using the URL */
  MPMoviePlayerController *newMoviePlayer =
  [[MPMoviePlayerController alloc] initWithContentURL:url];
  self.moviePlayer = newMoviePlayer;
  [newMoviePlayer release];

  if (self.moviePlayer != nil){

    /* Listen for the notification that the movie player
     sends us whenever it finishes playing a video file */
    [[NSNotificationCenter defaultCenter]
     addObserver:self
     selector:@selector(videoHasFinishedPlaying:)
     name:MPMoviePlayerPlaybackDidFinishNotification
     object:self.moviePlayer];
```

```
[[NSNotificationCenter defaultCenter]
 addObserver:self
 selector:@selector(videoThumbnailIsAvailable:)
 name:MPMoviePlayerThumbnailImageRequestDidFinishNotification
 object:self.moviePlayer];

NSLog(@"Successfully instantiated the movie player.");

/* Scale the movie player to fit the aspect ratio */
self.moviePlayer.scalingMode = MPMovieScalingModeAspectFit;

/* Get a reference to the view of the movie player */
UIView *playerView = self.moviePlayer.view;

/* We want to display the movie on our view */
[self.view addSubview:playerView];

/* Let's start playing the video in full screen mode */
[self.moviePlayer play];

[self.moviePlayer setFullscreen:YES
                       animated:YES];

/* Capture the frame at the third second into the movie */
NSNumber *thirdSecondThumbnail = [NSNumber numberWithFloat:3.0f];

/* We can ask to capture as many frames as we
   want. But for now, we are just asking to capture one frame */
NSArray  *requestedThumbnails =
[NSArray arrayWithObject:thirdSecondThumbnail];

/* Ask the movie player to capture this frame for us */
[self.moviePlayer
 requestThumbnailImagesAtTimes:requestedThumbnails
 timeOption:MPMovieTimeOptionExact];

} else {
  NSLog(@"Failed to instantiate the movie player.");
}

}
```

You can see that we are asking the movie player to capture the frame at the third second into the movie. Once this task is completed, the videoThumbnailIsAvailable: instance method of our view controller will be called. Here is how we can access the captured image:

```
- (void) videoThumbnailIsAvailable:(NSNotification *)paramNotification{

MPMoviePlayerController *controller = [paramNotification object];

if (controller != nil &&
    [controller isEqual:self.moviePlayer] == YES){

  NSLog(@"Screenshot is available");
```

```
    /* Now get the thumbnail out of the user info dictionary */
    UIImage *thumbnail =
      [paramNotification.userInfo
       objectForKey:MPMoviePlayerThumbnailImageKey];

    if (thumbnail != nil){

      /* We got the thumbnail image. You can now use it here */

    }

  }

}
```

Since we started listening to the `MPMoviePlayerThumbnailImageRequestDidFinishNotifi`
`cation` notifications when we instantiated our movie player object in the `startPlaying`
`Video:` method, we must also stop listening for this notification whenever we stop the
movie player (or whenever you believe is appropriate depending on your application
architecture):

```
- (IBAction) stopPlayingVideo:(id)paramSender {

  if (self.moviePlayer != nil){

    [[NSNotificationCenter defaultCenter]
     removeObserver:self
     name:MPMoviePlayerPlaybackDidFinishNotification
     object:self.moviePlayer];

    [[NSNotificationCenter defaultCenter]
     removeObserver:self
     name:MPMoviePlayerThumbnailImageRequestDidFinishNotification
     object:self.moviePlayer];

    [self.moviePlayer stop];

    if (self.moviePlayer.view != nil &&
        self.moviePlayer.view.superview != nil &&
        [self.moviePlayer.view.superview isEqual:self.view] == YES){
      [self.moviePlayer.view removeFromSuperview];
    }

  }

}
```

When calling the `requestThumbnailImagesAtTimes:timeOption:` instance method of
`MPMoviePlayerController`, we can specify one of two values for `timeOption`: `MPMovie`
`TimeOptionExact` or `MPMovieTimeOptionNearestKeyFrame`. The former gives us the frame
playing at the exact point we requested in the timeline of our video, whereas the latter
is less exact, but also uses fewer system resources and offers an overall better

performance when capturing thumbnails from a video. MPMovieTimeOptionNearestKey Frame is usually adequate in terms of precision because it is just a couple of frames off.

8.8 Accessing the iPod Library in Response to a User Request

Problem

You want to access an item that your user picks from his iPod library.

Solution

Use the MPMediaPickerController class:

```
- (IBAction) displayMediaPicker:(id)paramSender{

  MPMediaPickerController *mediaPicker =
  [[MPMediaPickerController alloc]
   initWithMediaTypes:MPMediaTypeAny];

  if (mediaPicker != nil){

    NSLog(@"Successfully instantiated a media picker.");
    mediaPicker.delegate = self;
    mediaPicker.allowsPickingMultipleItems = NO;

    [self.navigationController presentModalViewController:mediaPicker
                                                animated:YES];
    [mediaPicker release];

  } else {
    NSLog(@"Could not instantiate a media picker.");
  }

}
```

Discussion

MPMediaPickerController is a view controller that the iPod application displays to the user. By instantiating MPMediaPickerController, you can present a standard view controller to your users to allow them to select whatever item they want from their library and then transfer the control to your application. This is particularly useful in games, for instance, where the user plays the game and can have your application play his favorite tracks in the background.

You can get information from the media picker controller by becoming its delegate (conforming to MPMediaPickerControllerDelegate):

```
#import <UIKit/UIKit.h>
#import <MediaPlayer/MediaPlayer.h>

@interface AudioAndVideoViewController : UIViewController
```

```
            <MPMediaPickerControllerDelegate>{
}

- (IBAction) displayMediaPicker:(id)paramSender;

@end
```

Inside your `displayMediaPicker:` selector, implement the code required to display an instance of the media picker controller and present it to the user as a modal view controller:

```
- (IBAction) displayMediaPicker:(id)paramSender{

  MPMediaPickerController *mediaPicker =
  [[MPMediaPickerController alloc]
   initWithMediaTypes:MPMediaTypeAny];

  if (mediaPicker != nil){

    NSLog(@"Successfully instantiated a media picker.");
    mediaPicker.delegate = self;
    mediaPicker.allowsPickingMultipleItems = NO;

    [self.navigationController presentModalViewController:mediaPicker
                                                animated:YES];
    [mediaPicker release];

  } else {
    NSLog(@"Could not instantiate a media picker.");
  }

}
```

The `allowsPickingMultipleItems` property of the media picker controller lets you specify whether users can pick more than one item from their library before dismissing the media picker controller. This takes a BOOL value, so for now we just set it to NO; we will later see what this looks like. Now let's implement the various delegate messages defined in the `MPMediaPickerControllerDelegate` protocol:

```
- (void) mediaPicker:(MPMediaPickerController *)mediaPicker
  didPickMediaItems:(MPMediaItemCollection *)mediaItemCollection{

  NSLog(@"Media Picker returned");

  for (MPMediaItem *thisItem in mediaItemCollection.items){

    NSURL     *itemURL =
    [thisItem valueForProperty:MPMediaItemPropertyAssetURL];

    NSString *itemTitle =
    [thisItem valueForProperty:MPMediaItemPropertyTitle];

    NSString *itemArtist =
    [thisItem valueForProperty:MPMediaItemPropertyArtist];
```

```
   MPMediaItemArtwork *itemArtwork =
     [thisItem valueForProperty:MPMediaItemPropertyArtwork];

   NSLog(@"Item URL = %@", itemURL);
   NSLog(@"Item Title = %@", itemTitle);
   NSLog(@"Item Artist = %@", itemArtist);
   NSLog(@"Item Artwork = %@", itemArtwork);
 }

 [mediaPicker dismissModalViewControllerAnimated:YES];

}


- (void) mediaPickerDidCancel:(MPMediaPickerController *)mediaPicker{

  /* The media picker was cancelled */
  NSLog(@"Media Picker was cancelled");

  [mediaPicker dismissModalViewControllerAnimated:YES];

}
```

You can access different properties of each selected item using the valueForProperty: instance method of MPMediaItem. Instances of this class will be returned to your application through the mediaItemCollection parameter of the mediaPicker:didPick MediaItems: delegate message.

Now let's write a program with a very simple GUI that allows us to ask the user to pick one music item from his iPod library. After he picks the music file, we will attempt to play it using an MPMusicPlayerController instance. Our GUI has two simple buttons: Pick and Play, and Stop Playing. The first button will ask the user to pick an item from his iPod library for us to play, and the second button will stop the audio playback (if we are already playing the song). We will start with the design of the UI of our application. Let's create it in a simple way, as shown in Figure 8-1.

Now let's go ahead and define two IBAction selectors in the .h file of our view controller that we can later connect to the two buttons on our UI:

```
#import <UIKit/UIKit.h>
#import <MediaPlayer/MediaPlayer.h>
#import <AVFoundation/AVFoundation.h>

@interface AudioAndVideoViewController : UIViewController
         <MPMediaPickerControllerDelegate>{
@public
  MPMusicPlayerController *myMusicPlayer;
}

@property (nonatomic, retain) MPMusicPlayerController *myMusicPlayer;

- (IBAction)  displayMediaPickerAndPlayItem:(id)paramSender;
- (IBAction)  stopPlayingAudio:(id)paramSender;

@end
```

Figure 8-1. A very simple UI for our media picker and AV Audio Player

Now use Interface Builder to connect the buttons to the IBAction selectors we just defined. Then you can implement your view controller (the *.m* file). The two most important methods are the displayMediaPickerAndPlayItem: and stopPlayingAudio: instance methods:

```
- (IBAction) stopPlayingAudio:(id)paramSender{

  if (self.myMusicPlayer != nil){

    [[NSNotificationCenter defaultCenter]
     removeObserver:self
     name:MPMusicPlayerControllerPlaybackStateDidChangeNotification
     object:self.myMusicPlayer];

    [[NSNotificationCenter defaultCenter]
     removeObserver:self
     name:MPMusicPlayerControllerNowPlayingItemDidChangeNotification
     object:self.myMusicPlayer];

    [[NSNotificationCenter defaultCenter]
     removeObserver:self
     name:MPMusicPlayerControllerVolumeDidChangeNotification
     object:self.myMusicPlayer];

    [self.myMusicPlayer stop];
  }
```

```
}

- (IBAction) displayMediaPickerAndPlayItem:(id)paramSender{

  MPMediaPickerController *mediaPicker =
  [[MPMediaPickerController alloc]
   initWithMediaTypes:MPMediaTypeMusic];

  if (mediaPicker != nil){

    NSLog(@"Successfully instantiated a media picker.");
    mediaPicker.delegate = self;
    mediaPicker.allowsPickingMultipleItems = YES;

    [self.navigationController presentModalViewController:mediaPicker
                                                animated:YES];
    [mediaPicker release];

  } else {
    NSLog(@"Could not instantiate a media picker.");
  }

}
```

When our media picker controller succeeds, the `mediaPicker:didPickMediaItems` message will be called in the delegate object (in this case, our view controller). On the other hand, if the user cancels the media player, we'll get the `mediaPicker:mediaPickerDidCancel` message. The following code implements the method that will be called in each case:

```
- (void) mediaPicker:(MPMediaPickerController *)mediaPicker
    didPickMediaItems:(MPMediaItemCollection *)mediaItemCollection{

  NSLog(@"Media Picker returned");

  /* First, if we have already created a music player, let's
   deallocate it */
  self.myMusicPlayer = nil;

  /* Now make sure the music player has been deallocated
   and set to nil */
  if (self.myMusicPlayer == nil){
    MPMusicPlayerController *newMusicPlayer =
    [[MPMusicPlayerController alloc] init];

    self.myMusicPlayer = newMusicPlayer;
    [newMusicPlayer release];

    [self.myMusicPlayer beginGeneratingPlaybackNotifications];

    /* Get notified when the state of the playback changes */
    [[NSNotificationCenter defaultCenter]
     addObserver:self
     selector:@selector(musicPlayerStateChanged:)
```

```
    name:MPMusicPlayerControllerPlaybackStateDidChangeNotification
    object:self.myMusicPlayer];

  /* Get notified when the playback moves from one item
     to the other. In this recipe, we are only going to allow
     our user to pick one music file */
  [[NSNotificationCenter defaultCenter]
   addObserver:self
   selector:@selector(nowPlayingItemIsChanged:)
   name:MPMusicPlayerControllerNowPlayingItemDidChangeNotification
   object:self.myMusicPlayer];

  /* And also get notified when the volume of the
     music player is changed */
  [[NSNotificationCenter defaultCenter]
   addObserver:self
   selector:@selector(volumeIsChanged:)
   name:MPMusicPlayerControllerVolumeDidChangeNotification
   object:self.myMusicPlayer];

}

/* Start playing the items in the collection */
[self.myMusicPlayer setQueueWithItemCollection:mediaItemCollection];
[self.myMusicPlayer play];

/* Finally dismiss the media picker controller */
[mediaPicker dismissModalViewControllerAnimated:YES];

}

- (void) mediaPickerDidCancel:(MPMediaPickerController *)mediaPicker{

  /* The media picker was cancelled */
  NSLog(@"Media Picker was cancelled");

  [mediaPicker dismissModalViewControllerAnimated:YES];

}
```

We are listening for the events our music player generates through the notifications that it sends. Here are the three methods that are going to be responsible for handling the notifications we are listening to for the music player:

```
- (void) musicPlayerStateChanged:(NSNotification *)paramNotification{

  NSLog(@"Player State Changed");

  /* Let's get the state of the player */
  NSNumber *stateAsObject =
  [paramNotification.userInfo
   objectForKey:@"MPMusicPlayerControllerPlaybackStateKey"];

  NSInteger state = [stateAsObject integerValue];
```

```
      /* Make your decision based on the state of the player */
      switch (state){
        case MPMusicPlaybackStateStopped:{
          /* Here the media player has stopped playing the queue. */
          break;
        }
        case MPMusicPlaybackStatePlaying:{
          /* The media player is playing the queue. Perhaps you
            can reduce some processing that your application
            is using to give more processing power
            to the media player */
          break;
        }
        case MPMusicPlaybackStatePaused:{
          /* The media playback is paused here. You might want
            to indicate by showing graphics to the user */
          break;
        }
        case MPMusicPlaybackStateInterrupted:{
          /* An interruption stopped the playback of the media queue */
          break;
        }
        case MPMusicPlaybackStateSeekingForward:{
          /* The user is seeking forward in the queue */
          break;
        }
        case MPMusicPlaybackStateSeekingBackward:{
          /* The user is seeking backward in the queue */
          break;
        }
      } /* switch (State){ */

}

- (void) nowPlayingItemIsChanged:(NSNotification *)paramNotification{

  NSLog(@"Playing Item Is Changed");

  NSString *persistentID =
  [paramNotification.userInfo
   objectForKey:@"MPMusicPlayerControllerNowPlayingItemPersistentIDKey"];

  /* Do something with Persistent ID */

  NSLog(@"Persistent ID = %@", persistentID);

}

- (void) volumeIsChanged:(NSNotification *)paramNotification{

  NSLog(@"Volume Is Changed");

  /* The userInfo dictionary of this notification is normally empty */

}
```

We will also implement some common instance methods of our view controller to make sure we won't leave any memory leaking:

```
- (void)viewDidLoad {
    [super viewDidLoad];

    [self.navigationController setNavigationBarHidden:NO
                                            animated:NO];

}

- (void) viewDidUnload{
    [super viewDidUnload];

    [self stopPlayingAudio:nil];
    self.myMusicPlayer = nil;
}

- (BOOL)shouldAutorotateToInterfaceOrientation:
    (UIInterfaceOrientation)interfaceOrientation {
    return(YES);
}

- (void)dealloc {

    [self stopPlayingAudio:nil];
    [myMusicPlayer release];

    [super dealloc];
}
```

By running our application and pressing the Pick and Play button on our view controller, we will be presented with the media picker controller. Once the picker view controller is displayed, the same iPod UI will be presented to our user. After the user picks an item (or cancels the whole dialog), we will get appropriate delegate messages called in our view controller (since our view controller is the delegate of our media picker). After the items are picked (we only allow one item in this recipe, though), we will start our music player and start playing the whole collection.

If you want to allow your users to pick more than one item at a time, simply set the allowsPickingMultipleItems property of your media picker controller to YES:

```
mediaPicker.allowsPickingMultipleItems = YES;
```

 Sometimes when working with the media picker controller (MPMedia PickerController), the "MPMediaPicker: Lost connection to iPod library" message will be printed out to the console screen. This is because the media picker has been interrupted by an event such as syncing with iTunes while the picker was being displayed to the user. Immediately, your mediaPickerDidCancel: delegate message will be called as well.

CHAPTER 9

Address Book

9.0 Introduction

On an iOS device, the Contacts application allows users to add to, remove from, and manipulate their address book. The address book can be a collection of people and groups. Each person can have properties such as first name, last name, phone number, and email address assigned to her. Some properties can have a single value and some can have multiple values. For instance, the first name of a person is one value but the phone number can be multiple values (e.g., if the user has two home phone numbers).

The AddressBook.framework framework in the iOS SDK allows us to interact with the address book database on the device. We can get the array of all entities in the user's address book, insert and change values, and much more.

To use the address-book-related functions in your application, follow these steps to first add AddressBook.framework to your application:

1. Find your current target in Xcode.
2. Right-click on your target and choose Add→Existing Frameworks.
3. From the list, choose AddressBook.framework, as shown in Figure 9-1.
4. Click Add.

After you've added the framework to your application, whenever you want to use address-book-related functions you must include the main header file of the framework in your header (*.h*) or implementation (*.m*) file, like so:

```
#import <UIKit/UIKit.h>
#import <AddressBook/AddressBook.h>

@interface RootViewController : UIViewController {
@public
}

@end
```

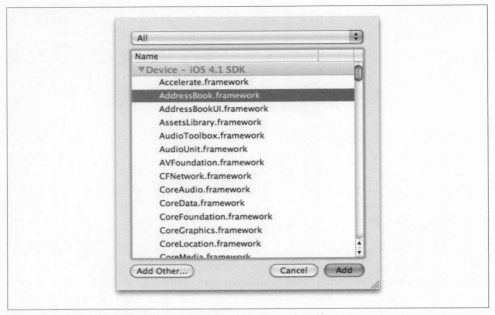

Figure 9-1. Adding AddressBook.framework to our target application in Xcode

 You can use the Address Book framework on iPhone Simulator, but the Contacts database on the simulator is empty by default. If you want to run the examples in this chapter on iPhone Simulator, first populate your address book (on the simulator) using the Contacts application.

I have populated my iPhone Simulator's contacts database with three entries, as shown in Figure 9-2.

I also suggest that you populate the address book of your iPhone Simulator with as many values as possible: multiple phone numbers for work and home, different addresses, and so forth. Only through such diversity can you correctly test the Address Book framework's functions.

9.1 Accessing the Address Book

Problem

You would like to get a reference to the user's address book database.

Figure 9-2. iPhone Simulator's sample contact entries

Solution

Use the `ABAddressBookCreate` function in the Address Book framework:

```
- (void)viewDidLoad {
[super viewDidLoad];

ABAddressBookRef addressBook = ABAddressBookCreate();

if (addressBook != nil){
  NSLog(@"Successfully accessed the address book.");

  /* Work with the address book here */

  /* Let's see if we have made any changes to the
   address book or not, before attempting to save it */

  if (ABAddressBookHasUnsavedChanges(addressBook) == YES){
    /* Now decide if you want to save the changes to
     the address book */
    NSLog(@"Changes were found in the address book.");

    BOOL doYouWantToSaveChanges = YES;

    /* We can make a decision to save or revert the
     address book back to how it was before */
    if (doYouWantToSaveChanges == YES){

      CFErrorRef saveError = NULL;

      if (ABAddressBookSave(addressBook, &saveError) == YES){
        /* We successfully saved our changes to the
         address book */
```

```
      } else {
        /* We failed to save the changes. You can now
           access the [saveError] variable to find out
           what the error is */
      }

    } else {

      /* We did NOT want to save the changes to the address
         book so let's revert it to how it was before */
      ABAddressBookRevert(addressBook);

    } /* if (doYouWantToSaveChanges == YES){ */

  } else {
    /* We have not made any changes to the address book */
    NSLog(@"No changes to the address book.");
  }

  CFRelease(addressBook);

} else {
  NSLog(@"Could not access the address book.");
}

}
```

I created the doYouWantToSaveChanges local variable and set it to YES just to demonstrate that we can, if we need to, revert an address book whose contents have been changed (reversion is done through the ABAddress BookRevert procedure). You can add code, for instance, asking the user if he wants the changes to be saved or not, and if not, you can revert the address book to its original state.

For more information about importing the Address Book framework into your application, please refer to this chapter's Introduction.

Discussion

To get a reference to the user's address book database, we must use the ABAddressBook Create function. This function returns a value of type ABAddressBookRef that will be nil if the address book cannot be accessed. You must check for nil values before accessing the address book reference returned by this function. Attempting to modify a nil address book will terminate your application with a runtime error.

After retrieving a reference to the user's address book, you can start making changes to the contacts, reading the entries, and so on. If you have made any changes to the address book, the ABAddressBookHasUnsavedChanges function will tell you by returning the value YES.

An instance of the address book database returned by the `ABAddress BookCreate` function must be released when you are finished working with it, using the `CFRelease` Core Foundation method, as demonstrated in our example code.

After determining whether changes were made to the address book database, you can either save or discard these changes using the `ABAddressBookSave` or `ABAddressBook Revert` procedure, respectively.

9.2 Retrieving All the People in the Address Book

Problem

You want to retrieve all contacts in the user's address book.

Solution

Use the `ABAddressBookCopyArrayOfAllPeople` function to retrieve an array of all contacts:

```
- (void)viewDidLoad {
[super viewDidLoad];

ABAddressBookRef addressBook = ABAddressBookCreate();

if (addressBook != nil){
  NSLog(@"Successfully accessed the address book.");

  CFArrayRef arrayOfAllPeople =
  ABAddressBookCopyArrayOfAllPeople(addressBook);

  if (arrayOfAllPeople != nil){

    NSUInteger peopleCounter = 0;
    for (peopleCounter = 0;
         peopleCounter < CFArrayGetCount(arrayOfAllPeople);
         peopleCounter++){

      ABRecordRef thisPerson =
      CFArrayGetValueAtIndex(arrayOfAllPeople,
                             peopleCounter);

      NSLog(@"%@", thisPerson);

      /* Use the [thisPerson] address book record */

    }

    CFRelease(arrayOfAllPeople);
  } /* if (allPeople != nil){ */
```

```
        CFRelease(addressBook);

    } /* if (addressBook != nil){ */

}
```

Discussion

After accessing the user's address book database, we can call the `ABAddressBookCopy`
`ArrayOfAllPeople` function to retrieve an array of all the contacts in that address book.
The return value of this function is an immutable array of type `CFArrayRef`. We cannot
work with this type of array the way we work with instances of `NSArray`, but here are a
few functions that will let us traverse an array of type `CFArrayRef`:

CFArrayGetCount
> Call this function to get the number of items in an instance of `CFArrayRef`. This is
> similar to the count instance method of an `NSArray`.

CFArrayGetValueAtIndex
> Call this function to retrieve an item at a specific location of an instance of
> `CFArrayRef`. This is similar to the `objectAtIndex:` instance method of an `NSArray`.

The items that are put in an array of all people, retrieved by calling the `ABAddressBook`
`CopyArrayOfAllPeople` function, are of type `ABRecordRef`. In Recipe 9.3, you will see how
to access different properties of the entries, such as a person's entry, in the address book
database.

See Also

Recipe 9.1

9.3 Retrieving Properties of Address Book Entries

Problem

You have retrieved a reference to an item in the address book, such as a person's entry,
and you want to retrieve that person's properties, such as first and last names.

Solution

Use the `ABRecordCopyValue` function:

```
- (void)viewDidLoad {
    [super viewDidLoad];

    ABAddressBookRef addressBook = ABAddressBookCreate();

    if (addressBook != nil){
        NSLog(@"Successfully accessed the address book.");
```

```
CFArrayRef allPeople =
ABAddressBookCopyArrayOfAllPeople(addressBook);

if (allPeople != nil){

  NSUInteger peopleCounter = 0;
  for (peopleCounter = 0;
       peopleCounter < CFArrayGetCount(allPeople);
       peopleCounter++){

    ABRecordRef thisPerson =
    CFArrayGetValueAtIndex(allPeople,
                           peopleCounter);

    NSString *firstName =
    (NSString *)ABRecordCopyValue(thisPerson,
                               kABPersonFirstNameProperty);

    NSString *lastName = (NSString *)
    ABRecordCopyValue(thisPerson,
                      kABPersonLastNameProperty);

    NSString *address = (NSString *)
    ABRecordCopyValue(thisPerson,
                      kABPersonEmailProperty);

    [self logContactEmailAddresses:thisPerson];

    NSLog(@"First Name = %@", firstName);
    NSLog(@"Last Name = %@", lastName);
    NSLog(@"Address = %@", address);

    [firstName release];
    [lastName release];
    [address release];

    /* Use the [thisPerson] address book record */

  } /* for (peopleCounter = 0; ... */

  CFRelease(allPeople);
} /* if (allPeople != nil){ */

CFRelease(addressBook);

} /* if (addressBook != nil){ */

}
```

Discussion

The records in the address book database are of type **ABRecordRef**. Each record could be either a group or a person. We have not discussed groups yet, so let's focus on people. Each person could have various types of information assigned to him, such as his first name, last name, email address, and so on. Bear in mind that many of these values are optional, and at the time of creating a new contact in the address book database, the user can simply leave out fields such as phone number, middle name, email address, URL, and so forth.

ABRecordCopyValue accepts an address book record and the property that has to be retrieved as its two parameters. The second parameter is the property of the record that we want to retrieve. Here are some of the common properties (all of these properties are defined as constant values in the *ABPerson.h* header file):

kABPersonFirstNameProperty
> This value will retrieve the first name of the given person. The return value is of type CFStringRef, which can be cast to NSString easily.

kABPersonLastNameProperty
> This value will retrieve the last name of the given person. Like the first name property, the return value will be of type CFString.

kABPersonMiddleNameProperty
> This value will retrieve the middle name of the given person. Like the first name and the last name, the return value will be of type CFString.

kABPersonEmailProperty
> This will retrieve the given person's email address. The return value in this case will be of type ABMultiValueRef. This type of data will be discussed next.

Some of the values that we retrieve from the ABRecordCopyValue function are straightforward, generic types, such as CFStringRef. This function can return complicated values, such as the email of a contact. The email could be further broken down into home email, work email, and so on. These values are called *multivalues* in the Address Book framework. Various functions allow us to work with multiple values (which are of type ABMultiValueRef):

ABMultiValueGetCount
> Returns the number of value/label pairs that are inside the multivalue.

ABMultiValueCopyLabelAtIndex
> Returns the label associated with a multivalue item at a specific index. For instance, if the user has three emails, such as work, home, and test emails, the index of the work email in the email multivalue would be 1. This function will then retrieve the label associated with that email (in this example, *work*). Please bear in mind that multivalues do not necessarily have to have labels. Make sure you check for NULL values.

ABMultiValueCopyValueAtIndex
> Returns the string value associated with a multivalue item at a specific index. I know this sounds complicated, but imagine the user has work, home, and test emails. If we provide the index 0 to this function, it will retrieve the given contact's work email.

Now let's go ahead and write a simple method that can log (to the console window) all the email addresses that are associated with a given contact from the address book:

```objc
- (void) logContactEmailAddresses:(ABRecordRef)paramContact{

  if (paramContact == nil){
    return;
  }

  ABMultiValueRef emails = ABRecordCopyValue(paramContact,
                                        kABPersonEmailProperty);

  /* This contact does not have an email property */
  if (emails == nil){
    return;
  }

  /* Go through all the emails */
  NSUInteger emailCounter = 0;
  for (emailCounter = 0;
       emailCounter < ABMultiValueGetCount(emails);
       emailCounter++){

    /* Get the label of the email (if any) */
    NSString *emailLabel = (NSString *)
    ABMultiValueCopyLabelAtIndex(emails, emailCounter);

    NSString *localizedEmailLabel = (NSString *)
    ABAddressBookCopyLocalizedLabel((CFStringRef)emailLabel);

    /* And then get the email address itself */
    NSString *email = (NSString *)
    ABMultiValueCopyValueAtIndex(emails, emailCounter);

    NSLog(@"Label = %@, Localized Label = %@, Email = %@",
          emailLabel, localizedEmailLabel, email);

    [email release];
    [localizedEmailLabel release];
    [emailLabel release];
  }

  CFRelease(emails);

}
```

 Calling the CFRelease procedure on a NULL value will crash your application. Make sure you check for NULL values before calling this Core Foundation procedure.

Label values returned by the ABMultiValueCopyLabelAtIndex function are rather cryptic and hard to read. Examples are "_$!<Other>!$_" and "_$!<Home>!$_" that might be set for email addresses with labels of Other and Home. However, if you want to retrieve a plain and readable version of these labels, you can first copy the label using the `ABMultiValueCopyLabelAtIndex` function and pass the returned value of this function to the `ABAddressBookCopyLocalizedLabel` function.

See Also

Recipe 9.1; Recipe 9.2

9.4 Inserting a Person Entry in the User's Address Book

Problem

You want to create a new person contact and insert it into the user's address book.

Solution

Use the `ABPersonCreate` function to create a new person. Set the person's properties using the `ABRecordSetValue` function and add the person to the address book using the `ABAddressBookAddRecord` function:

```
- (ABRecordRef) createNewPersonWithFirstName:(NSString *)paramFirstName
                lastName:(NSString *)paramLastName
                inAddressBook:(ABAddressBookRef)paramAddressBook{

    ABRecordRef result = nil;

    /* Check the address book parameter */
    if (paramAddressBook == nil){
      NSLog(@"The address book is nil.");
      return(nil);
    }

    /* Just create an empty person entry */
    ABRecordRef person = ABPersonCreate();

    if (person == nil){
      NSLog(@"Failed to create a new person.");
      return(nil);
    }

    BOOL couldSetFirstName = NO;
    BOOL couldSetLastName = NO;
    CFErrorRef error = nil;

    /* Set the first name of the person */
    couldSetFirstName =
    ABRecordSetValue(person,
                    kABPersonFirstNameProperty,
                    paramFirstName,
                    &error);
```

```objc
/* Set the last name of the person */
couldSetLastName =
ABRecordSetValue(person,
                 kABPersonLastNameProperty,
                 paramLastName,
                 &error);

if (couldSetFirstName == YES &&
    couldSetLastName == YES){
  NSLog(@"Successfully set the first and the last name.");

  /* Add the person record to the address book. We have NOT saved yet */
  if (ABAddressBookAddRecord(paramAddressBook,
                             person,
                             &error) == YES){
    NSLog(@"Successfully added the new person.");

    /* Now save the address book if there are any unsaved changes */
    if (ABAddressBookHasUnsavedChanges(paramAddressBook) == YES &&
        ABAddressBookSave(paramAddressBook, &error) == YES){

      NSLog(@"Successfully saved the address book.");

      result = person;

    } else {
      /* Either no unsaved changes were there or we could not save the
       address book */
    }

  } else {
    NSLog(@"Could not add a new person.");
  }

} else {
  NSLog(@"Either the first or the last name could not be set.");
}

return(result);

}

- (void)viewDidLoad {
  [super viewDidLoad];

  ABAddressBookRef addressBook = ABAddressBookCreate();

  if (addressBook != nil){
    NSLog(@"Successfully accessed the address book.");

    ABRecordRef newPerson =
    [self createNewPersonWithFirstName:@"Leonardo"
                             lastName:@"DiCaprio"
                        inAddressBook:addressBook];
```

```
  if (newPerson != nil){
    NSLog(@"Successfully added the new person");

    CFRelease(newPerson);

  } else {
    NSLog(@"Insertion failed");
  }

  CFRelease(addressBook);

} /* if (addressBook != nil){ */

}
```

Discussion

After accessing the address book database using the `ABAddressBookCreate` function, you can start inserting new group and person records into the database. In this recipe, we will concentrate on inserting new person records. For information about inserting new groups into the address book, please refer to Recipe 9.5.

We must use the `ABPersonCreate` function in order to create a new person record. Bear in mind that calling this function is not enough to add the person record to the address book. You must save the address book for your record to appear in the database.

By calling the `ABPersonCreate` function, you get a Core Foundation reference to a value of type `ABRecordRef`. Now you can call the `ABRecordSetValue` function to set the various properties of a new person entry. Once you are done, you must add the new person record to the database. You can do this using the `ABAddressBookAddRecord` function. After doing this, you must also save any unsaved changes to the address book database in order to truly save your new person record. Do this by using the `ABAddressBook Save` function.

The `createNewPersonWithFirstName:lastName:inAddressBook:` method that we implemented creates a new person entry in the address book database. After invoking this function as shown in the `viewDidLoad` method in this recipe's Solution, you will see the results, as shown in Figure 9-3, in the Contacts application on iPhone Simulator.

Memory management on Core Foundation is quite different from what you might be used to when writing applications for Cocoa Touch. As this topic is beyond the scope of this book, please make sure you read the "Memory Management Programming Guide for Core Foundation" documentation on Apple's website:

http://developer.apple.com/iphone/library/documentation/corefounda tion/Conceptual/CFMemoryMgmt/CFMemoryMgmt.html

Figure 9-3. A new person record added to the address book database

9.5 Inserting a Group Entry in the User's Address Book

Problem

You want to categorize your contacts into groups.

Solution

Use the `ABGroupCreate` function:

```
- (ABRecordRef) createNewGroupWithName:(NSString *)paramGroupName
                inAddressBook:(ABAddressBookRef)paramAddressBook{

    ABRecordRef result = nil;

    if (paramAddressBook == nil){
      NSLog(@"The address book is nil.");
    }

    ABRecordRef group = ABGroupCreate();

    if (group == nil){
      NSLog(@"Failed to create a new group.");
      return(nil);
    }

    BOOL couldSetGroupName = NO;
    CFErrorRef error = nil;

    couldSetGroupName =
    ABRecordSetValue(group,
                     kABGroupNameProperty,
```

```objc
                    paramGroupName,
                    &error);

  if (couldSetGroupName == YES){

    if (ABAddressBookAddRecord(paramAddressBook,
                               group,
                               &error) == YES){
      NSLog(@"Successfully added the new group.");

      if (ABAddressBookHasUnsavedChanges(paramAddressBook) == YES &&
          ABAddressBookSave(paramAddressBook, &error) == YES){

        NSLog(@"Successfully saved the address book.");

        result = group;

      } else {
        /* Either no unsaved changes were there or
           we could not save the address book */
      }

    } else {
      NSLog(@"Could not add a new group.");
    }

  } else {
    NSLog(@"Failed to set the name of the group.");
  }

  return(result);

}

- (void)viewDidLoad {
  [super viewDidLoad];

  ABAddressBookRef addressBook = ABAddressBookCreate();

  if (addressBook != nil){
    NSLog(@"Successfully accessed the address book.");

    ABRecordRef hollywoodGroup =
    [self createNewGroupWithName:@"Hollywood"
                  inAddressBook:addressBook];

    if (hollywoodGroup != nil){
      NSLog(@"Successfully created the group.");

      CFRelease(hollywoodGroup);

    } else {
      NSLog(@"Could not create the group.");
    }
```

```
        CFRelease(addressBook);

    }

}
```

Bear in mind that, as mentioned before, Core Foundation memory management is more complex than what Xcode's static analyzer could process. Therefore, attempting to compile this code with the LLVM compiler with static analysis turned on will give you a lot of warnings. You can ignore these warnings and test the code with Instruments to make sure your code does not leak. I encourage you to familiarize yourself with memory management in Core Foundation by reading Apple's "Memory Management Programming Guide for Core Foundation" document, as mentioned in the previous note.

Discussion

After retrieving the reference to the address book database, you can call the ABGroup Create function to create a new group entry. However, you must perform a few more operations before you can insert this group into the address book operation. The first thing you have to do is to set the name of this group using the ABRecordSetValue function with the kABGroupNameProperty property, as shown in the example code.

After the name of the group is set, you need to add it to the address book database just like you add a new person's entry, using the ABAddressBookAddRecord function. For more information about adding a new person's entry to the address book database, please read Recipe 9.4.

 Inserting a new group with a name that already exists in the address book database will create a new group with the same name but with no group members. In later recipes, we will learn how to avoid doing this by first finding the groups in the database and making sure a group with that name doesn't already exist.

After adding the group to the address book, you also need to save the address book's contents using the ABAddressBookSave function. After running your code, you will see the results shown in Figure 9-4 (you might have created other groups before, so your address book might not look exactly like that shown in the figure).

9.6 Adding Persons to Groups

Problem

You want to assign a person entry in the address book to a group.

Figure 9-4. A new group created in the address book database

Solution

Use the `ABGroupAddMember` function, like so:

```
- (BOOL)      addPerson:(ABRecordRef)paramPerson
               toGroup:(ABRecordRef)paramGroup
    saveToAddressBook:(ABAddressBookRef)paramAddressBook{

    BOOL result = NO;

    if (paramPerson == nil ||
        paramGroup == nil ||
        paramAddressBook == nil){
      NSLog(@"Invalid parameters are given.");
      return(NO);
    }

    CFErrorRef error = nil;

    /* Now attempt to add the person entry to the group */
    result = ABGroupAddMember(paramGroup,
                              paramPerson,
                              &error);

    if (result == NO){
      NSLog(@"Could not add the person to the group.");
      return(NO);
    }

    /* Make sure we save any unsaved changes */
    if (ABAddressBookHasUnsavedChanges(paramAddressBook) == YES &&
        ABAddressBookSave(paramAddressBook, &error) == YES){
      NSLog(@"Successfully added the person to the group.");
```

```
    result = YES;
  } else {
    NSLog(@"No changes were saved.");
  }

  return(result);

}
```

Discussion

We learned to insert into the address book database both person entries (in Recipe 9.4) and group entries (in Recipe 9.5). We implemented two custom methods named createNewPersonWithFirstName:lastName:inAddressBook: and createNewGroupWithName:inAddressBook: in the two aforementioned recipes. Now we want to add the person entry to the group we created and save the information to the address book database. Combining these three recipes, we can use the following code to achieve our goal:

```
- (void)viewDidLoad {
  [super viewDidLoad];

  ABAddressBookRef addressBook = ABAddressBookCreate();

  if (addressBook != nil){
    NSLog(@"Successfully accessed the address book.");

    ABRecordRef leonardoDiCaprio =
    [self createNewPersonWithFirstName:@"Leonardo"
                             lastName:@"DiCaprio"
                        inAddressBook:addressBook];

    if (leonardoDiCaprio != nil){

      ABRecordRef hollywoodGroup =
      [self createNewGroupWithName:@"Hollywood"
                    inAddressBook:addressBook];

      if (hollywoodGroup != nil){

        if ([self addPerson:leonardoDiCaprio
                    toGroup:hollywoodGroup
          saveToAddressBook:addressBook] == YES){
          NSLog(@"Successfully added the person to the group.");
        } else {
          NSLog(@"Could not add the user to the group.");
        }

        CFRelease(hollywoodGroup);

      } else {
        NSLog(@"Could not create the group.");
      }
```

```
      CFRelease(leonardoDiCaprio);

    } else {
      NSLog(@"The new person was not created.");
    }

    CFRelease(addressBook);

  } /* if (addressBook != nil){ */

}
```

We can see that the person entry we added to the "Hollywood" group and to the database is, in fact, now inside this group in the address book, as shown in Figure 9-5.

Figure 9-5. Adding a new person entry to a group

See Also

Recipe 9.5

9.7 Searching in the Address Book

Problem

You want to be able to find a specific person or group in the address book database.

Solution

Use the `ABAddressBookCopyArrayOfAllPeople` and `ABAddressBookCopyArrayOfAllGroups`
functions to find all people and groups in the address book. Traverse the returned arrays

to find the information you are looking for. Alternatively, you can use the `ABAddress BookCopyPeopleWithName` function to find person entries with a given name.

Discussion

Up to this point, we have been inserting group and person entries into the address book without checking if such a group or person already exists. We can use the `ABAddress BookCopyArrayOfAllPeople` and `ABAddressBookCopyArrayOfAllGroups` functions to get the array of all people and groups in the address book and search in the array to see if the person or group entries we are about to insert into the address book already exist. Here are two methods that will make use of these functions and that can also be used in other recipes:

```
- (BOOL) doesPersonExistWithFirstName:(NSString *)paramFirstName
         lastName:(NSString *)paramLastName
         inAddressBook:(ABAddressBookRef)paramAddressBook{

  BOOL result = NO;

  if (paramAddressBook == nil){
    NSLog(@"The address book is nil.");
    return(NO);
  }

  /* First we get all the people in the address book */
  CFArrayRef allPeople =
  ABAddressBookCopyArrayOfAllPeople(paramAddressBook);

  /* Were we successful? */
  if (allPeople != nil){

    /* Go through all of them one by one */
    NSUInteger peopleCounter = 0;
    for (peopleCounter = 0;
         peopleCounter < CFArrayGetCount(allPeople);
         peopleCounter++){

      /* Get the current person as we are going to need this person's
         address book details (first name and last name) */
      ABRecordRef person = CFArrayGetValueAtIndex(allPeople,
                                                  peopleCounter);

      /* Get the person's first name here */
      NSString *firstName = (NSString *)
      ABRecordCopyValue(person,
                   kABPersonFirstNameProperty);

      /* And then the last name */
      NSString *lastName = (NSString *)
      ABRecordCopyValue(person,
                   kABPersonLastNameProperty);

      /* This person has to pass a couple of checks
```

```
       before it can be identified as the person that we
       are looking for */
    BOOL firstNameIsEqual = NO;
    BOOL lastNameIsEqual = NO;

    /* Check the possible combinations of the current
     person's first name and the given first name parameter to
     detect if this person is the person we are looking for.
     Here we are checking the first name property */

    if (firstName == nil &&
        paramFirstName == nil){
      firstNameIsEqual = YES;
    }
    else if ([firstName isEqualToString:
              paramFirstName] == YES){
      firstNameIsEqual = YES;
    }

    /* And then check the last name property */
    if (lastName == nil &&
        paramLastName == nil){
      lastNameIsEqual = YES;
    }
    else if ([lastName isEqualToString:
              paramLastName] == YES){
      lastNameIsEqual = YES;
    }

    /* Release our resources. Note that we should
     NOT "break" before we have released these resources.
     This is a common programmming mistake */
    [firstName release];
    [lastName  release];

    if (firstNameIsEqual == YES &&
        lastNameIsEqual == YES){
      result = YES;
      break;
    }

  } /* for (peopleCounter = 0; ... */

} /* if (allPeople != nil){ */

allPeople == nil ? /* Do Nothing */ : CFRelease(allPeople);

return(result);


}
```

Similarly, we can check the existence of a group by first retrieving the array of all the groups in the address book database, using the `ABAddressBookCopyArrayOfAllGroups` function:

```
- (BOOL) doesGroupExistWithGroupName:(NSString *)paramGroupName
        inAddressBook:(ABAddressBookRef)paramAddressBook{

BOOL result = NO;

if (paramAddressBook == nil){
  NSLog(@"The address book is nil.");
  return(NO);
}

/* First we get all the groups in the address book */
CFArrayRef allGroups =
ABAddressBookCopyArrayOfAllGroups(paramAddressBook);

/* Were we successful? */
if (allGroups != nil){

  /* Go through all of them one by one */
  NSUInteger groupCounter = 0;
  for (groupCounter = 0;
       groupCounter < CFArrayGetCount(allGroups);
       groupCounter++){

    ABRecordRef group =
    CFArrayGetValueAtIndex(allGroups,
                           groupCounter);

    NSString *groupName = (NSString *)
    ABRecordCopyValue(group,
                      kABGroupNameProperty);

    BOOL groupNameIsEqual = NO;

    if (groupName == nil &&
        paramGroupName == nil){
      groupNameIsEqual = YES;
    }
    else if ([groupName isEqualToString:
              paramGroupName] == YES){
      groupNameIsEqual = YES;
    }

    [groupName release];

    if (groupNameIsEqual == YES){
      result = YES;
      break;
    }

  }
} /* if (allGroups != nil){ */

allGroups == nil ? /* Do Nothing */ : CFRelease(allGroups);

return(result);
```

```
}
```

 Attempting to create a group with the name equal to @", (an empty string) or nil or NULL will create a new group with the name "Contacts" in the address book database. Please try to avoid creating groups with empty names or names equal to nil or NULL.

We can use the doesGroupExistWithGroupName:inAddressBook: method in this way:

```
- (void)viewDidLoad {
  [super viewDidLoad];

  ABAddressBookRef addressBook = ABAddressBookCreate();

  if (addressBook != nil){
    NSLog(@"Successfully accessed the address book.");

    if ([self doesGroupExistWithGroupName:@"O'Reilly"
                          inAddressBook:addressBook] == YES){

      NSLog(@"This group already exists.");

    } else {

      NSLog(@"This group does not exist.");

      ABRecordRef newGroup =
      [self createNewGroupWithName:@"O'Reilly"
                    inAddressBook:addressBook];

      if (newGroup != nil){
        NSLog(@"The new group is created successfully.");
        CFRelease(newGroup);
      } else {
        NSLog(@"Failed to create the new group.");
      }

    }

    CFRelease(addressBook);

  } /* if (addressBook != nil){ */

}
```

For the implementation of the createNewGroupWithName:inAddressBook: method, please refer to Recipe 9.5.

As we saw earlier, we have two ways of finding a person in the address book database:

• Retrieving the array of all people in the address book, using the ABAddressBook CopyArrayOfAllPeople function. We then get each record inside the array and compare the first and last name properties of each person with the strings we are looking for. This function helps you find the exact match you are looking for. You can search in any of the properties assigned to that person in the address book, including first name, last name, email, phone number, and so on.

- Asking the Address Book framework to perform the search based on a composite name for us. This is done using the `ABAddressBookCopyPeopleWithName` function.

Here is an example of using the `ABAddressBookCopyPeopleWithName` function to search for a contact with a specific name:

```
- (BOOL) doesPersonExistWithName:(NSString *)paramName
       inAddressBook:(ABAddressBookRef)paramAddressBook{

  BOOL result = NO;

  if (paramAddressBook == nil){
    NSLog(@"The address book is nil.");
    return(NO);
  }

  CFArrayRef people =
  ABAddressBookCopyPeopleWithName(paramAddressBook,
                                (CFStringRef)paramName);

  if (people != nil){
    /* Is there at least one person in the array with the name
     we were looking for? */
    if (CFArrayGetCount(people) > 0){
      result = YES;
    }
  }

  people == nil ? /* Do Nothing */ : CFRelease(people);

  return(result);

}
```

Here is how we can use this method that we just created:

```
- (void)viewDidLoad {
  [super viewDidLoad];

  ABAddressBookRef addressBook = ABAddressBookCreate();

  if (addressBook != nil){
    NSLog(@"Successfully accessed the address book.");

    if ([self doesPersonExistWithName:@"leonardo d"
                       inAddressBook:addressBook] == YES){

      NSLog(@"This person already exists.");

    } else {
```

```
    NSLog(@"This person does not exist. Creating...");

    ABRecordRef newPerson =
    [self createNewPersonWithFirstName:@"Leonardo"
                             lastName:@"DiCaprio"
                        inAddressBook:addressBook];

    if (newPerson != nil){
      NSLog(@"The new person object is created successfully.");
      CFRelease(newPerson);
    } else {
      NSLog(@"Failed to create the new person object.");
    }

  }

  CFRelease(addressBook);

} /* if (addressBook != nil){ */

}
```

Using this function, we won't have to know the full name to be able to find a contact in the address book. We can just pass a part of the name—for instance, just the first name—in order to find all the contacts with that specific first name.

The search performed by the ABAddressBookCopyPeopleWithName function is case-insensitive.

9.8 Retrieving and Setting a Person's Address Book Image

Problem

You want to be able to retrieve and set the images of address book people entries.

Solution

Use one of the following functions:

ABPersonHasImageData
    Use this function to find out if an address book entry has an image set.

ABPersonCopyImageData
    Use this function to retrieve the image data (if any).

ABPersonSetImageData
    Use this function to set the image data for an entry.

Discussion

As mentioned in this recipe's Solution, we can use the `ABPersonCopyImageData` function to retrieve the data associated with an image of a person entry in the address book. We can use this function in a method of our own to make it more convenient to use:

```
- (UIImage *) getPersonImage:(ABRecordRef)paramPerson{

  UIImage *result = nil;

  if (paramPerson == nil){
    NSLog(@"The person is nil.");
    return(nil);
  }

  NSData *imageData = (NSData *)ABPersonCopyImageData(paramPerson);
  if (imageData != nil){
    UIImage *image = [UIImage imageWithData:imageData];
    result = image;
  }

  [imageData release];

  return(result);

}
```

The `ABPersonSetImageData` function sets the image data for a person entry in the address book. Since this function uses data, not the image itself, we need to get `NSData` from `UIImage`. If we want the data pertaining to a PNG image, we can use the `UIImagePNG Representation` function to retrieve the PNG `NSData` representation of the image of type `UIImage`. To retrieve JPEG image data from an instance of `UIImage`, use the `UIImage JPEGRepresentation` function. Here is the method that will allow you to set the image of a person entry in the address book database:

```
- (BOOL) setPersonImage:(ABRecordRef)paramPerson
        inAddressBook:(ABAddressBookRef)paramAddressBook
        withImageData:(NSData *)paramImageData{

  BOOL result = NO;

  if (paramPerson == nil){
    NSLog(@"The person is nil.");
    return(NO);
  }

  if (paramAddressBook == nil){
    NSLog(@"The address book is nil.");
    return(NO);
  }

  CFErrorRef error = nil;

  if (ABPersonSetImageData(paramPerson,
```

```
                              (CFDataRef)paramImageData,
                              &error) == YES){

     NSLog(@"Successfully set the person's image. Saving...");

     if (ABAddressBookHasUnsavedChanges(paramAddressBook) == YES &&
         ABAddressBookSave(paramAddressBook, &error) == YES){
       NSLog(@"Successfully saved the changes.");
       result = YES;
     } else {
       NSLog(@"Failed to save the changes.");
     }

   } else {
     NSLog(@"Could not set the image data.");
   }

   return(result);

 }
```

Now let's write a simple application to demonstrate use of these methods. In this example code, we want to achieve the following:

- Create a simple view controller with an XIB file with two labels and two image views, as shown in Figure 9-6. Connect these image views to two corresponding outlets in the *.h* file of our view controller.

- Attempt to retrieve a contact with the first name "Leonardo" and the last name "DiCaprio" from our address book. If this contact doesn't exist, we will create it.

- Retrieve the previous image (if any) of the contact and display it in the first image view (the top image view, as shown in Figure 9-6).

- Set a new image for the contact, retrieved from our application bundle, and display the new image in the second image view (the bottom image view, as shown in Figure 9-6).

Let's get started. Here is the *.h* file of our view controller:

```
#import <UIKit/UIKit.h>
#import <AddressBook/AddressBook.h>

@interface RootViewController : UIViewController {
@public
  UIImageView *imageViewOldImage;
  UIImageView *imageViewNewImage;
}

@property (nonatomic, retain)
IBOutlet UIImageView *imageViewOldImage;

@property (nonatomic, retain)
IBOutlet UIImageView *imageViewNewImage;
```

@end

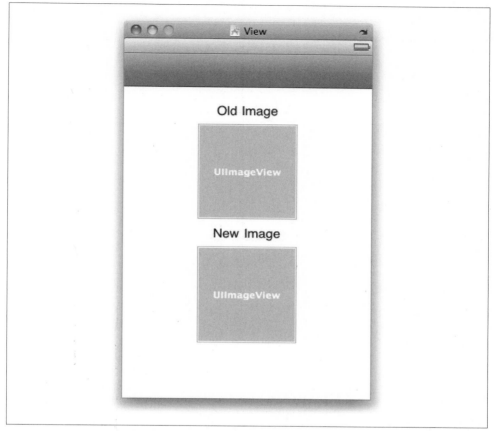

Figure 9-6. The view on an XIB file with two labels and two image views

In this recipe, we will use the copyFirstPersonWithName:inAddressBook: method that we implemented in Recipe 9.7. We are also using the createNewPersonWithFirst Name:lastName:inAddressBook: method that we implemented in Recipe 9.4. Here are the viewDidLoad and viewDidUnload instance methods of our view controller:

```
- (void)viewDidLoad {
[super viewDidLoad];

ABAddressBookRef addressBook = ABAddressBookCreate();

if (addressBook != nil){
  NSLog(@"Successfully accessed the address book.");

  ABRecordRef person = nil;
```

```objc
    person =
    [self copyFirstPersonWithName:@"Leonardo DiCaprio"
                    inAddressBook:addressBook];

    if (person != nil){
      NSLog(@"This person already exists in the address book.");
    } else {
      NSLog(@"This person doesn't exist. Creating...");
      person =
      [self createNewPersonWithFirstName:@"Leonardo"
                               lastName:@"DiCaprio"
                          inAddressBook:addressBook];
    }

    if (person != nil){

      UIImage *oldImage = [self getPersonImage:person];
      self.imageViewOldImage.image = oldImage;

      UIImage *newImage =
      [UIImage imageNamed:@"LeonardoDiCaprio.png"];

      NSData *newImageData =
      UIImagePNGRepresentation(newImage);

      if ([self setPersonImage:person
                 inAddressBook:addressBook
                 withImageData:newImageData] == YES){
        NSLog(@"Successfully set the new image.");
        self.imageViewNewImage.image = newImage;
      } else {
        NSLog(@"Failed to set the new image.");
      }

      CFRelease(person);

    } else {
      NSLog(@"The person cannot be found.");
    }


    CFRelease(addressBook);

  } /* if (addressBook != nil){ */

}

- (void) viewDidUnload{
  [super viewDidUnload];

  self.imageViewNewImage = nil;
  self.imageViewOldImage = nil;
}
```

For our example code to run, we will need an image named *Leonardo-DiCaprio.png* to be placed inside our application bundle. You can simply go to *http://images.google.com* and find a simple PNG image and place it inside your application bundle. If the filename differs from what we have in the source code, you must either rename the file or use the file's correct name in the source code.

After running our example, we can go to the Contacts application and see if our code was run successfully, as depicted in Figure 9-7.

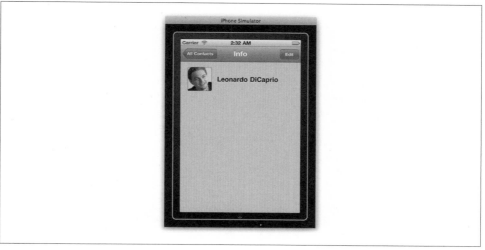

Figure 9-7. An image assigned to a contact in our address book, through code

CHAPTER 10

Camera and the Photo Library

10.0 Introduction

Devices running iOS, such as the iPhone, are equipped with cameras—for instance, the iPhone 4 has two cameras and the iPhone 3G and 3GS each have one. Some iOS devices, such as the first generation of the iPad, do not have cameras. The UIImage PickerController class (in iOS SDK 2.0 and later) allows programmers to display the familiar Camera interface to their users and ask them to take a photo or shoot a video. The photos taken or the videos shot by the user with the UIImagePickerController class then become accessible to the programmer.

In this chapter, you will learn how to let users take photos and shoot videos from inside applications, access these photos and videos, and access the photos and videos that are placed inside the Photo Library on an iOS device such as the iPod Touch and iPad.

 iPhone Simulator does not support the Camera interface. Please test and debug all your applications that require a Camera interface on a real device. All the examples in this chapter are run on three devices: the iPhone 3GS, iPhone 4, and iPad (the iPad does not have a camera, but we are including it in this chapter to see if our code really can detect the availability of a camera on an iOS device).

In this chapter, first we will attempt to determine if a camera is available on the iOS device that is running our application. You can also determine whether the camera allows you (the programmer) to capture videos, images, or both. To do this, make sure you have added the MobileCoreServices.framework framework to your target by following these steps:

1. Find the Frameworks group in Xcode.
2. Right-click on the Frameworks group and choose Add→Existing Frameworks.
3. Choose MobileCoreServices.framework from the list.

4. Click Add.

We will then move to other topics, such as accessing videos and photos from different albums on an iOS device. These are the same albums that are accessible through the Photos application built into iOS.

Accessing photos inside albums is more straightforward than accessing videos, however. For photos, we will be given the address of the photo and we can simply load the data of the image either in an instance of NSData or directly into an instance of UIImage. For videos, we won't be given a file address on the filesystem from which to load the data of the video. Instead, we will be given an address such as this:

```
assets-library://asset/asset.MOV?id=1000000004&ext=MOV
```

For addresses such as this, we need to use the Assets Library framework available in iOS SDK 4.0 and later. The Assets Library framework allows us to access the contents accessible through the Photos application, such as videos and photos shot by the user. You can also use the Assets Library framework to save images and videos on the device. These photos and videos will then become accessible by the Photo Library as well as other applications that wish to access these contents.

To make sure the recipes in this chapter compile correctly, follow these steps to add the Assets Library framework to your target:

1. Find your target in Xcode.
2. Right-click on the Frameworks group and choose Add→Existing Frameworks.
3. Choose AssetsLibrary.framework from the list.
4. Click Add.

To access the data of an asset given the URL to the asset, follow these steps:

1. Allocate and initialize an object of type ALAssetsLibrary. The Assets Library object facilitates the bridge that you need in order to access the videos and photos accessible by the Photos application.
2. Use the assetForURL:resultBlock:failureBlock instance method of the Assets Library object (allocated and initialized in step 1) to access the asset. An asset could be an image, a video, or any other resource available to the Photo Library. This method works with block objects, and a full discussion of block objects is beyond the scope of this book. Please consult Apple's "Blocks Programming Topics" documentation at *http://developer.apple.com/library/ios/#documentation/Cocoa/Con ceptual/Blocks/Articles/00_Introduction.html* for more information.
3. Release the Assets Library object allocated and initialized in step 1.

At this point, you might be wondering: how do I access the data for the asset? The resultBlock parameter of the assetForURL:resultBlock:failureBlock instance method of the Assets Library object will need to point to a block object that accepts a single parameter of type ALAsset. ALAsset is a class provided by the Assets Library that encapsulates an asset available to Photos and any other iOS application that wishes to use these assets. For more information about storing photos and videos in the Photo Library, please refer to Recipe 10.4 and Recipe 10.5. If you want to learn more about retrieving photos and videos from the Photo Library and the Assets Library, please refer to Recipe 10.6 and Recipe 10.7.

10.1 Detecting and Probing the Camera

Problem

You want to know whether the iOS device running your application has a camera that you can access. This is an important check to make before attempting to use the camera, unless you are sure your application will never run on a device that lacks one.

Solution

Use the isSourceTypeAvailable: class method of UIImagePickerController with the UIImagePickerControllerSourceTypeCamera value, like so:

```
- (BOOL)  isCameraAvailable{

  return([UIImagePickerController
          isSourceTypeAvailable:
          UIImagePickerControllerSourceTypeCamera]);

}

- (void)viewDidLoad {
  [super viewDidLoad];

  if ([self isCameraAvailable] == YES){
    NSLog(@"The camera is available.");
  } else {
    NSLog(@"The camera is not available.");
  }
}
```

Discussion

Before attempting to display an instance of UIImagePickerController to your user for taking photos or shooting videos, you must detect whether the device supports that interface. The isSourceTypeAvailable: class method allows you to determine three sources of data:

- The camera, by passing the UIImagePickerControllerSourceTypeCamera value to this method.
- The Photo Library, by passing the UIImagePickerControllerSourceTypePhoto Library value to this method. This browses the root folder of the *Photos* directory on the device.

- The camera roll folder in the *Photos* directory, by passing the `UIImagePicker ControllerSourceTypeSavedPhotosAlbum` value to this method.

If you want to check the availability of any of these facilities on an iOS device, you must pass these values to the `isSourceTypeAvailable:` class method of `UIImagePickerControl ler` before attempting to present the interfaces to the user.

Now in the *.h* or the *.m* file of your object(s) (depending on your requirements), import the main header file of the framework we just added. Here I will import this framework into the *.h* file of my main view controller:

```
#import <UIKit/UIKit.h>
#import <MobileCoreServices/MobileCoreServices.h>

@interface RootViewController : UIViewController {

}

@end
```

Now we can use the `isSourceTypeAvailable:` and `availableMediaTypesForSource Type:` class methods of `UIImagePickerController` to determine first if a media source is available (camera, Photo Library, etc.), and if so, whether media types such as image and video are available on that media source:

```
- (BOOL) doesCameraSupportMediaType:(NSString *)paramMediaType
  onSourceType:(UIImagePickerControllerSourceType)paramSourceType{

  BOOL result = NO;

  if (paramMediaType == nil ||
      [paramMediaType length] == 0){
    return(NO);
  }

  if ([UIImagePickerController
       isSourceTypeAvailable:paramSourceType] == NO){
    return(NO);
  }

  NSArray *mediaTypes =
  [UIImagePickerController
   availableMediaTypesForSourceType:paramSourceType];

  if (mediaTypes == nil){
    return(NO);
  }

  for (NSString *mediaType in mediaTypes){

    if ([mediaType isEqualToString:paramMediaType] == YES){
      return(YES);
    }
```

```
    }

    return(result);

}

- (BOOL)  doesCameraSupportShootingVideos{

    BOOL result = NO;

    result =
    [self doesCameraSupportMediaType:(NSString *)kUTTypeMovie
            onSourceType:UIImagePickerControllerSourceTypeCamera];

    return(result);

}

- (BOOL)  doesCameraSupportTakingPhotos{

    BOOL result = NO;

    result =
    [self doesCameraSupportMediaType:(NSString *)kUTTypeImage
            onSourceType:UIImagePickerControllerSourceTypeCamera];

    return(result);

}

- (void)viewDidLoad {
    [super viewDidLoad];

    if ([self doesCameraSupportTakingPhotos] == YES){
      NSLog(@"The camera supports taking photos.");
    } else {
      NSLog(@"The camera does not support taking photos");
    }

    if ([self doesCameraSupportShootingVideos] == YES){
      NSLog(@"The camera supports shooting videos.");
    } else {
      NSLog(@"The camera does not support shooting videos.");
    }

}
```

Some iOS devices, such as the iPhone 4, could have more than one camera. The two cameras on the iPhone 4 are called the front and the rear cameras. To determine whether these cameras are available, use the isCameraDeviceAvailable: class method of UIImagePickerController, like so:

```
- (BOOL)  isFrontCameraAvailable{

    BOOL result = NO;
```

```
#ifdef __IPHONE_4_0
  #if (__IPHONE_OS_VERSION_MAX_ALLOWED >= __IPHONE_4_0)
    result = [UIImagePickerController
              isCameraDeviceAvailable:
              UIImagePickerControllerCameraDeviceFront];
  #endif
#endif

return(result);

}

- (BOOL) isRearCameraAvailable{

  BOOL result = NO;

  #ifdef __IPHONE_4_0
    #if (__IPHONE_OS_VERSION_MAX_ALLOWED >= __IPHONE_4_0)
      result = [UIImagePickerController
                isCameraDeviceAvailable:
                UIImagePickerControllerCameraDeviceRear];
    #endif
  #endif

  return(result);

}
```

 The `UIImagePickerControllerCameraDeviceFront` and `UIImagePicker` `ControllerCameraDeviceRear` values are available only on iOS SDK 4.0 and later, so to write a universal application that can target an iPad and other devices with an SDK earlier than iOS SDK 4.0, you need to check the version in preprocessor directives.

By calling these methods on an iPhone 3GS, you will see that the `isFrontCameraAvail` `able` method returns `NO` and the `isRearCameraAvailable` method returns `YES`. Running the code on an iPhone 4 will prove that both methods will return `YES`, as iPhone 4 devices are equipped with both front- and rear-facing cameras.

If detecting which camera is present on a device isn't enough for your application, you can retrieve other settings using the `UIImagePickerController` class. One such setting is whether flash capability is available for a camera on the device. You can use the `isFlashAvailableForCameraDevice:` class method of `UIImagePickerController` to determine the availability of a flash capability on the rear or front camera. Please bear in mind that the `isFlashAvailableForCameraDevice:` class method of `UIImagePickerCon` `troller` checks the availability of the given camera device first, before checking the availability of a flash capability on that camera. Therefore, you can run the methods we will implement on devices that do not have front or rear cameras without a need to first check if the camera is available.

```objc
- (BOOL)  isFlashAvailableOnRearCamera{

  BOOL result = NO;

  #ifdef __IPHONE_4_0
    #if (__IPHONE_OS_VERSION_MAX_ALLOWED >= __IPHONE_4_0)
      result = [UIImagePickerController
                 isFlashAvailableForCameraDevice:
                 UIImagePickerControllerCameraDeviceRear];
    #endif
  #endif

  return(result);

}

- (BOOL)  isFlashAvailableOnFrontCamera{

  BOOL result = NO;

  #ifdef __IPHONE_4_0
    #if (__IPHONE_OS_VERSION_MAX_ALLOWED >= __IPHONE_4_0)
      result = [UIImagePickerController
                 isFlashAvailableForCameraDevice:
                 UIImagePickerControllerCameraDeviceFront];
    #endif
  #endif

  return(result);

}
```

Now if we take advantage of all the methods that we wrote in this recipe and test them in the `viewDidLoad` (for example) instance method of a view controller, we can see the results on different devices:

```objc
- (void)viewDidLoad {
  [super viewDidLoad];

  if ([self isFrontCameraAvailable] == YES){
    NSLog(@"The front camera is available.");
    if ([self isFlashAvailableOnFrontCamera] == YES){
      NSLog(@"The front camera is equipped with a flash");
    } else {
      NSLog(@"The front camera is not equipped with a flash");
    }
  } else {
    NSLog(@"The front camera is not available.");
  }

  if ([self isRearCameraAvailable] == YES){
    NSLog(@"The rear camera is available.");
    if ([self isFlashAvailableOnRearCamera] == YES){
      NSLog(@"The rear camera is equipped with a flash");
    } else {
```

```
      NSLog(@"The rear camera is not equipped with a flash");
    }
  } else {
    NSLog(@"The rear camera is not available.");
  }

  if ([self doesCameraSupportTakingPhotos] == YES){
    NSLog(@"The camera supports taking photos.");
  } else {
    NSLog(@"The camera does not support taking photos");
  }

  if ([self doesCameraSupportShootingVideos] == YES){
    NSLog(@"The camera supports shooting videos.");
  } else {
    NSLog(@"The camera does not support shooting videos.");
  }

}
```

Here are the results when we run the application on an iPhone 4:

```
The front camera is available.
The front camera is not equipped with a flash
The rear camera is available.
The rear camera is equipped with a flash
The camera supports taking photos.
The camera supports shooting videos.
```

Here is the output of the same code when run on an iPhone 3GS:

```
The front camera is not available.
The rear camera is available.
The rear camera is not equipped with a flash
The camera supports taking photos.
The camera supports shooting videos.
```

Running the same code on an iPad will result in this output in the console window:

```
The front camera is not available.
The rear camera is not available.
The camera does not support taking photos
The camera does not support shooting videos.
```

10.2 Taking Photos with the Camera

Problem

You want to ask the user to take a photo with the camera on his iOS device, and you want to access that photo once the user is done.

Solution

Instantiate an object of type `UIImagePickerController` and present it as a modal view controller on your current view controller. Here is the *.h* file of our view controller:

```
#import <UIKit/UIKit.h>
#import <MobileCoreServices/MobileCoreServices.h>

@interface RootViewController : UIViewController
        <UINavigationControllerDelegate,
         UIImagePickerControllerDelegate> {

}

@end
```

The delegate of an instance of `UIImagePickerController` must conform to the `UINavigationControllerDelegate` and `UIImagePickerControllerDelegate` protocols. If you forget to include them in the *.h* file of your delegate object, you'll get warnings from the compiler when assigning a value to the delegate property of your image picker controller. Please bear in mind that you can still assign an object to the delegate property of an instance of `UIImagePickerController` where that object does not explicitly conform to the `UIImagePickerControllerDelegate` and `UINavigationControllerDelegate` protocols, but implements the required methods in these protocols. I, however, suggest that you give a hint to the compiler that the delegate object does, in fact, conform to the aforementioned protocols in order to avoid getting compiler warnings.

In the implementation of our view controller (*.m* file), we will attempt to display an image picker controller as a modal view controller, like so:

```
- (void)viewDidLoad {
  [super viewDidLoad];

  if ([self isCameraAvailable] == YES &&
      [self doesCameraSupportTakingPhotos] == YES){

    UIImagePickerController *imagePicker =
    [[UIImagePickerController alloc] init];

    imagePicker.sourceType = UIImagePickerControllerSourceTypeCamera;

    NSString *requiredMediaType = (NSString *)kUTTypeImage;

    imagePicker.mediaTypes =
    [NSArray arrayWithObject:requiredMediaType];

    imagePicker.allowsEditing = YES;

    imagePicker.delegate = self;

    [self presentModalViewController:imagePicker
                            animated:YES];
```

```
    [imagePicker release];

  } else {
    NSLog(@"The camera is not available.");
  }

}
```

 We are using the isCameraAvailable and doesCameraSupportTaking Photos methods in this example. These methods are implemented and explained in Recipe 10.1.

In this example, we are allowing the user to take photos using the image picker. You must have noticed that we are setting the delegate property of the image picker to self, which refers to our view controller. For this, we have to make sure we have implemented the methods defined in the UIImagePickerControllerDelegate protocol, like so:

```
- (void)    imagePickerController:(UIImagePickerController *)picker
    didFinishPickingMediaWithInfo:(NSDictionary *)info{

  NSLog(@"Picker returned successfully.");

  NSString     *mediaType = [info objectForKey:
                          UIImagePickerControllerMediaType];

  if ([mediaType isEqualToString:(NSString *)kUTTypeMovie] == YES){

    NSURL *urlOfVideo =
    [info objectForKey:UIImagePickerControllerMediaURL];

    NSLog(@"Video URL = %@", urlOfVideo);

  }

  else if ([mediaType isEqualToString:(NSString *)kUTTypeImage] == YES){

    /* Let's get the metadata. This is only for
     images. Not videos */

    NSDictionary *metadata =
    [info objectForKey:
     UIImagePickerControllerMediaMetadata];

    UIImage *theImage =
    [info objectForKey:
     UIImagePickerControllerOriginalImage];

    NSLog(@"Image Metadata = %@", metadata);
    NSLog(@"Image = %@", theImage);
```

```
    }

    [picker dismissModalViewControllerAnimated:YES];


}

- (void)    imagePickerControllerDidCancel:(UIImagePickerController *)picker{

    NSLog(@"Picker was cancelled");

    [picker dismissModalViewControllerAnimated:YES];

}
```

Discussion

There are a couple of important things that you must keep in mind about the image picker controller's delegate. First, two delegate messages are called on the delegate object of the image picker controller. The `imagePickerController:didFinishPicking MediaWithInfo:` method gets called when the user finishes execution of the image picker (e.g., takes a photo and presses a button at the end), whereas the `imagePickerControl lerDidCancel:` method gets called when the image picker's operation is cancelled.

Also, the `imagePickerController:didFinishPickingMediaWithInfo:` delegate method contains information about the item that was captured by the user, be it an image or a video. The `didFinishPickingMediaWithInfo` parameter is a dictionary of values that tell you what the image picker has captured and the metadata of that item, along with other useful information. The first thing you have to do in this method is to read the value of the `UIImagePickerControllerMediaType` key in this dictionary. The object for this key is an instance of `NSString` that could be one of these values:

kUTTypeImage
>   For a photo that was shot by the camera

kUTTypeMovie
>   For a movie/video that was shot by the camera

 The kUTTypeImage and kUTTypeMovie values are available in the Mobile Core Services framework and are of type `CFStringRef`. You can simply typecast these values to `NSString` if needed.

After determining the type of resource created by the camera (video or photo), you can access that resource's properties using the `didFinishPickingMediaWithInfo` dictionary parameter again.

For images (kUTTypeImage), you can access these keys:

UIImagePickerControllerMediaMetadata
This key's value is an object of type NSDictionary. This dictionary contains a lot of useful information about the image that was shot by the user. A complete discussion of the values inside this dictionary is beyond the scope of this chapter.

UIImagePickerControllerOriginalImage
This key's value is an object of type UIImage containing the image that was shot by the user.

UIImagePickerControllerCropRect
If editing is enabled (using the allowsEditing property of UIImagePickerController), the object of this key will contain the rectangle of the cropped area.

UIImagePickerControllerEditedImage
If editing is enabled (using the allowsEditing property of UIImagePickerController), this key's value will contain the edited (resized and scaled) image.

For videos (kUTTypeMovie) that are shot by the user, you can access the UIImagePickerControllerMediaURL key in the didFinishPickingMediaWithInfo dictionary parameter of the imagePickerController:didFinishPickingMediaWithInfo: method. The value of this key is an object of type NSURL containing the URL of the video that was shot by the user.

After you get a reference to the UIImage instance that the user took with the camera, you can simply use that instance within your application.

 The images that are shot by the image picker controller within your application are not saved to the camera roll by default.

See Also

Recipe 10.1

10.3 Taking Videos with the Camera

Problem

You want to allow your users to shoot a video using their iOS device, and you would like to be able to use that video from inside your application.

Solution

Use UIImagePickerController with the UIImagePickerControllerSourceTypeCamera source type and the kUTTypeMovie media type:

```
- (void)viewDidLoad {
  [super viewDidLoad];

  if ([self isCameraAvailable] == YES &&
      [self doesCameraSupportShootingVideos] == YES){

    UIImagePickerController *imagePicker =
    [[UIImagePickerController alloc] init];

    imagePicker.sourceType =
    UIImagePickerControllerSourceTypeCamera;

    NSString *requiredMediaType = (NSString *)kUTTypeMovie;

    imagePicker.mediaTypes =
    [NSArray arrayWithObject:requiredMediaType];

    imagePicker.allowsEditing = YES;

    imagePicker.delegate = self;

    [self presentModalViewController:imagePicker
                            animated:YES];

    [imagePicker release];

  } else {
    NSLog(@"The camera is not available.");
  }

}
```

The isCameraAvailable and doesCameraSupportShootingVideos methods used in this sample code are implemented and discussed in Recipe 10.1.

We will implement the delegate methods of our image picker controller like so:

```
- (void) imagePickerController
        :(UIImagePickerController *)picker
        didFinishPickingMediaWithInfo:(NSDictionary *)info{

  NSLog(@"Picker returned successfully.");

  NSString     *mediaType = [info objectForKey:
                             UIImagePickerControllerMediaType];

  if ([mediaType isEqualToString:(NSString *)kUTTypeMovie] == YES){

    NSURL *urlOfVideo =
    [info objectForKey:UIImagePickerControllerMediaURL];

    NSLog(@"Video URL = %@", urlOfVideo);
```

```
    NSError *dataReadingError = nil;

    NSData *videoData =
    [NSData dataWithContentsOfURL:urlOfVideo
                         options:NSDataReadingMapped
                           error:&dataReadingError];

    if (videoData != nil){
      /* We were able to read the data */
      NSLog(@"Successfully loaded the data.");
    } else {
      /* We failed to read the data. Use the dataReadingError
       variable to determine what the error is */
      NSLog(@"Failed to load the data with error = %@",
            dataReadingError);
    }

  }

  [picker dismissModalViewControllerAnimated:YES];


}

- (void) imagePickerControllerDidCancel
        :(UIImagePickerController *)picker{

  NSLog(@"Picker was cancelled");

  [picker dismissModalViewControllerAnimated:YES];

}
```

Discussion

Once you detect that the iOS device your application is running on supports video recording, you can bring up the image picker controller with the UIImagePickerControl lerSourceTypeCamera source type and kUTTypeMovie media type to allow the users of your application to shoot videos. Once they are done, the imagePickerController:did FinishPickingMediaWithInfo: delegate method will get called and you can use the didFinishPickingMediaWithInfo dictionary parameter to find out more about the captured video (the values that can be placed inside this dictionary are thoroughly explained in Recipe 10.2).

When the user shoots a video using the image picker controller, the video will be saved in a temporary folder inside your application's bundle, not inside the camera roll. An example of such a URL is:

file://localhost/private/var/mobile/Applications/<APPID>/tmp/capture-T0x104e20.tmp.TQ9UTr/capturedvideo.MOV

The value *APPID* in the URL represents your application's unique identifier, and will clearly be different depending on your application.

As the programmer, not only can you allow your users to shoot videos from inside your application, but also you can modify how the videos are captured. You can change two important properties of the UIImagePickerController class in order to modify the default behavior of video recording:

videoQuality

    This property specifies the quality of the video. You can choose a value such as UIImagePickerControllerQualityTypeHigh or UIImagePickerControllerQuality TypeMedium for the value of this property.

videoMaximumDuration

    This property specifies the maximum duration of the video. This value is measured in seconds.

For instance, if we were to allow our users to record high-quality videos for up to 30 seconds, we could simply modify the values of the aforementioned properties of our instance of UIImagePickerController like so:

```
- (void)viewDidLoad {
  [super viewDidLoad];

  if ([self isCameraAvailable] == YES &&
      [self doesCameraSupportShootingVideos] == YES){

    UIImagePickerController *imagePicker =
    [[UIImagePickerController alloc] init];

    imagePicker.sourceType =
    UIImagePickerControllerSourceTypeCamera;

    /* We will only allow the user to shoot videos.
     Note that we must first detect if the camera on the
     user's iOS device allows us to shoot videos. We determined
     this using the doesCameraSupportShootingVideos method */
    NSString *requiredMediaType = (NSString *)kUTTypeMovie;

    imagePicker.mediaTypes =
    [NSArray arrayWithObject:requiredMediaType];

    /* Allow the user to trim/edit the video once they are done */
    imagePicker.allowsEditing = YES;

    /* Record in high quality */
    imagePicker.videoQuality =
    UIImagePickerControllerQualityTypeHigh;

    /* Only allow 30 seconds of recording */
```

```
    imagePicker.videoMaximumDuration = 30.0f;

    /* The current view controller will receive the
     delegate messages */
    imagePicker.delegate = self;

    /* Present the image picker as a modal view controller */
    [self presentModalViewController:imagePicker
                            animated:YES];

    [imagePicker release];

  } else {
    NSLog(@"The camera is not available.");
  }

}
```

See Also

Recipe 10.1

10.4 Storing Photos in the Photo Library

Problem

You want to be able to store a photo in the user's photo library.

Solution

Use the UIImageWriteToSavedPhotosAlbum procedure:

```
- (void) imageWasSavedSuccessfully:(UIImage *)paramImage
          didFinishSavingWithError:(NSError *)paramError
                       contextInfo:(void *)paramContextInfo{

  if (paramError == nil){
    NSLog(@"Image was saved successfully.");
  } else {
    NSLog(@"An error happened while saving the image.");
    NSLog(@"Error = %@", paramError);
  }

}

- (void)    imagePickerController:(UIImagePickerController *)picker
    didFinishPickingMediaWithInfo:(NSDictionary *)info{

  NSLog(@"Picker returned successfully.");

  NSString    *mediaType = [info objectForKey:
                            UIImagePickerControllerMediaType];
```

```objectivec
    if ([mediaType isEqualToString:(NSString *)kUTTypeImage] == YES){

      UIImage *theImage = nil;

      if (picker.allowsEditing == YES){

        theImage =
        [info
         objectForKey:UIImagePickerControllerEditedImage];

      } else {

        theImage =
        [info
         objectForKey:UIImagePickerControllerOriginalImage];

      }

      /* To avoid a very long line of text, we will break the
       selector's name into two lines using the backslash */
      NSString *targetSelectorAsString =
      @"imageWasSavedSuccessfully:didFinishSavingWithError:\
      contextInfo:";

      SEL targetSelector =
      NSSelectorFromString(targetSelectorAsString);

      NSLog(@"Saving the photo...");
      UIImageWriteToSavedPhotosAlbum(theImage,
                                     self,
                                     targetSelector,
                                     nil);

    }

    [picker dismissModalViewControllerAnimated:YES];


}

- (void) imagePickerControllerDidCancel
          :(UIImagePickerController *)picker{

  NSLog(@"Picker was cancelled");

  [picker dismissModalViewControllerAnimated:YES];

}

- (void)viewDidLoad {
  [super viewDidLoad];

  if ([self isCameraAvailable] == YES &&
      [self doesCameraSupportTakingPhotos] == YES){
```

```
        UIImagePickerController *imagePicker =
        [[UIImagePickerController alloc] init];

        imagePicker.sourceType =
        UIImagePickerControllerSourceTypeCamera;

        imagePicker.mediaTypes =
        [UIImagePickerController
         availableMediaTypesForSourceType:
         imagePicker.sourceType];

        imagePicker.allowsEditing = YES;

        imagePicker.delegate = self;

        [self presentModalViewController:imagePicker
                                animated:YES];

        [imagePicker release];

    } else {
      NSLog(@"The camera is not available.");
    }

}
```

 The isCameraAvailable and doesCameraSupportTakingPhotos methods used in this example are thoroughly explained in Recipe 10.1.

Discussion

Usually after a user is done taking a photo with her iOS device, she expects the photo to be saved into her photo library. However, applications that are not originally shipped with iOS can ask the user to take a photo, using the UIImagePickerController class, and then process that image. In this case, the user will understand that the application provided by us might not save the photo to her photo library and might simply use it internally. For instance, if an instant messaging application allows users to transfer their photos to each other's devices, the user will understand that a photo she takes inside the application will not be saved to her photo library, but will instead be transferred over the Internet to the other user.

However, if you decide you want to store an instance of UIImage to the photo library on the user's device, you can use the UIImageWriteToSavedPhotosAlbum function. This function accepts four parameters:

1. The image
2. The object that will get notified whenever the image is fully saved

---

3. A parameter that specifies the selector that has to be called on the target object (specified by the second parameter) when the save operation finishes

4. A context value that will get passed to the specified selector once the operation is done

Providing the second, third, and fourth parameters to this procedure is optional. If you do provide the second and third parameters, the fourth parameter still remains optional. The selector that gets passed to this procedure must follow this structure:

```
- (void) <name is up to you>:(UIImage *)<name up to you>
     didFinishSavingWithError:(NSError *)<name up to you>
              contextInfo:(void *)<name up to you>{

}
```

For instance, this is the selector we have chosen in our example:

```
- (void) imageWasSavedSuccessfully:(UIImage *)paramImage
          didFinishSavingWithError:(NSError *)paramError
                    contextInfo:(void *)paramContextInfo{

  if (paramError == nil){
    NSLog(@"Image was saved successfully.");
  } else {
    NSLog(@"An error happened while saving the image.");
    NSLog(@"Error = %@", paramError);
  }

}
```

If the error parameter that you receive in this selector is equal to nil, that means the image was saved in the user's photo library successfully. Otherwise, you can retrieve the value of this parameter to determine what the issue was.

10.5 Storing Videos in the Photo Library

Problem

You want to store a video accessible through a URL, such as a video in your application bundle, to the Photo Library.

Solution

Use the writeVideoAtPathToSavedPhotosAlbum:completionBlock: instance method of ALAssetsLibrary:

```
- (void)viewDidLoad {
  [super viewDidLoad];
```

```
/* First get the URL of the video in the application bundle */
NSBundle *mainBundle = [NSBundle mainBundle];

NSURL *videoFileURL = [mainBundle URLForResource:@"MyVideo"
                                    withExtension:@"MOV"];

if (videoFileURL != nil){

  /* Now attempt to save the video to the photo library using
   the assets library object */
  [self.assetsLibrary
   writeVideoAtPathToSavedPhotosAlbum:videoFileURL
   completionBlock:handleVideoWasSavedSuccessfully];

} else {
  NSLog(@"Could not find the MyVideo.MOV file in the app bundle");
}

}
```

In our example, assetsLibrary is a property, of type ALAssetsLibrary, of the current UIView. Although this example allocates and initializes this property in a view, you do not necessarily have to use Assets Library objects in view objects. You can allocate, initialize, and use them anywhere in your application that you find most appropriate. This is how to allocate and initialize our Assets Library object:

```
- (id) initWithNibName:(NSString *)nibNameOrNil
                bundle:(NSBundle *)nibBundleOrNil{

self = [super initWithNibName:nibNameOrNil
                       bundle:nibBundleOrNil];

if (self != nil){

  ALAssetsLibrary *newLibrary = [[ALAssetsLibrary alloc] init];
  assetsLibrary = [newLibrary retain];
  [newLibrary release];

}

return(self);

}
```

Discussion

The Assets Library framework, as mentioned in this chapter's Introduction, appears in the iOS SDK starting from iOS SDK 4.0. This framework is a convenient bridge between developers and the Photo Library. As mentioned in Recipe 10.6, the iOS SDK provides you with built-in GUI components that you can use to access the contents of the Photo Library. However, you might sometimes require direct access to these contents. In such instances, you can use the Assets Library framework.

After allocating and initializing the Assets Library object of type `ALAssetsLibrary`, you can use the `writeVideoAtPathToSavedPhotosAlbum:completionBlock:` instance method of this object to write a video from a URL to the Photo Library. All you have to do is provide the URL of the video in `NSURL` form and a block object whose code will be called when the video is saved. The block object must accept two parameters of type `NSURL` and `NSError`, respectively:

```
void (^handleVideoWasSavedSuccessfully)(NSURL *, NSError *) =
^(NSURL *assetURL, NSError *error){

  if (error == nil){
    NSLog(@"no errors happened");
  } else {
    NSLog(@"Error happened while saving the video.");
    NSLog(@"The error is = %@", error);
  }

};
```

If the `error` parameter is `nil`, the save process went well and you don't have to worry about anything. One of the common errors that iOS could return to you is similar to this:

```
Error Domain=ALAssetsLibraryErrorDomain Code=-3302 "Invalid data"
UserInfo=0x7923590 {NSLocalizedFailureReason=
There was a problem writing this asset because
the data is invalid and cannot be viewed or played.,
NSLocalizedRecoverySuggestion=Try with different data,
NSLocalizedDescription=Invalid data}
```

You will get this error message if you attempt to pass a URL that is not inside your application bundle. If you are testing your application on iPhone Simulator, occasionally you might also get this error message:

```
Error Domain=ALAssetsLibraryErrorDomain Code=-3310 "Data unavailable"
UserInfo=0x6456810 {NSLocalizedRecoverySuggestion=
Launch the Photos application, NSLocalizedDescription=Data unavailable}
```

If so, please open the Photos application in iPhone Simulator once, and then launch your application.

The first parameter passed to the block object provided to the `writeVideoAtPathTo SavedPhotosAlbum:completionBlock:` method will point to the Assets Library URL of the stored video. A sample URL of this kind will look like this:

```
assets-library://asset/asset.MOV?id=1000000002&ext=MOV
```

In Recipe 10.7, we will learn how to use such a URL to load the data for the video file into memory.

10.6 Retrieving Photos and Videos from the Photo Library

Problem

You want users to be able to pick a photo or a video from their photo library and use it in your application.

Solution

Use the UIImagePickerControllerSourceTypePhotoLibrary value for the source type of your UIImagePickerController and the kUTTypeImage or kUTTypeMovie value, or both, for the media type, like so:

```
- (BOOL) isPhotoLibraryAvailable{

    return ([UIImagePickerController
            isSourceTypeAvailable:
            UIImagePickerControllerSourceTypePhotoLibrary]);

}

- (BOOL) canUserPickVideosFromPhotoLibrary{

    BOOL result = NO;

    result =
    [self
     doesCameraSupportMediaType:(NSString *)kUTTypeMovie
     onSourceType:UIImagePickerControllerSourceTypePhotoLibrary];

    return(result);

}

- (BOOL) canUserPickPhotosFromPhotoLibrary{

    BOOL result = NO;

    result =
    [self
     doesCameraSupportMediaType:(NSString *)kUTTypeImage
     onSourceType:UIImagePickerControllerSourceTypePhotoLibrary];

    return(result);

}

- (void)viewDidLoad {
    [super viewDidLoad];

    if ([self isPhotoLibraryAvailable] == YES){

        UIImagePickerController *imagePicker =
```

```
[[UIImagePickerController alloc] init];

imagePicker.sourceType =
UIImagePickerControllerSourceTypePhotoLibrary;

NSMutableArray *mediaTypes = [[NSMutableArray alloc] init];

if ([self canUserPickPhotosFromPhotoLibrary] == YES){
  [mediaTypes addObject:(NSString *)kUTTypeImage];
}

if ([self canUserPickVideosFromPhotoLibrary] == YES){
  [mediaTypes addObject:(NSString *)kUTTypeMovie];
}

imagePicker.mediaTypes = mediaTypes;

imagePicker.delegate = self;

[self presentModalViewController:imagePicker
                        animated:YES];

[imagePicker release];
[mediaTypes release];

}

}
```

For the implementation of the doesCameraSupportMediaType:onSourceType: method we are using in this example, please refer to Recipe 10.1.

Discussion

To allow your users to pick photos or videos from their photo library, you must set the sourceType property of an instance of UIImagePickerController to UIImagePickerCon trollerSourceTypePhotoLibrary before presenting the image picker to them. In addition, if you want to filter the videos or photos out of the items presented to your users once the image picker is shown, exclude the kUTTypeMovie or kUTTypeImage value from the array of media types of the image picker (in the mediaTypes property), respectively.

Bear in mind that setting the mediaTypes property of an image picker controller to nil or an empty array will result in a runtime error similar to this:

```
Terminating app due to uncaught exception
'NSInvalidArgumentException', reason: 'No available types for source 0'
```

After the user is done picking the image, you will get the usual delegate messages through the UIImagePickerControllerDelegate protocol. For more information on how you can implement the methods defined in this protocol for processing images, please refer to Recipe 10.2.

See Also

Recipe 10.7

10.7 Retrieving Assets from the Assets Library

Problem

You want to directly retrieve photos or videos from the Photo Library without the help of any built-in GUI components.

Solution

Use the Assets Library framework. Follow these steps:

1. Allocate and initialize an object of type ALAssetsLibrary.

2. Provide two block objects to the enumerateGroupsWithTypes:usingBlock:failure Block: instance method of the Assets Library object. The first block will retrieve all the groups associated with the type that we passed to this method. The groups will be of type ALAssetsGroup. The second block returns an error in case of failure.

3. Use the enumerateAssetsUsingBlock: instance method of each group object to enumerate the assets available in each group. This method takes a single parameter, a block that retrieves information on a single asset. The block that you pass as a parameter must accept three parameters, of which the first must be of type ALAsset.

4. After retrieving the ALAsset objects available in each group, you can retrieve various properties of each asset, such as their type, available URLs, and so on. Retrieve these properties using the valueForProperty: instance method of each asset of type ALAsset. The return value of this method, depending on the property passed to it, could be NSDictionary, NSString, or any other object type. We will see a few common properties that we can retrieve from each asset soon.

5. Invoke the defaultRepresentation instance method of each object of type ALAsset to retrieve its representation object of type ALAssetRepresentation. Each asset in the Assets Library can have more than one representation. For instance, a photo might have a PNG representation by default, but a JPEG representation as well. Using the defaultRepresentation method of each asset of type ALAsset, you can retrieve the ALAssetRepresentation object, and then use that to retrieve different representations (if available) of each asset.

6. Use the size and the getBytes:fromOffset:length:error: instance methods of each asset representation to load the asset's representation data. You can then write the read bytes into an NSData object or do whatever else you need to do in your application. Additionally, for photos, you can use the fullResolutionImage, fullScreen Image, and CGImageWithOptions: instance methods of each representation to

---

retrieve images of type `CGImageRef`. You can then construct a `UIImage` from `CGImageRef` using the `imageWithCGImage:` class method of `UIImage`.

```
void (^groupAssetEnumerationBlock)(ALAsset *, NSUInteger, BOOL *) =
^(ALAsset *result, NSUInteger index, BOOL *stop){

  /* Get the asset type */
  NSString *assetType = [result valueForProperty:ALAssetPropertyType];

  if ([assetType isEqualToString:ALAssetTypePhoto] == YES){
    NSLog(@"This is a photo asset");
  }

  else if ([assetType isEqualToString:ALAssetTypeVideo] == YES){
    NSLog(@"This is a video asset");
  }

  else if ([assetType isEqualToString:ALAssetTypeUnknown] == YES){
    NSLog(@"This is an unknown asset");
  }

  /* Get the URLs for the asset */
  NSDictionary  *assetURLs =
  [result valueForProperty:ALAssetPropertyURLs];

  NSUInteger    assetCounter = 0;
  for (NSString *assetURLKey in assetURLs){
    assetCounter++;
    NSLog(@"Asset URL %lu = %@",
          (unsigned long)assetCounter,
          [assetURLs valueForKey:assetURLKey]);
  }

  /* Get the asset's representation object */
  ALAssetRepresentation *assetRepresentation =
  [result defaultRepresentation];

  NSLog(@"Representation Size = %lld", [assetRepresentation size]);

};

void (^assetGroupEnumerationBlock)(ALAssetsGroup *, BOOL *) =
^(ALAssetsGroup *group, BOOL *stop){

  [group enumerateAssetsUsingBlock:groupAssetEnumerationBlock];

};

void (^assetGroupEnumerationFailedBlock)(NSError *) =
     ^(NSError *error){

  NSLog(@"Error = %@", error);

};
```

```
- (void)viewDidLoad {
  [super viewDidLoad];

  [self.assetsLibrary
   enumerateGroupsWithTypes:ALAssetsGroupAll
   usingBlock:assetGroupEnumerationBlock
   failureBlock:assetGroupEnumerationFailedBlock];

}
```

Discussion

The Assets Library is broken down into groups. Each group contains assets and each asset has properties such as URLs and representation objects.

As demonstrated in this recipe's Solution, you can retrieve all assets of all types from the Assets Library using the ALAssetsGroupAll constant passed to the enumerateGroups WithTypes parameter of the enumerateGroupsWithTypes:usingBlock:failureBlock: instance method of the Assets Library object. Here is a list of values you can pass to this parameter to enumerate different groups of assets:

ALAssetsGroupAlbum

> Groups representing albums that have been stored on an iOS device through iTunes.

ALAssetsGroupFaces

> Groups representing albums that contain face assets that were stored on an iOS device through iTunes.

ALAssetsGroupSavedPhotos

> Groups representing the saved photos in the Photo Library. These are accessible to an iOS device through the Photos application as well.

ALAssetsGroupAll

> All available groups in the Assets Library.

Now let's go ahead and write a simple application that retrieves the data for the first image found in the Assets Library, creates a UIImageView out of it, and adds the image view to the view of the current view controller. This way, we will learn how to read the contents of an asset using its representation.

Let's start with the declaration of our view controller (.h file):

```
#import <UIKit/UIKit.h>
#import <AssetsLibrary/AssetsLibrary.h>

#define NOTIFICATION_ASSET_IMAGE_RETRIEVED @"ASSET_IMG_RETRIEVED" ❶

@interface RootViewController : UIViewController {
@protected
  ALAssetsLibrary *assetsLibrary; ❷
  UIImageView     *imageView; ❸
}
```

```
@property (nonatomic, retain) ALAssetsLibrary *assetsLibrary;
@property (nonatomic, retain) UIImageView      *imageView;

@end
```

❶ NOTIFICATION_ASSET_IMAGE_RETRIEVED is the name of the notification that we will send from a block object to tell our view controller that we have been able to construct a UIImage out of the first image we retrieved from the Assets Library.

❷ assetsLibrary is the Assets Library instance variable for our view controller. We will allocate and initialize it when our view controller is initialized and dispose of it in the dealloc method of the view controller.

❸ imageView is the image view that will display the image of the first photo asset in the Assets Library.

We will implement the initialization and deallocation of our view controller so that we can have access to the Assets Library:

```
- (id) initWithNibName:(NSString *)nibNameOrNil
              bundle:(NSBundle *)nibBundleOrNil{

  self = [super initWithNibName:nibNameOrNil
                        bundle:nibBundleOrNil];

  if (self != nil){

    ALAssetsLibrary *newLibrary = [[ALAssetsLibrary alloc] init];
    assetsLibrary = [newLibrary retain];
    [newLibrary release];

  }

  return(self);

}

- (void)dealloc {
  [assetsLibrary release];
  [imageView release];
  [super dealloc];
}
```

When the viewDidLoad method of our view controller gets called, we will start enumerating all available groups in the Assets Library, searching for the first image asset we can find:

```
- (void)viewDidLoad {
  [super viewDidLoad];

  [[NSNotificationCenter defaultCenter]
    addObserver:self
    selector:@selector(createImageViewOutOfImage:)
    name:NOTIFICATION_ASSET_IMAGE_RETRIEVED
```

```
    object:nil];

    [self.assetsLibrary
     enumerateGroupsWithTypes:ALAssetsGroupAll
     usingBlock:assetGroupEnumerationBlock
     failureBlock:assetGroupEnumerationFailedBlock];

}

- (void)viewDidUnload {
    [super viewDidUnload];
    self.imageView = nil;

    [[NSNotificationCenter defaultCenter]
     removeObserver:self
     name:NOTIFICATION_ASSET_IMAGE_RETRIEVED
     object:nil];

}
```

As you can see, we have written a method called createImageViewOutOfImage: that will
be invoked as the result of the NOTIFICATION_ASSET_IMAGE_RETRIEVED notification sent
from the block object enumerating the assets in every group in the Assets Library. This
method will simply take a notification object that is supposed to carry an instance of
UIImage and will create a UIImageView out of the image. It will eventually place the image
view on the view of the current view controller:

```
- (void) createImageViewOutOfImage:(NSNotification *)paramNotification{

    UIImage *image = (UIImage *)[paramNotification object];

    if (image == nil){
      NSLog(@"The given image is nil.");
      return;
    }

    /* Create the image view with the view's bounds */
    UIImageView *newImageView = [[UIImageView alloc]
                                 initWithFrame:self.view.bounds];
    self.imageView = [newImageView retain];
    [newImageView release];

    /* Make sure the image gets scaled properly in the image view */
    [self.imageView setContentMode:UIViewContentModeScaleAspectFit];
    [self.imageView setImage:image];

    /* And add the image view to the view */
    [self.view addSubview:self.imageView];

}
```

In the viewDidLoad method of our view controller, we used the assetGroupEnumeration
Block block object to process the groups that are enumerated in the Assets Library. The
code in this block object will be called as soon as an asset group is enumerated:

```
void (^assetGroupEnumerationBlock)(ALAssetsGroup *, BOOL *) =
^(ALAssetsGroup *group, BOOL *stop){

  [group enumerateAssetsUsingBlock:groupAssetEnumerationBlock];

};

void (^assetGroupEnumerationFailedBlock)(NSError *) =
^(NSError *error){

  NSLog(@"Group enumeration Error = %@", error);

};
```

This block object itself calls the `groupAssetEnumerationBlock` block object to enumerate the assets inside the current group. The implementation of this block object is the most vital part of the whole process:

```
void (^groupAssetEnumerationBlock)(ALAsset *, NSUInteger, BOOL *) =
^(ALAsset *result, NSUInteger index, BOOL *stop){

  static BOOL firstImageIsFound = NO;

  /* We don't want to get all the images. As soon as we get
   the first image, we want to end the enumeration */
  if (firstImageIsFound == YES){
    *stop = YES;
    return;
  }

  /* Get the asset type */
  NSString *assetType = [result valueForProperty:ALAssetPropertyType];

  if ([assetType isEqualToString:ALAssetTypePhoto] == YES){
    NSLog(@"This is a photo asset");
    firstImageIsFound = YES;

    /* Get the asset's representation object */
    ALAssetRepresentation *assetRepresentation =
    [result defaultRepresentation];

    /* We need the scale and orientation to be able to
     construct a properly oriented and scaled UIImage out of the
     representation object */
    CGFloat      imageScale = [assetRepresentation scale];

    UIImageOrientation  imageOrientation =
    [assetRepresentation orientation];

    CGImageRef  imageReference =
    [assetRepresentation fullResolutionImage];

    /* Construct the image now */
    UIImage      *image =
    [[UIImage alloc] initWithCGImage:imageReference
```

```
                          scale:imageScale
                    orientation:imageOrientation];

    if (image != nil){

      /* Send a notification saying that we have constructed
       the image */
      [[NSNotificationCenter defaultCenter]
       postNotificationName:NOTIFICATION_ASSET_IMAGE_RETRIEVED
       object:image];

    } else {
      NSLog(@"Failed to create the image.");
    }

    [image release];

  }

};
```

You can see that we are following the exact same steps mentioned in this recipe's Solution. We enumerate the groups and every asset in the groups. Then we find the first photo asset and retrieve its representation. Using the representation, we construct a UIImage, and from the UIImage, we construct a UIImageView to display that image on our view. Quite simple, isn't it?

For video files, we are dealing with a slightly different issue as the ALAssetRepresenta tion class does not have any methods that could return an object that encapsulates our video files. For this reason, we have to read the contents of a video asset into a buffer and perhaps save it to our *Documents* folder where it is easier for us to access later. Of course, the requirements depend on your application, but in this example code, we will go ahead and find the first video in the Assets Library and store it in our application's *Documents* folder under the name *Temp.MOV*:

```
void (^groupAssetEnumerationBlock)(ALAsset *, NSUInteger, BOOL *) =
^(ALAsset *result, NSUInteger index, BOOL *stop){

  static BOOL firstVideoIsFound = NO;

  /* We don't want to get all the videos. As soon as we get
   the first video, we want to end the enumeration */
  if (firstVideoIsFound == YES){
    *stop = YES;
    return;
  }

  /* Get the asset type */
  NSString *assetType = [result valueForProperty:ALAssetPropertyType];

  if ([assetType isEqualToString:ALAssetTypeVideo] == YES){
    NSLog(@"This is a video asset");
    firstVideoIsFound = YES;
```

```
/* Get the asset's representation object */
ALAssetRepresentation *assetRepresentation =
[result defaultRepresentation];

const NSUInteger BufferSize = 1024;
uint8_t     buffer[BufferSize];
NSUInteger  bytesRead = 0;
long long   currentOffset = 0;
NSError     *readingError = nil;

/* Find the documents folder (an array) */
NSArray *documents =
NSSearchPathForDirectoriesInDomains(NSDocumentDirectory,
                                    NSUserDomainMask,
                                    YES);

/* Retrieve the one documents folder that we need */
NSString *documentsFolder = [documents objectAtIndex:0];

/* Construct the path where the video has to be saved */
NSString *videoPath =
[documentsFolder
 stringByAppendingPathComponent:@"Temp.MOV"];

NSFileManager *fileManager = [[NSFileManager alloc] init];

/* Create the file if it doesn't exist already */
if ([fileManager fileExistsAtPath:videoPath] == NO){
  [fileManager createFileAtPath:videoPath
                       contents:nil
                     attributes:nil];
}

[fileManager release];

/* We will use this file handle to write the contents
 of the media assets to the disk */
NSFileHandle *fileHandle =
[NSFileHandle
 fileHandleForWritingAtPath:videoPath];

do{

  /* Read as many bytes as we can put in the buffer */
  bytesRead = [assetRepresentation getBytes:(uint8_t *)&buffer
                                 fromOffset:currentOffset
                                     length:BufferSize
                                      error:&readingError];

  /* If we couldn't read anything, we will exit this loop */
  if (bytesRead == 0){
    break;
  }
```

```
        /* Keep the offset up to date */
        currentOffset += bytesRead;

        /* Put the buffer into an NSData */
        NSData *readData = [[NSData alloc]
                            initWithBytes:(const void *)buffer
                            length:bytesRead];

        /* And write the data to file */
        [fileHandle writeData:readData];
        [readData release];

    } while (bytesRead > 0);

    NSLog(@"Finished reading and storing the \
        video in the documents folder");

    }

};
```

This is what's happening in our sample code:

- We get the default representation of the first video asset that we find in the Assets Library.
- We create a file called *Temp.MOV* in our application's *Documents* folder to save the contents of the video asset.
- We create a loop that runs so long as there is still data in the asset representation waiting to be read. The `getBytes:fromOffset:length:error:` instance method of our asset representation object reads as many bytes as we can fit into our buffer for as many times as necessary until we get to the end of the representation data.
- After reading the data into our buffer, we encapsulate the data into an object of type `NSData` using the `initWithBytes:length:` initialization method of `NSData`. We then write this data to the file we created previously using the `writeData:` instance method of `NSFileHandle`.

Go ahead and give this a try. You can test this example code on either iPhone/iPad Simulator or a real iOS device.

10.8 Editing Videos on an iOS Device

Problem

You want the user of your application to be able to edit videos straight from your application.

Solution

Use the `UIVideoEditorController` class. In this example, we will use this class in conjunction with an image picker controller. First we will ask the user to pick a video from her photo library. After she does, we will display an instance of the video editor controller and allow the user to edit the video she picked. Here is the *.h* file of our view controller:

```objc
#import <UIKit/UIKit.h>
#import <MobileCoreServices/MobileCoreServices.h>

@interface RootViewController : UIViewController
                                <UINavigationControllerDelegate,
                                UIImagePickerControllerDelegate,
                                UIVideoEditorControllerDelegate> {
@public
  NSURL *videoURLToEdit;
}

@property (nonatomic, retain) NSURL *videoURLToEdit;

@end
```

Here is the implementation (*.m* file) of the view controller:

```objc
#import "RootViewController.h"

@implementation RootViewController

@synthesize videoURLToEdit;

- (BOOL) doesCameraSupportMediaType:(NSString *)paramMediaType
  onSourceType:(UIImagePickerControllerSourceType)paramSourceType{

  BOOL result = NO;

  if (paramMediaType == nil ||
      [paramMediaType length] == 0){
    return(NO);
  }

  if ([UIImagePickerController
       isSourceTypeAvailable:paramSourceType] == NO){
    return(NO);
  }

  NSArray *mediaTypes =
  [UIImagePickerController
   availableMediaTypesForSourceType:paramSourceType];

  if (mediaTypes == nil){
    return(NO);
  }

  for (NSString *mediaType in mediaTypes){
```

```objc
      if ([mediaType isEqualToString:paramMediaType] == YES){
        return(YES);
      }

    }

    return(result);

}

- (void)videoEditorController:(UIVideoEditorController *)editor
      didSaveEditedVideoToPath:(NSString *)editedVideoPath{

  NSLog(@"The video editor finished saving video");

  NSLog(@"The edited video path is at = %@", editedVideoPath);

  [editor dismissModalViewControllerAnimated:YES];
}

- (void)videoEditorController:(UIVideoEditorController *)editor
             didFailWithError:(NSError *)error{

  NSLog(@"Video editor error occurred = %@", error);

  [editor dismissModalViewControllerAnimated:YES];
}

- (void)videoEditorControllerDidCancel
        :(UIVideoEditorController *)editor{

  NSLog(@"The video editor was cancelled");

  [editor dismissModalViewControllerAnimated:YES];
}

void (^handleFailedToRetrieveAsset)(NSError *) = ^(NSError *error){
  NSLog(@"Failed to retrieve the asset with error = %@", error);
};

void (^saveVideoAssetToDisk)(ALAsset *) = ^(ALAsset *asset){

  /* Get the asset type */
  NSString *assetType = [asset valueForProperty:ALAssetPropertyType];

  if ([assetType isEqualToString:ALAssetTypeVideo] == YES){
    NSLog(@"This is a video asset");

    /* Get the asset's representation object */
    ALAssetRepresentation *assetRepresentation =
    [asset defaultRepresentation];

    const NSUInteger BufferSize = 1024;
    uint8_t      buffer[BufferSize];
```

```objc
NSUInteger  bytesRead = 0;
long long   currentOffset = 0;
NSError     *readingError = nil;

/* Find the documents folder (an array) */
NSArray *documents =
NSSearchPathForDirectoriesInDomains(NSDocumentDirectory,
                                    NSUserDomainMask,
                                    YES);

/* Retrieve the one documents folder that we need */
NSString *documentsFolder = [documents objectAtIndex:0];

/* Construct the path where the video has to be saved */
NSString *videoPath =
[documentsFolder
 stringByAppendingPathComponent:@"Temp.MOV"];

NSFileManager *fileManager = [[NSFileManager alloc] init];

/* Create the file if it doesn't exist already */
if ([fileManager fileExistsAtPath:videoPath] == NO){
  [fileManager createFileAtPath:videoPath
                       contents:nil
                     attributes:nil];
}

[fileManager release];

/* We will use this file handle to write the contents
 of the media assets to the disk */
NSFileHandle *fileHandle =
[NSFileHandle
 fileHandleForWritingAtPath:videoPath];

do{

  /* Read as many bytes as we can put in the buffer */
  bytesRead = [assetRepresentation getBytes:(uint8_t *)&buffer
                                 fromOffset:currentOffset
                                     length:BufferSize
                                      error:&readingError];

  /* If we couldn't read anything, we will exit this loop */
  if (bytesRead == 0){
    break;
  }

  /* Keep the offset up to date */
  currentOffset += bytesRead;

  /* Put the buffer into an NSData */
  NSData *readData = [[NSData alloc
                      initWithBytes:(const void *)buffer
                      length:bytesRead];
```

```
        /* And write the data to file */
        [fileHandle writeData:readData];
        [readData release];

    } while (bytesRead > 0);

    NSLog(@"Finished reading and storing the video \
            in the documents folder");

  }

};

- (void)    imagePickerController:(UIImagePickerController *)picker
    didFinishPickingMediaWithInfo:(NSDictionary *)info{

  NSLog(@"Picker returned successfully.");

  NSString    *mediaType = [info objectForKey:
                            UIImagePickerControllerMediaType];

  if ([mediaType isEqualToString:(NSString *)kUTTypeMovie] == YES){

    /* We now have to save the video into the documents folder of our
     application and then show the video editor based on the new path */

    NSURL *urlOfVideo =
    [info objectForKey:UIImagePickerControllerReferenceURL];

    ALAssetsLibrary *assetLibrary = [[ALAssetsLibrary alloc] init];

    [assetLibrary assetForURL:urlOfVideo
                  resultBlock:saveVideoAssetToDisk
                 failureBlock:handleFailedToRetrieveAsset];

    [assetLibrary release];

    /* Find the documents folder (an array) */
    NSArray *documents =
    NSSearchPathForDirectoriesInDomains(NSDocumentDirectory,
                                        NSUserDomainMask,
                                        YES);

    /* Retrieve the one documents folder that we need */
    NSString *documentsFolder = [documents objectAtIndex:0];

    /* Construct the path where the video has to be saved */
    NSString *videoPath =
    [documentsFolder
     stringByAppendingPathComponent:@"Temp.MOV"];

    self.videoURLToEdit = [NSURL URLWithString:videoPath];

  }
```

```
    [picker dismissModalViewControllerAnimated:YES];

}

- (void) imagePickerControllerDidCancel
            :(UIImagePickerController *)picker{

   NSLog(@"Picker was cancelled");

   self.videoURLToEdit = nil;

   [picker dismissModalViewControllerAnimated:YES];

}

- (BOOL) canUserPickVideosFromPhotoLibrary{

   BOOL result = NO;

   result =
   [self
    doesCameraSupportMediaType:(NSString *)kUTTypeMovie
    onSourceType:UIImagePickerControllerSourceTypePhotoLibrary];

   return(result);

}

- (BOOL) isPhotoLibraryAvailable{

   return ([UIImagePickerController
           isSourceTypeAvailable:
           UIImagePickerControllerSourceTypePhotoLibrary]);

}

- (void)viewDidLoad {
   [super viewDidLoad];

   /* First see if we are allowed to access the photo library and
    if the user is able to pick videos from the photo library */
   if ([self isPhotoLibraryAvailable] == YES &&
       [self canUserPickVideosFromPhotoLibrary] == YES){

     UIImagePickerController *imagePicker =
     [[UIImagePickerController alloc] init];

     /* Set the source type to photo library */
     imagePicker.sourceType =
     UIImagePickerControllerSourceTypePhotoLibrary;

     /* And we want our user to be able to pick movies from the library */
     NSArray  *mediaTypes =
     [NSArray arrayWithObject:(NSString *)kUTTypeMovie];
```

```
        imagePicker.mediaTypes = mediaTypes;

        /* Set the delegate to the current view controller */
        imagePicker.delegate = self;

        /* Present our image picker */
        [self presentModalViewController:imagePicker
                                animated:YES];

        [imagePicker release];

    }

}

- (void) viewDidAppear:(BOOL)animated{
    [super viewDidAppear:animated];

    if (self.videoURLToEdit != nil){

        NSString *videoPath = [self.videoURLToEdit absoluteString];

        /* First let's make sure the video editor is able to edit the
         video at the path in our documents folder */
        if ([UIVideoEditorController canEditVideoAtPath:videoPath] == YES){

            /* Instantiate the video editor */
            UIVideoEditorController *videoEditor =
            [[UIVideoEditorController alloc] init];

            /* We become the delegate of the video editor */
            videoEditor.delegate = self;

            /* Make sure to set the path of the video */
            videoEditor.videoPath = videoPath;

            /* And present the video editor */
            [self presentModalViewController:videoEditor
                                    animated:YES];

            [videoEditor release];

            self.videoURLToEdit = nil;

        } else {
            NSLog(@"Cannot edit the video at this path");
        }

    }

}

- (void)viewDidUnload {
    [super viewDidUnload];
```

```
    }

    - (BOOL)shouldAutorotateToInterfaceOrientation:
    (UIInterfaceOrientation)interfaceOrientation {
      return (YES);
    }

    - (void)didReceiveMemoryWarning {
      [super didReceiveMemoryWarning];
    }

    - (void)dealloc {
      [videoURLToEdit release];
      [super dealloc];
    }

    @end
```

The UIVideoEditorController is not designed to work in landscape mode. Even if the view controller that displays an instance of the video editor supports all orientations, the video editor will be shown in portrait mode only.

Discussion

The UIVideoEditorController in the iOS SDK allows programmers to display a video editor interface to the users of their applications. All you have to do is to provide the URL of the video that needs to be edited and then present the video editor controller as a modal view. You should not overlay the view of this controller with any other views and you should not modify this view.

Calling the presentModalViewController:animated: method immediately after calling the dismissModalViewControllerAnimated: method of a view controller will terminate your application with a runtime error. You must wait for the first view controller to be dismissed and then present the second view controller. You can take advantage of the viewDidAppear: instance method of your view controllers to detect when your view is displayed. You know at this point that any modal view controllers must have disappeared.

In our example, the user is allowed to pick any video from the photo library. Once she does, we will save the video to our application's *Documents* folder and then display the video editor controller by providing the path of the video we just saved into the *Documents* folder of our application. The GUI of the video editor controller is similar to that shown in Figure 10-1.

Figure 10-1. The video editor controller editing a simple video shot of the desktop of a MacBook Pro

The video editor controller's delegate gets important messages about the state of the video editor. This delegate object must conform to the `UIVideoEditorControllerDele gate` and `UINavigationControllerDelegate` protocols. In our example, we chose our view controller to become the delegate of our video editor. Once the editing is done, the delegate object receives the `videoEditorController:didSaveEditedVideoToPath:` delegate method from the video editor controller. The path of the edited video will be passed through the `didSaveEditedVideoToPath` parameter.

 It is very important to bear in mind that the path set as the `videoPath` property of an instance of `UIVideoEditorController` must be a local path inside the application. This path cannot be in the *temp* directory inside the application. Usually the best place is the *Documents* folder. For instance, if you allow users to record a video using `UIImagePickerControl ler`, save the video to the *Documents* folder inside your application and pass the path of the video inside the *Documents* folder to `UIVideoEdi torController`. The path that `UIImagePickerController` gives to your application (whether it is in the `UIImagePickerControllerReferenceURL` or `UIImagePickerControllerMediaURL` key of the `didFinishPickingMedia WithInfo` dictionary parameter in the `imagePickerController:didFinish PickingMediaWithInfo:` delegate message) will not work with `UIVideoEditorController`.

Before attempting to display the interface of the video editor to your users, you must call the `canEditVideoAtPath:` class method of `UIVideoEditorController` to make sure the path you are trying to edit is editable by the controller. If the return value of this class method is `YES`, proceed to configuring and displaying the video editor's interface. If not, take a separate path, perhaps displaying an alert to your user.

See Also

Recipe 10.6; Recipe 10.7

CHAPTER 11

Multitasking

11.0 Introduction

iOS introduced support for multitasking in version 4. Multitasking in iOS is straightforward, with simple rules that give the user a great experience while allowing applications to multitask with ease. Not all iOS devices support multitasking, as you will learn in Recipe 11.1.

Multitasking enables *background execution*, which means the application can keep working as usual—running tasks, spawning new threads, listening for notifications, and reacting to events—but simply does not display anything on the screen or have any way to interact with the user. When the user presses the Home button on the device, which in previous versions of the iPhone and iPad would terminate the application, the application is now sent into the background.

An application running on an iOS version that supports multitasking is, by default, opted into background execution. If you link your application against iOS SDK 4.0 and later, you can opt out of background execution, as you will see in Recipe 11.11. If you do, your application will be terminated when the user presses the Home button, as before.

When our application moves to the background (such as when the user presses the Home button) and then back to the foreground (when the user selects the application again), various messages are sent by the system and are expected to be received by an object we designate as our application delegate. For instance, when our application is sent to the background, our application delegate will receive the `applicationDidEnterBackground:` method, and as the application comes back to the foreground by the user, the application delegate will receive the `applicationWillEnterForeground:` delegate message.

In addition to these delegate messages, iOS also sends notifications to the running application when it transitions the application to the background and from the background to the foreground. The notification that gets sent when the application is moved to the background is `UIApplicationDidEnterBackgroundNotification`, and the notification that gets sent when an application transitions from the background to the foreground is UIApplicationWillEnterForegroundNotification. You can use the default notification center to register for these notifications, like so:

```
- (void) handleEnteringBackground:(NSNotification *)paramNotification{

  /* We have entered background */
  NSLog(@"Going to background.");

}

- (void) handleEnteringForeground:(NSNotification *)paramNotification{

  /* We have entered foreground */
  NSLog(@"Coming to foreground");

}

- (void)viewDidLoad {
  [super viewDidLoad];

  if ([self isMultitaskingSupported] == YES){

    [[NSNotificationCenter defaultCenter]
     addObserver:self
       selector:@selector(handleEnteringBackground:)
           name:UIApplicationDidEnterBackgroundNotification
         object:nil];

    [[NSNotificationCenter defaultCenter]
     addObserver:self
       selector:@selector(handleEnteringForeground:)
           name:UIApplicationWillEnterForegroundNotification
         object:nil];

  } else {
    NSLog(@"Multitasking is not enabled.");
  }
}

- (void)viewDidUnload {
  [super viewDidUnload];

  if ([self isMultitaskingSupported] == YES){

    [[NSNotificationCenter defaultCenter]
     removeObserver:self
               name:UIApplicationDidEnterBackgroundNotification
             object:nil];

    [[NSNotificationCenter defaultCenter]
     removeObserver:self
               name:UIApplicationWillEnterForegroundNotification
             object:nil];
```

```
    }
}
```

We are using the `isMultitaskingSupported` method in this code. This method is explained in Recipe 11.1.

11.1 Detecting the Availability of Multitasking

Problem

You want to find out whether the iOS device running your application supports multitasking.

Solution

Call the `isMultitaskingSupported` instance method of `UIDevice`, like so:

```
- (BOOL) isMultitaskingSupported{

    BOOL result = NO;

    UIDevice *device = [UIDevice currentDevice];

    if (device != nil){
      if ([device respondsToSelector:
           @selector(isMultitaskingSupported)] == YES){
        /* Make sure this only gets compiled on iOS SDK 4.0 and
        later so we won't get any compile-time warnings */
        #ifdef __IPHONE_4_0
          #if (__IPHONE_OS_VERSION_MAX_ALLOWED >= __IPHONE_4_0)
            result = [device isMultitaskingSupported];
          #endif
        #endif
      }
    }

    return(result);

}
```

The `isMultitaskingSupported` instance method of `UIDevice` is available in iOS SDK 4.0 and later. If you plan to switch from one SDK to the other, make sure you conditionally compile this line, as we are doing in our code:

```
result = [device isMultitaskingSupported];
```

If you unconditionally compile this code, you will get warnings from SDKs earlier than iOS SDK 4.0 since this instance method is available in iOS SDK 4.0 and later.

Discussion

Your application, depending on the iOS devices it targets, can be run and executed on a variety of devices on different operating systems. For instance, if you compile your application with iOS SDK 4.0 and your deployment target OS is 4.0, your application can be run on the iPhone 3G, iPhone 3GS, iPhone 4, and iPod Touch (second and third generations), provided that the iOS on these devices has been updated to iOS 4.0. Furthermore, a device could have iOS 4.0 or later installed on it, but the underlying hardware might not be strong enough for multitasking to be supported. Because of this, your application must be aware of whether multitasking is enabled on that specific hardware and on that specific iOS before attempting to act like a multitasking application.

11.2 Completing a Long-Running Task in the Background

Problem

You want to borrow some time from iOS to complete a long-running task when your application is being sent to the background.

Solution

Use the `beginBackgroundTaskWithExpirationHandler:` instance method of `UIApplica tion`. After you have finished the task, call the `endBackgroundTask:` instance method of `UIApplication`, as shown in the following code. Here is the *.h* file of our sample application delegate:

```
#import <UIKit/UIKit.h>

@interface MultitaskingAppDelegate : NSObject <UIApplicationDelegate> {
    UIWindow                *window;
    UIBackgroundTaskIdentifier  backgroundTaskIdentifier;
}

@property (nonatomic, retain) IBOutlet UIWindow *window;
@property (nonatomic, assign) UIBackgroundTaskIdentifier
                              backgroundTaskIdentifier;

@end
```

Here is the *.m* file of our application delegate:

```
#import "MultitaskingAppDelegate.h"

@implementation MultitaskingAppDelegate

@synthesize window;
@synthesize backgroundTaskIdentifier;

- (BOOL) isMultitaskingSupported{
```

```
    BOOL result = NO;

    UIDevice *device = [UIDevice currentDevice];

    if (device != nil){
      if ([device respondsToSelector:
            @selector(isMultitaskingSupported)] == YES){
        /* Make sure this only gets compiled on iOS SDK 4.0 and
         later so we won't get any compile-time warnings */
        #ifdef __IPHONE_4_0
          #if (__IPHONE_OS_VERSION_MAX_ALLOWED >= __IPHONE_4_0)
            result = [device isMultitaskingSupported];
          #endif
        #endif
      }
    }

    return(result);

}

- (void) myThread{

    NSAutoreleasePool *pool = [[NSAutoreleasePool alloc] init];

    /* Just run for as long as nobody has cancelled this thread */
    while ([[NSThread currentThread] isCancelled] == NO){

      [NSThread sleepForTimeInterval:1.0f];

      if ([[NSThread currentThread] isCancelled] == NO){

        /* Print out how much time we have left */
        if ([[UIApplication sharedApplication]
              backgroundTimeRemaining] == DBL_MAX){
          NSLog(@"Remaining time = Infinite");
        } else {
          NSLog(@"Remaining time = %02.02F seconds",
                [[UIApplication sharedApplication]
                  backgroundTimeRemaining]);
        }

      }

    }

    [pool release];

}

- (void) endBackgroundTask:(NSNumber *)paramIdentifier{

    /* We are asked to end the given background task
     by its identifier */
```

```objc
    UIBackgroundTaskIdentifier identifier =
    [paramIdentifier integerValue];

    [[UIApplication sharedApplication]
     endBackgroundTask:identifier];

}

- (BOOL)              application:(UIApplication *)application
  didFinishLaunchingWithOptions:(NSDictionary *)launchOptions {

    /* Start a new thread that prints out the number of
     seconds our application has left to execute code
     before it is sent to the background */

    [NSThread detachNewThreadSelector:@selector(myThread)
                             toTarget:self
                           withObject:nil];

    [window makeKeyAndVisible];

    return(YES);
}

- (void)applicationDidEnterBackground:(UIApplication *)application {

    /* Start a long-running task */

    if ([self isMultitaskingSupported] == NO){
      return;
    }

    self.backgroundTaskIdentifier =
    [[UIApplication sharedApplication]
     beginBackgroundTaskWithExpirationHandler:^{
       /* Make sure our identifier is a valid one */
       if (self.backgroundTaskIdentifier != UIBackgroundTaskInvalid){

         /* Now attempt to end the task whenever the system asks us */

         NSNumber *taskIdentifier = [NSNumber numberWithInteger:
                                     self.backgroundTaskIdentifier];

         [self performSelectorOnMainThread:@selector(endBackgroundTask:)
                                withObject:taskIdentifier
                             waitUntilDone:YES];

         /* Mark this task as done/invalid */
         self.backgroundTaskIdentifier = UIBackgroundTaskInvalid;
       }
     }];

}
```

```
- (void)applicationWillEnterForeground:(UIApplication *)application {

    /* When the application comes to the foreground again, we HAVE to
       check if we had previously borrowed some time from iOS. If so, we
       have to mark that task as done */

    if (self.backgroundTaskIdentifier != UIBackgroundTaskInvalid){
      [[UIApplication sharedApplication]
        endBackgroundTask:self.backgroundTaskIdentifier];
      self.backgroundTaskIdentifier = UIBackgroundTaskInvalid;
    }

}

- (void)dealloc {
  [window release];
  [super dealloc];
}

@end
```

 The `beginBackgroundTaskWithExpirationHandler:` instance method of `UIApplication` requires a block object as its parameter. The discussion of block objects is beyond the scope of this chapter. The caret character (^) marks the beginning of a block object, followed by an open and closed curly brace. Please refer to the "Blocks Programming Topics" article on Apple's website (*http://developer.apple.com/library/ios/#docu mentation/Cocoa/Conceptual/Blocks/Articles/00_Introduction.html*) for more information.

Discussion

When an iOS application is sent to the background, its main thread is paused. The threads you create within your application using the `detachNewThreadSelector:toTar get:withObject:` class method of `NSThread` are also suspended. If you are attempting to finish a long-running task when your application is being sent to the background, you must call the `beginBackgroundTaskWithExpirationHandler:` instance method of `UIAppli cation` to borrow some time from iOS. The `backgroundTimeRemaining` property of `UIAp plication` contains the number of seconds the application has to finish its job. If the application doesn't finish the long-running task before this time expires, iOS will terminate the application. Every call to the `beginBackgroundTaskWithExpirationHandler:` method must have a corresponding call to `endBackgroundTask:` (another instance method of `UIApplication`). In other words, if you ask for more time from iOS to complete a task, you must tell iOS when you are done with that task. Once this is done and no more tasks are requested to be running in the background, your application will be fully put into the background with all threads paused.

When your application is in the foreground, the `backgroundTimeRemaining` property of `UIApplication` is equal to the `DBL_MAX` constant, which is the largest value a value of type `double` can contain (the integer equivalent of this value is normally equal to –1 in this case). After iOS is asked for more time before the application is fully suspended, this property will indicate the number of seconds the application has before it finishes running its task(s).

You can call the `beginBackgroundTaskWithExpirationHandler:` method as many times as you wish inside your application. The important thing to keep in mind is that whenever iOS returns a token or a task identifier to your application with this method, you must call the `endBackgroundTask:` method to mark the end of that task once you are finished running the task. Failing to do so might cause iOS to terminate your application.

While in the background, applications are not supposed to be fully functioning and processing heavy data. They are indeed only supposed to *finish* a long-running task. An example could be an application that is calling a web service API and has not yet received the response of that API from the server. During this time, if the application is sent to the background, the application can request more time until it receives a response from the server. Once the response is received, the application must save its state and mark that task as finished by calling the `endBackgroundTask:` instance method of `UIApplication`.

In our example, whenever the application is put into the background, we ask for more time to finish a long-running task (in this case, for instance, our thread's code). In our thread, we constantly read the value of the `backgroundTimeRemaining` property of `UIApplication`'s instance and print that value out to the console. In the `beginBackgroundTaskWithExpirationHandler:` instance method of `UIApplication`, we provided the code that will be executed just before our application's extra time to execute a long-running task finishes (usually about 5 to 10 seconds before the expiration of the task). In here, we can simply end the task by calling the `endBackgroundTask:` instance method of `UIApplication`.

When an application is sent to the background and the application has requested more execution time from iOS, before the execution time is finished, the application could be revived and brought to the foreground by the user again. If you had previously asked for a long-running task to be executed in the background when the application was being sent to the background, you must end the long-running task using the `endBackgroundTask:` instance method of `UIApplication`.

See Also

Recipe 11.1

11.3 Receiving Local Notifications in the Background

Problem

You want to present an alert to your user even when your application is not running. You want to create this alert locally inside your application without using push notifications.

Solution

Instantiate an object of type `UILocalNotification` and schedule it using the `schedule LocalNotification:` instance method of `UIApplication`:

```
- (BOOL) localNotificationWithMessage
        :(NSString *)paramMessage
        actionButtonTitle:(NSString *)paramActionButtonTitle
        launchImage:(NSString *)paramLaunchImage
        applicationBadge:(NSInteger)paramApplicationBadge
        howManySecondsFromNow:(NSTimeInterval)paramHowManySecondsFromNow
        userInfo:(NSDictionary *)paramUserInfo{

  UILocalNotification *notification =
  [[UILocalNotification alloc] init];

  notification.alertBody = paramMessage;

  notification.alertAction = paramActionButtonTitle;

  if (paramActionButtonTitle != nil &&
      [paramActionButtonTitle length] > 0){
    /* Make sure we have the action button for the user to press
     to open our application */
    notification.hasAction = YES;
  } else {
    notification.hasAction = NO;
  }

  /* Here you have a chance to change the launch
   image of your application when the notification's
   action is viewed by the user */
  notification.alertLaunchImage = paramLaunchImage;

  /* Change the badge number of the application once the
   notification is presented to the user. Even if the user
   dismisses the notification, the badge number of the
   application will change */
  notification.applicationIconBadgeNumber = paramApplicationBadge;

  /* This dictionary will get passed to your application
   later if and when the user decides to view this notification */
  notification.userInfo = paramUserInfo;

  /* We need to get the system time zone so that the alert view
```

```
        will adjust its fire date if the user's time zone changes */
    NSTimeZone *timeZone = [NSTimeZone systemTimeZone];
    notification.timeZone = timeZone;

    /* Schedule the delivery of this notification 10 seconds from
     now */
    NSDate *fireDate = [NSDate date];

    fireDate =
    [fireDate dateByAddingTimeInterval:paramHowManySecondsFromNow];

    NSCalendar *calendar = [NSCalendar autoupdatingCurrentCalendar];

    NSUInteger dateComponents = NSYearCalendarUnit |
                               NSMonthCalendarUnit |
                               NSDayCalendarUnit |
                               NSHourCalendarUnit |
                               NSMinuteCalendarUnit |
                               NSSecondCalendarUnit;

    NSDateComponents *components = [calendar components:dateComponents
                                             fromDate:fireDate];

    /* Here you have a chance to change these components. That's why we
     retrieved the components of the date in the first place. */

    fireDate = [calendar dateFromComponents:components];

    /* Finally set the schedule date for this notification */
    notification.fireDate = fireDate;

    [[UIApplication sharedApplication]
     cancelAllLocalNotifications];

    [[UIApplication sharedApplication]
     scheduleLocalNotification:notification];

    [notification release];

    return(YES);

}
```

 The UILocalNotification class is available on iOS SDK 4.0 and later. So, this recipe's Solution uses conditional compilation, as in other recipes in this book using iOS 4 features, because you might build your application with SDKs earlier than SDK 4.0.

Discussion

Local notifications were introduced with iOS SDK 4.0. A *local notification* is an alert view (an object of type UIAlertView) that gets presented to the user if your application is running in the background or not running at all. You can schedule the delivery of a local notification using the `scheduleLocalNotification:` instance method of `UIApplica` `tion`. The `cancelAllLocalNotifications` instance method cancels the delivery of all pending local notifications.

You can ask iOS to deliver a local notification to the user in the future when your application is not even running. These notifications could also be recurring—for instance, every week at a certain time. However, extra care must be taken when you are specifying the *fire date* for your notifications.

For instance, let's say the time is now 13:00 in London, the time zone is GMT+0, and your application is currently running on a user's device. You want to be able to deliver a notification at 14:00 to your user even if your application is not running at that time. Now your user is on a plane at London's Gatwick Airport and plans to fly to Stockholm where the time zone is GMT+1. If the flight takes 30 minutes, the user will be in Stockholm at 13:30 GMT+0. However, when he lands, the iOS device will detect the change in the time zone and will change the user's device time to 14:30. Your notification was supposed to occur at 14:00 (GMT+0), so as soon as the time zone is changed, iOS detects that the notification is due to be displayed (30 minutes late, in fact, with the new time zone) and will display your notification.

The issue is that your notification was supposed to be displayed at 14:00 GMT+0 or 15:00 GMT+1, and not 14:30 GMT+1. To deal with occasions such as this (which may be more common than you think, with modern travel habits), when specifying a date and time for your local notifications to be fired, you should also specify the time zone of the date and time you are specifying.

The previous code did not include the alert view that you need to write in order to have something to display to the user. Let's go ahead and add that code in our application and see what happens on iPhone Simulator in different scenarios. Here is the *.h* file of our application delegate:

```
#import <UIKit/UIKit.h>

@interface MultitaskingAppDelegate : NSObject <UIApplicationDelegate> {
  UIWindow                      *window;
}

@property (nonatomic, retain) IBOutlet UIWindow *window;

@end
```

Here is the implementation of our application delegate, excluding the code that already appeared in the method listed in this recipe's Solution:

```
- (void) displayAlertWithTitle:(NSString *)paramTitle
                       message:(NSString *)paramMessage{

    UIAlertView *alertView = [[[UIAlertView alloc]
                               initWithTitle:paramTitle
                               message:paramMessage
```

```objc
                            delegate:self
                            cancelButtonTitle:@"OK"
                            otherButtonTitles:nil] autorelease];

    [alertView show];

}

- (BOOL)            application:(UIApplication *)application
  didFinishLaunchingWithOptions:(NSDictionary *)launchOptions {

    id      scheduledLocalNotification = nil;

    scheduledLocalNotification =
    [launchOptions
     valueForKey:UIApplicationLaunchOptionsLocalNotificationKey];

    if (scheduledLocalNotification != nil){

      /* We received a local notification while
       our application wasn't running. You can now typecase the
       ScheduledLocalNotification variable to UILocalNotification and
       use it in your application */

      [self displayAlertWithTitle:@"Notification"
                          message:@"Local Notification Woke Us Up"];

    } else {

      NSString *message =
      NSLocalizedString(@"A new instant message is available. \
                        Would you like to read this message?", nil);

      /* If a local notification didn't start our application,
       then we start a new local notification */
      [self localNotificationWithMessage:message
                        actionButtonTitle:nil
                              launchImage:nil
                          applicationBadge:1
                   howManySecondsFromNow:10
                                 userInfo:nil];

      [self displayAlertWithTitle:@"Set Up"
                          message:@"A new Local Notification is set up \
                                  to be displayed 10 seconds from now"];

    }

    [window makeKeyAndVisible];

    return YES;
}

- (void)           application:(UIApplication *)application
  didReceiveLocalNotification:(UILocalNotification *)notification{
```

```
/* We will receive this message whenever our application
 is running in the background when the local notification
 is delivered. If the application is terminated and the
 local notification is viewed by the user, the
 application:didFinishLaunchingWithOptions: method will be
 called and the notification will be passed via the
 didFinishLaunchingWithOptions parameter */

[self displayAlertWithTitle:@"Local Notification"
                    message:@"The Local Notification is delivered."];

}

- (void)dealloc {
  [window release];
  [super dealloc];
}
```

Now let's test the code. Here is our first scenario: the user has just installed our application and will launch it for the first time. Figure 11-1 shows what he will see.

Figure 11-1. The alert that we display to the user as an indication of local notifications being set up

The user taps the OK button and stays in the application. Figure 11-2 depicts the message that will be shown to the user after the notification is delivered to our application.

When the application is running or even in the background (has not been terminated yet), iOS will call the application:didReceiveLocalNotification: method of our application delegate to let our application know a local notification has been delivered to us. If the user is inside the application, iOS will not do anything special and will not display a message automatically. However, iOS does display a notification message automatically when our application is running in the background.

Figure 11-2. A local notification delivered while our application is running

In scenario 2, the user opens our application for the first time, as shown in Figure 11-1, and immediately after pressing the OK button presses the Home button on his iOS device, sending our application to the background. Now when the notification is delivered, our user will see a message similar to that shown in Figure 11-3.

Figure 11-3. A local notification delivered to an application that either is in the background or has been terminated

Because we set the application badge number property of our local notification to 1 when we created the notification, our application's badge number is immediately set to 1 when the notification is delivered. The user doesn't have to close or accept the notification for the badge number to be changed. Now if the user presses the Yes button, iOS will launch the application associated with this local notification and the user will see a screen similar to that shown in Figure 11-2. Please note that in this scenario, our application has not been terminated but sent to the background.

Scenario 3 is when our application runs for the first time, as shown in Figure 11-1, and the user sends our application to the background. Then the user terminates our application manually by double-tapping the Home button and closing the application using the Close button that will appear on the application icon when the user presses and holds his finger on the icon for a few seconds, as shown in Figure 11-4.

Figure 11-4. The user attempting to terminate our application before the local notification that we have scheduled is displayed to him

Once our application is terminated, the local notification will be displayed to the user after a few seconds (10 seconds from the time we scheduled the notification). Once the notification is delivered, the user will see a screen similar to that shown in Figure 11-3. After the user presses the Yes button, iOS will relaunch our application and the user will see a screen similar to that shown in Figure 11-5.

So, you can visually see how local notifications work. When our application is running in the foreground or the background, iOS will deliver the local notification through the `application:didReceiveLocalNotification:` delegate method. However, if our application has been terminated either by the user or by iOS, we will receive the local notification (that is, if the user decides to view it) through the application's `didFinishLaunchingWithOptions:` method. We can retrieve the notification using the `UIApplicationLaunchOptionsLocalNotificationKey` key of the `didFinishLaunchingWith Options` parameter.

Figure 11-5. A local notification waking up the terminated application

A local notification does not necessarily have to be an action notification. Action notifications have two buttons. You can change the title of one button through the `alertAction` property of `UILocalNotification`. The other button is an OK button that simply dismisses the alert; you cannot change the title or action. If a notification is not an action notification (when the `hasAction` property of `UILocalNotification` is set to `NO`), the notification will simply have an OK button, and pressing this button will *not* relaunch your application (see Figure 11-6).

Figure 11-6. A local notification that is not an action notification

11.4 Playing Audio in the Background

Problem

You are writing an application that plays audio files (such as a music player) and you would like the audio files to be played even if your application is running in the background.

Solution

Create a new array key in your application's main *.plist* file. Set the name of the key to UIBackgroundModes. Add the value audio to this new key. Here is an example of the contents of a *.plist* file with the aforementioned key and value added:

```
<dict>
    <key>CFBundleDevelopmentRegion</key>
    <string>English</string>
    <key>CFBundleDisplayName</key>
    <string>${PRODUCT_NAME}</string>
    <key>CFBundleExecutable</key>
    <string>${EXECUTABLE_NAME}</string>
    <key>CFBundleIconFile</key>
    <string></string>
    <key>CFBundleIdentifier</key>
    <string>com.pixolity.multitasking</string>
    <key>CFBundleInfoDictionaryVersion</key>
    <string>6.0</string>
    <key>CFBundleName</key>
    <string>${PRODUCT_NAME}</string>
    <key>CFBundlePackageType</key>
    <string>APPL</string>
    <key>CFBundleSignature</key>
    <string>????</string>
    <key>CFBundleVersion</key>
    <string>1.0</string>
    <key>LSRequiresIPhoneOS</key>
    <true/>
    <key>NSMainNibFile</key>
    <string>MainWindow</string>
    <key>UIBackgroundModes</key>
    <array>
        <string>audio</string>
    </array>
</dict>
```

Now you can use the AV Foundation to play audio files, and your audio files will be played even if your application is in the background.

Discussion

In iOS 4, applications can request that their audio files continue playing even if the application is sent to the background. AV Foundation's AVAudioPlayer is an easy-to-use audio player that we will use in this recipe. Our mission is to start an audio player and play a simple song, and while the song is playing, send the application to the background by pressing the Home button. If we have included the UIBackground Modes key in our application's *.plist* file, iOS will continue playing the music the audio player in our application is playing, even in the background. While in the background, we should only play music and provide our music player with the data that is necessary for it to run. We should not be performing any other tasks, such as displaying new screens and such.

Here is the *.h* file of a simple view controller that starts an AVAudioPlayer:

```
#import <UIKit/UIKit.h>
#import <AVFoundation/AVFoundation.h>

@interface RootViewController : UIViewController
                              <AVAudioPlayerDelegate> {
@public
  AVAudioPlayer    *audioPlayer;
}

@property (nonatomic, retain) AVAudioPlayer    *audioPlayer;

@end
```

Here is the *.m* file of the same view controller:

```
#import "RootViewController.h"

@implementation RootViewController

@synthesize audioPlayer;

- (id)initWithNibName:(NSString *)nibNameOrNil
             bundle:(NSBundle *)nibBundleOrNil {
  if ((self = [super initWithNibName:nibNameOrNil
                             bundle:nibBundleOrNil])) {
    // Custom initialization
  }
  return self;
}

- (void)audioPlayerBeginInterruption:(AVAudioPlayer *)player{

  /* Audio Session is interrupted.
   The player will be paused here */

}

- (void)audioPlayerEndInterruption:(AVAudioPlayer *)player
                        withFlags:(NSUInteger)flags{

  /* Check the flags, if we can resume the audio,
   then we should do it here */
```

```
    if (flags == AVAudioSessionInterruptionFlags_ShouldResume){
      [player play];
    }

  }

- (void)audioPlayerDidFinishPlaying:(AVAudioPlayer *)player
                      successfully:(BOOL)flag{

  NSLog(@"Finished playing the song");

  /* The flag parameter tells us if the playback was successfully
   finished or not */

  if ([player isEqual:self.audioPlayer] == YES){
    self.audioPlayer = nil;
  } else {
    [player release];
  }

}

- (void) startPlayingAudio{

  NSAutoreleasePool *pool = [[NSAutoreleasePool alloc] init];

  NSError *audioSessionError = nil;
  AVAudioSession *audioSession = [AVAudioSession sharedInstance];
  if ([audioSession setCategory:AVAudioSessionCategoryPlayback
                        error:&audioSessionError] == YES){
    NSLog(@"Successfully set the audio session.");
  } else {
    NSLog(@"Could not set the audio session");
  }

  NSBundle *mainBundle = [NSBundle mainBundle];

  NSString *filePath = [mainBundle pathForResource:@"MySong"
                                          ofType:@"mp3"];

  NSData    *fileData = [NSData dataWithContentsOfFile:filePath];

  NSError   *error = nil;

  /* Start the audio player */
  AVAudioPlayer *newPlayer =
  [[AVAudioPlayer alloc] initWithData:fileData
                              error:&error];

  self.audioPlayer = newPlayer;
  [newPlayer release];

  /* Did we get an instance of AVAudioPlayer? */
  if (self.audioPlayer != nil){
    /* Set the delegate and start playing */
```

```
      self.audioPlayer.delegate = self;
      if ([self.audioPlayer prepareToPlay] == YES &&
          [self.audioPlayer play] == YES){
        /* Successfully started playing */
      } else {
        /* Failed to play */
      }
    } else {
      /* Failed to instantiate AVAudioPlayer */
    }

    [pool release];

}

- (void)viewDidLoad {
  [super viewDidLoad];

  [NSThread
   detachNewThreadSelector:@selector(startPlayingAudio)
   toTarget:self
   withObject:nil];

}

- (void) viewDidUnload{
  [super viewDidUnload];

  if([self.audioPlayer isPlaying] == YES){
    [self.audioPlayer stop];
  }
  self.audioPlayer = nil;
}

- (BOOL)shouldAutorotateToInterfaceOrientation
  :(UIInterfaceOrientation)interfaceOrientation {
  return (YES);
}

- (void)didReceiveMemoryWarning {
  [super didReceiveMemoryWarning];
}

- (void)dealloc {
  [audioPlayer release];
  [super dealloc];
}

@end
```

 Please bear in mind that playing audio in the background might not
work in iPhone Simulator. You need to test this recipe on a real device.
On the simulator, chances are that the audio will stop playing once your
application is sent to the background.

In this example code, we are using AV audio sessions to silence music playback from other applications (such as the iPod application) before starting to play the audio. For more information about audio sessions, please refer to Recipe 8.5. When in the background, you are not limited to playing only the current audio file. If the currently playing audio file (in the background) finishes playing, you can start another instance of AVAudioPlayer and play a completely new audio file. iOS will adjust the processing required for this, and there is no guarantee your application will be given permission to allocate enough memory while in the background to accommodate for the data of the new sound file that needs to be played.

Another important thing to keep in mind is that while your application is running an audio file in the background, the value returned by the backgroundTimeRemaining property of UIApplication will not be changed. In other words, an application that requests to play audio files in the background is not implicitly or explicitly asking iOS for extra execution time.

11.5 Handling Location Changes in the Background

Problem

You are writing an application whose main functionality is processing location changes, using Core Location. You want the application to retrieve the iOS device location changes even if the application is sent to the background.

Solution

Add the location value to the UIBackgroundModes key of your main application *.plist* file, like so:

```
<dict>
    <key>CFBundleDevelopmentRegion</key>
    <string>English</string>
    <key>CFBundleDisplayName</key>
    <string>${PRODUCT_NAME}</string>
    <key>CFBundleExecutable</key>
    <string>${EXECUTABLE_NAME}</string>
    <key>CFBundleIconFile</key>
    <string></string>
    <key>CFBundleIdentifier</key>
    <string>com.pixolity.multitasking</string>
    <key>CFBundleInfoDictionaryVersion</key>
    <string>6.0</string>
    <key>CFBundleName</key>
    <string>${PRODUCT_NAME}</string>
    <key>CFBundlePackageType</key>
    <string>APPL</string>
    <key>CFBundleSignature</key>
    <string>????</string>
    <key>CFBundleVersion</key>
```

```
    <string>1.0</string>
    <key>LSRequiresIPhoneOS</key>
    <true/>
    <key>NSMainNibFile</key>
    <string>MainWindow</string>
    <key>UIBackgroundModes</key>
    <array>
        <string>location</string>
    </array>
</dict>
```

Discussion

When your application is running in the foreground, you can receive delegate messages from an instance of `CLLocationManager` telling you when iOS detects that the device is at a new location. However, if your application is sent to the background and is no longer active, the location delegate messages will normally not be delivered to your application. They will instead be delivered in a batch when your application becomes the foreground application again.

If you still want to be able to receive changes in the location of the user's device while running in the background, you must add the `location` value to the `UIBackground Modes` key of your application's main *.plist* file, as shown in this recipe's Solution. Once in the background, your application will continue to receive the changes in the device's location. Let's test this in a simple view controller. Here is the *.h* file of our view controller:

```
#import <UIKit/UIKit.h>
#import <CoreLocation/CoreLocation.h>

@interface RootViewController : UIViewController
                                <CLLocationManagerDelegate> {
@public
  CLLocationManager      *myLocationManager;
@private
  BOOL                   isExecutingInBackground;
}

@property (nonatomic, retain)
CLLocationManager      *myLocationManager;

@property (nonatomic, assign)
BOOL                   isExecutingInBackground;

@end
```

We will register for notifications when the application goes to the background and comes to the foreground, and we will set the `isExecutingInBackground` property's value accordingly. Now let's implement the view controller (*.m* file):

```
#import "RootViewController.h"

@implementation RootViewController
```

```objc
@synthesize myLocationManager;
@synthesize isExecutingInBackground;

- (void)locationManager:(CLLocationManager *)manager
    didUpdateToLocation:(CLLocation *)newLocation
           fromLocation:(CLLocation *)oldLocation{

  NSLog(@"New Location arrived...");

  if (self.isExecutingInBackground == YES){

    /* Just process the location and do not do any
       heavy processing here */

  } else {

    /* Display messages, alerts, etc. if needed because
       we are not in the background */

  }

}

- (void)locationManager:(CLLocationManager *)manager
       didFailWithError:(NSError *)error{

  NSLog(@"Failed to get location.");

  if (self.isExecutingInBackground == YES){

    /* Simply note this error and do not display any
       alerts and messages and more importantly, do not
       do any heavy processing here */

  } else {

    /* maybe display a message to the user */

  }

}

- (id)initWithNibName:(NSString *)nibNameOrNil
               bundle:(NSBundle *)nibBundleOrNil {
  if ((self = [super initWithNibName:nibNameOrNil
                              bundle:nibBundleOrNil])) {
    // Custom initialization
  }
  return self;
}

- (BOOL) isMultitaskingSupported{

  BOOL result = NO;
```

```
    UIDevice *device = [UIDevice currentDevice];

  if (device != nil){
    if ([device respondsToSelector:
        @selector(isMultitaskingSupported)] == YES){
      /* Make sure this only gets compiled on iOS SDK 4.0 and
       later so we won't get any compile-time warnings */
      #ifdef __IPHONE_4_0
        #if (__IPHONE_OS_VERSION_MAX_ALLOWED >= __IPHONE_4_0)
          result = [device isMultitaskingSupported];
        #endif
      #endif
    }
  }

  return(result);

}

- (void) handleEnteringBackground:(NSNotification *)paramNotification{

  /* We have entered background */
  NSLog(@"Going to background.");

  self.isExecutingInBackground = YES;

  if (self.myLocationManager != nil){
    /* If we are going to the background, let's reduce the accuracy
     of the location manager so that we use less system resources */
    self.myLocationManager.desiredAccuracy =
    kCLLocationAccuracyHundredMeters;
  }

}

- (void) handleEnteringForeground:(NSNotification *)paramNotification{

  /* We have entered foreground */
  NSLog(@"Coming to foreground");

  self.isExecutingInBackground = NO;

  if (self.myLocationManager != nil){
    /* Now that we are in the foreground, we can increase the accuracy
     of the location manager */
    self.myLocationManager.desiredAccuracy =
    kCLLocationAccuracyBest;
  }

}

- (void)viewDidLoad {
  [super viewDidLoad];
```

```
    /* Listen for the notifications sent to our application when the
     user puts the application in the background or when
     the application is brought to the foreground  */
    if ([self isMultitaskingSupported] == YES){

      [[NSNotificationCenter defaultCenter]
        addObserver:self
        selector:@selector(handleEnteringBackground:)
        name:UIApplicationDidEnterBackgroundNotification
        object:nil];

      [[NSNotificationCenter defaultCenter]
        addObserver:self
        selector:@selector(handleEnteringForeground:)
        name:UIApplicationWillEnterForegroundNotification
        object:nil];

    } else {
      NSLog(@"Multitasking is not enabled.");
    }

    /* Now let's create the location manager and start getting
     location change messages */
    CLLocationManager *newManager = [[CLLocationManager alloc] init];
    self.myLocationManager = newManager;
    [newManager release];

    self.myLocationManager.delegate = self;
    self.myLocationManager.desiredAccuracy = kCLLocationAccuracyBest;
    [self.myLocationManager startUpdatingLocation];

}

- (void)viewDidUnload {
  [super viewDidUnload];

  if ([self isMultitaskingSupported] == YES){

    [[NSNotificationCenter defaultCenter]
      removeObserver:self
      name:UIApplicationDidEnterBackgroundNotification
      object:nil];

    [[NSNotificationCenter defaultCenter]
      removeObserver:self
      name:UIApplicationWillEnterForegroundNotification
      object:nil];

  }

  /* Get rid of the location manager in cases such as
   a low memory warning */
  if (self.myLocationManager != nil){
    [self.myLocationManager stopUpdatingLocation];
  }
```

```
    self.myLocationManager = nil;

}

- (BOOL)shouldAutorotateToInterfaceOrientation
  :(UIInterfaceOrientation)interfaceOrientation {
  return (YES);
}

- (void)didReceiveMemoryWarning {
  [super didReceiveMemoryWarning];
}

- (void)dealloc {

  /* make sure we also deallocate our location manager here */
  if (myLocationManager != nil){
    [myLocationManager stopUpdatingLocation];
  }
  [myLocationManager release];

  [super dealloc];
}

@end
```

In the `viewDidLoad` method, we are creating our location manager with the best available accuracy. However, when the application goes into the background, we decrease the accuracy of the location manager. This is part of being a responsible multitasking-aware application. The simple rule here is that if we are in the background, we should be using the smallest amount of memory and processing power to satisfy our application's needs. So, by decreasing the accuracy of the location manager while in the background, we are decreasing the amount of processing iOS has to do to deliver new locations to our application.

 Depending on the version of iPhone Simulator that you are testing your applications with, background location processing might not work for you. Please test your applications, including the source code in this recipe, on a real device.

11.6 Saving and Loading the State of a Multitasking iOS Application

Problem

You want the state of your iPhone application to be saved when it is sent to the background, and for the same state to resume when the application is brought to the foreground.

Solution

Use a combination of the `UIApplicationDelegate` protocol's messages sent to your application delegate and the notifications sent by iOS to preserve the state of your multitasking application.

Discussion

When an empty iOS application (an application with just one window and no code written for it) is run on iOS 4.0 (and later versions) for the first time (not from the background), the following `UIApplicationDelegate` messages will be sent to your application delegate, in this order:

1. `application:didFinishLaunchingWithOptions:`
2. `applicationDidBecomeActive:`

If the user presses the Home button on her iOS device, your application delegate will receive these messages, in this order:

1. `applicationWillResignActive:`
2. `applicationDidEnterBackground:`

Once the application is in the background, the user can press the Home button twice and select our application from the list of background applications. Once our application is brought to the foreground again, we will receive these messages in the application delegate, in this order:

1. `applicationWillEnterForeground:`
2. `applicationDidBecomeActive:`

In addition to these messages, as mentioned in this chapter's Introduction, we will also receive various notification messages from iOS when our application is sent to the background or brought to the foreground again.

Now going back to the main subject of this recipe, you need to think carefully about the tasks you need to pause when going into the background and then resume after the application is brought to the foreground. Let me give you an example. As mentioned in Recipe 11.7, network connections can be easily resumed by the system itself, so we might not need to do anything special in the case of downloading a file from the network. However, if you are writing a game, for instance, it is best to listen for the notifications that iOS sends when your application is being sent to the background, and to act accordingly. In such a scenario, you can simply put the game engine into a paused state. You can also put the state of the sound engine into a paused state if necessary.

After an application is sent to the background, it has about 10 seconds to save any unsaved data and prepare itself to be brought to the foreground at any moment by the user. You can optionally ask for extra execution time if required (further information about this is available in Recipe 11.2).

Let's demonstrate saving your state with an example. Suppose we are writing a game for iOS. When our game is sent to the background, we want to:

1. Put the game engine into a paused state.
2. Save the user's score to disk.
3. Save the current level's data to disk. This includes where the user is in the level, the physical aspects of the level, the camera position, and so on.

When the user opens the application again, bringing the application to the foreground, we want to:

1. Load the user's score from disk.
2. Load the level the user was playing the last time from disk.
3. Resume the game engine.

Now let's assume our game engine runs in a single view controller, and it is the only one we have in the application. Here is the *.h* file of the sample view controller:

```
#import <UIKit/UIKit.h>

@interface RootViewController : UIViewController{
@private
  BOOL  isExecutingInBackground;
}

@property (nonatomic, assign) BOOL  isExecutingInBackground;

@end
```

Here is the implementation (*.m* file) of the view controller:

```
#import "RootViewController.h"

@implementation RootViewController

@synthesize isExecutingInBackground;

- (id)initWithNibName:(NSString *)nibNameOrNil
              bundle:(NSBundle *)nibBundleOrNil {
  if ((self = [super initWithNibName:nibNameOrNil
                            bundle:nibBundleOrNil])) {
    // Custom initialization
  }
  return self;
}

- (BOOL) isMultitaskingSupported{

  BOOL result = NO;

  UIDevice *device = [UIDevice currentDevice];

  if (device != nil){
```

```
      if ([device respondsToSelector:
          @selector(isMultitaskingSupported)] == YES){
        /* Make sure this only gets compiled on iOS SDK 4.0 and
         later so we won't get any compile-time warnings or
         runtime errors */
        #ifdef __IPHONE_4_0
          #if (__IPHONE_OS_VERSION_MAX_ALLOWED >= __IPHONE_4_0)
            result = [device isMultitaskingSupported];
          #endif
        #endif
      }
  }

  return(result);

}

- (void) saveUserScoreToDisk{

  /* Save the score */
  NSLog(@"Saving user's score to the disk...");

}

- (void) loadUserScoreFromDisk{

  NSLog(@"Loading user's score from the disk...");

}

- (void) saveCurrentLevelDataToDisk{

  /* Save the current game level to disk. Where in the
   level the user is, how much she has advanced in this
   level, etc. */

  NSLog(@"Saving the current level's data to the disk...");

}

- (NSData *) loadCurrentLevelDataFromDisk{

  NSData *result = nil;

  /* Load the current level from the disk and put
   it in an NSData (result) if possible. The level doesn't
   necessarily have to be loaded into an NSData. This is just
   an example */

  NSLog(@"Loading the current level's data from the disk...");

  return(result);

}
```

```objc
- (void) pauseGameEngine{

  NSLog(@"Pausing the game engine...");

}

- (void) resumeGameEngine{

  NSLog(@"Resuming the game engine...");

}

- (void) saveGameState{

  [self pauseGameEngine];
  [self saveUserScoreToDisk];
  [self saveCurrentLevelDataToDisk];

}

- (void) loadGameState{

  [self loadUserScoreFromDisk];
  [self loadCurrentLevelDataFromDisk];
  [self resumeGameEngine];

}

- (void) handleEnteringBackground:(NSNotification *)paramNotification{

  /* We have entered background */
  NSLog(@"Going to background.");

  self.isExecutingInBackground = YES;

  [self saveGameState];

}

- (void) handleEnteringForeground:(NSNotification *)paramNotification{

  /* We have entered foreground */
  NSLog(@"Coming to foreground");

  self.isExecutingInBackground = NO;

  [self loadGameState];

}

- (void)viewDidLoad {
  [super viewDidLoad];

  /* Listen for the notifications sent to our application when the
   user puts the application in the background or when
```

```objc
    the application is brought to the foreground  */
    if ([self isMultitaskingSupported] == YES){

      [[NSNotificationCenter defaultCenter]
       addObserver:self
       selector:@selector(handleEnteringBackground:)
       name:UIApplicationDidEnterBackgroundNotification
       object:nil];

      [[NSNotificationCenter defaultCenter]
       addObserver:self
       selector:@selector(handleEnteringForeground:)
       name:UIApplicationWillEnterForegroundNotification
       object:nil];

    } else {
      NSLog(@"Multitasking is not enabled.");
    }

}

- (void)viewDidUnload {
  [super viewDidUnload];

  if ([self isMultitaskingSupported] == YES){

    [[NSNotificationCenter defaultCenter]
     removeObserver:self
     name:UIApplicationDidEnterBackgroundNotification
     object:nil];

    [[NSNotificationCenter defaultCenter]
     removeObserver:self
     name:UIApplicationWillEnterForegroundNotification
     object:nil];

  }

}

- (BOOL)shouldAutorotateToInterfaceOrientation
  :(UIInterfaceOrientation)interfaceOrientation {
  return (YES);
}

- (void)didReceiveMemoryWarning {
  [super didReceiveMemoryWarning];
}

- (void)dealloc {
  [super dealloc];
}

@end
```

You can see we are listening for notifications that iOS sends to our application when it is getting sent to the background or being brought back to the foreground. Using these notifications, we can stop the game engine when we are getting sent to the background and resume it once we are in the foreground again.

Not every application is a game. However, you can use this technique to load and save the state of your application in the multitasking environment of iOS.

See Also

Recipe 11.2

11.7 Handling Network Connections in the Background

Problem

You are using instances of `NSURLConnection` to send and receive data to and from a web server and are wondering how you can allow your application to work in the multitasking environment of iOS without connection failures.

Solution

Make sure you support connection failures in the `connection:didFailWithError:` delegate method of `NSURLConnection` in case iOS tears down the network sockets of your application.

Discussion

For applications that use `NSURLConnection` but do not borrow extra time from iOS when they are sent to the background, connection handling is truly simple. Let's go through an example to see how an asynchronous connection will act in case the application is sent to the background and brought to the foreground again. Here is the .h file of our application delegate:

```
#import <UIKit/UIKit.h>

@interface MultitaskingAppDelegate : NSObject <UIApplicationDelegate> {
  UIWindow          *window;
  NSURLConnection   *downloadConnection;
  NSInteger         connectionDataLength;
}

@property (nonatomic, retain) IBOutlet UIWindow *window;
@property (nonatomic, retain) NSURLConnection   *downloadConnection;
@property (nonatomic, assign) NSInteger         connectionDataLength;

@end
```

We will implement the application delegate like so (the .m file):

```objc
#import "MultitaskingAppDelegate.h"
#import "RootViewController.h"

@implementation MultitaskingAppDelegate

@synthesize window;
@synthesize downloadConnection;
@synthesize connectionDataLength;

- (void) connection:(NSURLConnection *)connection
     didReceiveData:(NSData *)data{

  NSLog(@"%s", __FUNCTION__);

  self.connectionDataLength += [data length];

  NSLog(@"Data Length = %ld", (long)self.connectionDataLength);

}

- (void) connection:(NSURLConnection *)connection
 didReceiveResponse:(NSURLResponse *)response{

  NSLog(@"%s", __FUNCTION__);

  self.connectionDataLength = 0;

}

- (void) connection:(NSURLConnection *)connection
   didFailWithError:(NSError *)error{

  NSLog(@"%s", __FUNCTION__);

}

- (NSURLRequest *)connection:(NSURLConnection *)connection
             willSendRequest:(NSURLRequest *)request
            redirectResponse:(NSURLResponse *)redirectResponse{

  NSLog(@"%s", __FUNCTION__);

  return(request);
}

- (BOOL)            application:(UIApplication *)application
  didFinishLaunchingWithOptions:(NSDictionary *)launchOptions {

  self.connectionDataLength = 0;

  NSString *urlAsString =
  @"PUT THE REMOTE URL OF A RATHER BIG FILE HERE";

  NSURL *url = [NSURL URLWithString:urlAsString];
```

```objc
    NSURLRequest *request = [NSURLRequest requestWithURL:url];

    NSURLConnection *newConnection =
    [[NSURLConnection alloc] initWithRequest:request
                                    delegate:self
                             startImmediately:YES];

    self.downloadConnection = newConnection;

    [newConnection release];

    [window makeKeyAndVisible];

    return YES;
}

- (void)applicationWillResignActive:(UIApplication *)application {

    NSLog(@"%s", __FUNCTION__);

}

- (void)applicationDidEnterBackground:(UIApplication *)application {

    NSLog(@"%s", __FUNCTION__);

}

- (void)applicationWillEnterForeground:(UIApplication *)application {

    NSLog(@"%s", __FUNCTION__);

}

- (void)applicationDidBecomeActive:(UIApplication *)application {

    NSLog(@"%s", __FUNCTION__);

}

- (void)applicationWillTerminate:(UIApplication *)application {

    NSLog(@"%s", __FUNCTION__);

    if (self.downloadConnection != nil){
      [self.downloadConnection cancel];
    }

}

- (void)applicationDidReceiveMemoryWarning:(UIApplication *)application {

    NSLog(@"%s", __FUNCTION__);

}
```

```
- (void)dealloc {
  [downloadConnection release];
  [window release];
  [super dealloc];
}

@end
```

 The __FUNCTION__ macro is a GCC macro that will print the name of the current method. This is extremely useful when you want to find out what code is being executed in which method.

Before running this example, please make sure you have set the value of the urlAsString variable in the application:didFinishLaunchingWithOptions: method to the URL of the file you found on the Internet (preferably a large file so that you can test the application before the file is fully downloaded). You will see that in the foreground, our application will continue downloading the file. While downloading, the user can press the Home button and send the application to the background. What you will observe is true magic! iOS will automatically put the download process into a paused state for you. When the user brings your application to the foreground again, the downloading will resume without you writing a single line of code to handle multi-tasking.

In the example, we used an instance of NSURLConnection to download a file asynchronously. All delegate messages will be delivered to the delegate object on the thread that calls the initWithRequest:delegate:startImmediately: instance method of NSURLConnection. In our code, this is done on the main thread, so all delegate messages will be delivered on the main thread. Now let's see what happens with synchronous connections. We are going to download a very big file on the main thread (a very bad exercise, do not do this in a production application!) as soon as our application launches:

```
- (BOOL)            application:(UIApplication *)application
  didFinishLaunchingWithOptions:(NSDictionary *)launchOptions {

  NSString      *urlAsString =
  @"PUT THE REMOTE URL OF A VERY BIG FILE HERE";

  NSURL          *url = [NSURL URLWithString:urlAsString];
  NSURLRequest   *request = [NSURLRequest requestWithURL:url];
  NSError        *error = nil;
  NSURLResponse *response = nil;

  NSLog(@"Downloading started...");

  NSData *connectionData =
  [NSURLConnection sendSynchronousRequest:request
                        returningResponse:&response
```

```
                              error:&error];

    NSLog(@"Downloading finished");

    if (connectionData != nil){
      /* Do something here */
      NSLog(@"Successfully retrieved the data.");
    } else {
      /* Display an error */
      NSLog(@"Failed to retrieve the data.");
    }

    [window makeKeyAndVisible];

    return YES;
}
```

If you run this application and send it to the background, you will notice that the application's GUI is sent to the background, but the application's core is never sent to the background and the appropriate delegate messages—applicationWillResign Active: and applicationDidEnterBackground:—will never be received. I have conducted this test on an iPhone 4 and an iPhone 3GS.

The problem with this approach is that we are consuming the main thread's time slice by downloading files synchronously. We can fix this by either downloading the files asynchronously on the main thread, as mentioned before, or downloading them synchronously on separate threads.

Take the previous sample code, for example. If we download the same big file synchronously on a different thread, the connection will be paused when the application is sent to the background and will resume once it is brought to the foreground again:

```
- (void) downloadFile:(NSURL *)paramURL{

    NSAutoreleasePool *pool = [[NSAutoreleasePool alloc] init];

    NSURLRequest  *request = [NSURLRequest requestWithURL:paramURL];
    NSError       *error = nil;
    NSURLResponse *response = nil;

    NSLog(@"Downloading started...");

    NSData *connectionData =
    [NSURLConnection sendSynchronousRequest:request
                          returningResponse:&response
                                      error:&error];

    NSLog(@"Downloading finished");

    if (connectionData != nil){
      /* Do something here */
      NSLog(@"Successfully retrieved the data.");
    } else {
      /* Display an error */
```

```
        NSLog(@"Failed to retrieve the data.");
    }

    [pool release];

}

- (BOOL)              application:(UIApplication *)application
  didFinishLaunchingWithOptions:(NSDictionary *)launchOptions {

    NSString      *urlAsString =
    @"PUT THE REMOTE URL OF A VERY BIG FILE HERE";

    NSURL        *url = [NSURL URLWithString:urlAsString];

    [NSThread detachNewThreadSelector:@selector(downloadFile:)
                             toTarget:self
                           withObject:url];

    [window makeKeyAndVisible];

    return YES;
}
```

See Also

Recipe 11.2

11.8 Handling Notifications Delivered to a Waking Application

Problem

When your application is brought to the foreground, you want to be able to get notifications about important system changes such as the user's locale changes.

Solution

Simply listen to one of the many system notifications that iOS sends to waking applications. Some of these notifications are listed here:

NSCurrentLocaleDidChangeNotification
: This notification is delivered to applications when the user changes her locale: for instance, if the user switches her iOS device's language from English to Spanish in the Settings page of the device.

NSUserDefaultsDidChangeNotification
: This notification is fired when the user changes the application's settings in the Settings page of the iOS device (if any setting is provided to the user).

`UIDeviceBatteryStateDidChangeNotification`

This notification gets sent whenever the state of the battery of the iOS device is changed. For instance, if the device is plugged into a computer when the application is in the foreground and then unplugged when in the background, the application will receive this notification (if the application has registered for this notification). The state can then be read using the `batteryState` property of an instance of `UIDevice`.

`UIDeviceProximityStateDidChangeNotification`

This notification gets sent whenever the state of the proximity sensor changes. The last state is available through the `proximityState` property of an instance of `UIDevice`.

Discussion

When your application is in the background, a lot of things could happen! For instance, the user might suddenly change the locale of her iOS device through the Settings page from English to Spanish. Applications can register themselves for such notifications. These notifications will be coalesced and then delivered to a waking application. Let me explain what I mean by the term *coalesced*. Suppose your application is in the foreground and you have registered for `UIDeviceOrientationDidChangeNotification` notifications. Now the user presses the Home button and your application gets sent to the background. The user then rotates the device from portrait, to landscape right, to portrait, and then to landscape left. When the user brings your application to the foreground, you will receive only one notification of type `UIDeviceOrientationDidChange Notification`. This is coalescing. All the other orientations that happened along the way before your application opens are not important (since your application isn't on the screen) and the system will not deliver them to your application. However, the system will deliver you at least one notification for each aspect of the system, such as orientation, and you can then detect the most up-to-date orientation of the device.

Here is the *.h* file of our view controller:

```
#import <UIKit/UIKit.h>

@interface RootViewController : UIViewController {

}

@end
```

Here is the implementation of the same view controller (the *.m* file):

```
#import "RootViewController.h"

@implementation RootViewController

- (id)initWithNibName:(NSString *)nibNameOrNil
                bundle:(NSBundle *)nibBundleOrNil {
    if ((self = [super initWithNibName:nibNameOrNil
```

```
                              bundle:nibBundleOrNil])) {
    // Custom initialization
  }
  return self;
}

- (BOOL) isMultitaskingSupported{

  BOOL result = NO;

  UIDevice *device = [UIDevice currentDevice];

  if (device != nil){
    if ([device respondsToSelector:
        @selector(isMultitaskingSupported)] == YES){
      /* Make sure this only gets compiled on iOS SDK 4.0 and
       later so we won't get any compile-time warnings */
      #ifdef __IPHONE_4_0
        #if (__IPHONE_OS_VERSION_MAX_ALLOWED >= __IPHONE_4_0)
          result = [device isMultitaskingSupported];
        #endif
      #endif
    }
  }

  return(result);

}

- (void) orientationChanged:(NSNotification *)paramNotification{

  NSLog(@"Orientation Changed");

  NSLog(@"%@", [paramNotification object]);

}

- (void)viewDidLoad {
  [super viewDidLoad];

  if ([self isMultitaskingSupported] == YES){

    [[NSNotificationCenter defaultCenter]
     addObserver:self
     selector:@selector(orientationChanged:)
     name:UIDeviceOrientationDidChangeNotification
     object:nil];

  }

}


- (void)viewDidUnload {
  [super viewDidUnload];
```

```
        if ([self isMultitaskingSupported] == YES){
          [[NSNotificationCenter defaultCenter]
            removeObserver:self
            name:UIDeviceOrientationDidChangeNotification
            object:nil];
        }

    }

    - (BOOL)shouldAutorotateToInterfaceOrientation
      :(UIInterfaceOrientation)interfaceOrientation {
      return (YES);
    }

    - (void)didReceiveMemoryWarning {
      [super didReceiveMemoryWarning];
    }

    - (void)dealloc {
      [super dealloc];
    }

    @end
```

Run the application on the device now. After the view controller is displayed on the screen, press the Home button to send the application to the background. Now try changing the orientation of the device a couple of times, and then relaunch the application. Observe the results, and you will see that initially when your application opens, at most one notification will be sent to the orientationChanged: method. You might get a second call, though, if your view hierarchy supports orientation changes.

11.9 Handling Locale Changes in the Background

Problem

When your application is in the background and the user changes the language or locale in the Settings page of an iOS device, you want to know about the change when you return to the foreground. This problem is similar to the one in Recipe 11.8, but deals only with locale specifically.

Solution

Register for the NSCurrentLocaleDidChangeNotification notification:

```
    - (BOOL) isMultitaskingSupported{

      BOOL result = NO;

      UIDevice *device = [UIDevice currentDevice];
```

```
    if (device != nil){
      if ([device respondsToSelector:
          @selector(isMultitaskingSupported)] == YES){
        /* Make sure this only gets compiled on iOS SDK 4.0 and
         later so we won't get any compile-time warnings */
        #ifdef __IPHONE_4_0
          #if (__IPHONE_OS_VERSION_MAX_ALLOWED >= __IPHONE_4_0)
            result = [device isMultitaskingSupported];
          #endif
        #endif
      }
    }

    return(result);

}

- (void) localeChanged:(NSNotification *)paramNotification{

    NSLog(@"Locale Changed");

    NSLog(@"%@", [paramNotification object]);

    NSLocale *currentLocale = [NSLocale autoupdatingCurrentLocale];

    NSLog(@"Current Locale = %@", currentLocale);

}

- (void)viewDidLoad {
    [super viewDidLoad];

    if ([self isMultitaskingSupported] == YES){

      [[NSNotificationCenter defaultCenter]
        addObserver:self
        selector:@selector(localeChanged:)
        name:NSCurrentLocaleDidChangeNotification
        object:nil];

    }

}

- (void)viewDidUnload {
    [super viewDidUnload];

    if ([self isMultitaskingSupported] == YES){
      [[NSNotificationCenter defaultCenter]
        removeObserver:self
        name:NSCurrentLocaleDidChangeNotification
        object:nil];
    }

}
```

Discussion

When your application is running in the background, the user might go to the Settings application on iOS and change the device's locale (Settings→General→International→Region Format). If you register for the NSCurrentLocaleDidChangeNotification notification, your application will be notified when these settings change. Multiple notifications will be coalesced and delivered to your application when it is brought to the foreground. Coalescing here means that if the user changes the Region Format (for instance) three times (from Canadian to Chinese, from Chinese to Dutch, and from Dutch to French) while your application is running in the background, the moment the user brings your application to the foreground, you will only get one notification of type NSCurrentLocaleDidChangeNotification. Right there, you can access the current user locale using the autoupdatingCurrentLocale class method of NSLocale to access the current locale (French).

11.10 Responding to Changes in an Application's Settings

Problem

Your application exposes a settings bundle to the user. You want to get notified of the changes the user has made to your application's settings (while the application was in the background) as soon as your application is brought to the foreground.

Solution

Register for the NSUserDefaultsDidChangeNotification notification:

```
- (BOOL) isMultitaskingSupported{

  BOOL result = NO;

  UIDevice *device = [UIDevice currentDevice];

  if (device != nil){
    if ([device respondsToSelector:
        @selector(isMultitaskingSupported)] == YES){
      /* Make sure this only gets compiled on iOS SDK 4.0 and
       later so we won't get any compile-time warnings */
      #ifdef __IPHONE_4_0
        #if (__IPHONE_OS_VERSION_MAX_ALLOWED >= __IPHONE_4_0)
          result = [device isMultitaskingSupported];
        #endif
      #endif
    }
  }

  return(result);

}
```

```
- (void) settingsChanged:(NSNotification *)paramNotification{

  NSLog(@"Settings Changed");

  /* The object is of type NSUserDefaults */
  NSLog(@"%@", [paramNotification object]);

}

- (void)viewDidLoad {
  [super viewDidLoad];

  if ([self isMultitaskingSupported] == YES){

    [[NSNotificationCenter defaultCenter]
      addObserver:self
      selector:@selector(settingsChanged:)
      name:NSUserDefaultsDidChangeNotification
      object:nil];

  }

}

- (void)viewDidUnload {
  [super viewDidUnload];

  if ([self isMultitaskingSupported] == YES){
    [[NSNotificationCenter defaultCenter]
      removeObserver:self
      name:NSUserDefaultsDidChangeNotification
      object:nil];
  }

}
```

Please refer to this recipe's Discussion for information on creating a settings bundle in order to test this code.

Discussion

Applications written for iOS can expose a bundle file for their settings. These settings will be available to users through the Settings application on their device. To get a better understanding of how this works, let's create a settings bundle:

1. In Xcode, choose File→New File.
2. Choose Resources as the template.
3. Choose Settings Bundle as the file type and click Next.
4. Set the filename as *Settings.bundle*.
5. Click Finish.

Now you have a file in Xcode named *Settings.bundle*. Leave this file as it is, without modifying it. Put the code in this recipe's Solution into your root view controller and run the application. Press the Home button on the device and go to the device's Settings application. If you have named your application "foo" you will see "Foo" in the Settings application, as shown in Figure 11-7 (the name of the sample application I created is "Multitasking").

Figure 11-7. The Settings application on an iPhone Simulator

Tap on your application's name to see the settings your application exposes to the user, as shown in Figure 11-8.

Now try to change some of these settings while your application is running in the background. After you are done, bring the application to the foreground and you will see that only one NSUserDefaultsDidChangeNotification notification will be delivered to your application. The object attached to this notification will be of type NSUserDefaults and will contain the contents of your application's settings bundle.

11.11 Opting Out of Background Execution

Problem

You are maintaining an application that was built before the introduction of iOS 4, or you are writing an application that is built for iOS 4 but for some reason you do not want your application to participate in multitasking.

Figure 11-8. Settings bundle of a simple application

Solution

Add the `UIApplicationExitsOnSuspend` key to your application's main *.plist* file and set the value to `true`:

```
<dict>
    <key>CFBundleDevelopmentRegion</key>
    <string>English</string>
    <key>CFBundleDisplayName</key>
    <string>${PRODUCT_NAME}</string>
    <key>CFBundleExecutable</key>
    <string>${EXECUTABLE_NAME}</string>
    <key>CFBundleIconFile</key>
    <string></string>
    <key>CFBundleIdentifier</key>
    <string>com.pixolity.multitasking</string>
    <key>CFBundleInfoDictionaryVersion</key>
    <string>6.0</string>
    <key>CFBundleName</key>
    <string>${PRODUCT_NAME}</string>
    <key>CFBundlePackageType</key>
    <string>APPL</string>
    <key>CFBundleSignature</key>
    <string>????</string>
    <key>CFBundleVersion</key>
    <string>1.0</string>
    <key>LSRequiresIPhoneOS</key>
    <true/>
    <key>NSMainNibFile</key>
    <string>MainWindow</string>
    <key>UIApplicationExitsOnSuspend</key>
    <true/>
</dict>
```

Discussion

In some circumstances, you might require that your iOS applications not be multi-tasking (although I strongly encourage you to develop your applications to be multitasking-aware). In such cases, you can add the UIApplicationExitsOnSuspend key to your application's main *.plist* file. Devices on iOS 4.0 and later understand this value and the OS will terminate an application with this key set to true in the application's *.plist* file. On iOS earlier than version 4.0, this value will have no meaning to the operating system and will be ignored, so you can add this value even to applications that run on iOS 2.0.

When such an application runs on iOS 4.0, the following application delegate messages will be posted to your application:

1. application:didFinishLaunchingWithOptions:
2. applicationDidBecomeActive:

If the user presses the Home button on the device, the following messages will be sent to your application delegate:

1. applicationDidEnterBackground:
2. applicationWillTerminate:

CHAPTER 12

Core Data

12.0 Introduction

Core Data is a powerful framework on the iOS SDK that allows programmers to store and manage data in an object-oriented way. Traditionally, programmers had to store their data on disk using the archiving capabilities of Objective-C or write their data to files and manage them manually. With the introduction of Core Data, programmers can simply interact with its object-oriented interface to manage their data efficiently. In this chapter, you will learn how to use Core Data to create the model of your application (in the Model-View-Controller software architecture).

Core Data interacts with a persistent store at a lower level that is not visible to the programmer. iOS decides how the low-level data management is implemented. All the programmer must know is the high-level API she is provided with. But understanding the structure of Core Data and how it works internally is very important. Let's create a Core Data application to understand this a bit better.

Core Data in an iOS application needs a bit of setting up. Fortunately, with Xcode, this process is very easy. You can simply create a Core Data application and leave the rest up to Xcode.

Follow these steps to create an application in Xcode to take advantage of Core Data:

1. In Xcode, choose File→New→New Project.
2. Choose iOS Application as the template.
3. Choose Window-based Application as the template object.
4. Choose Universal in the Product drop-down menu and make sure you choose "Use Core Data for storage," as shown in Figure 12-1.
5. Click Choose.
6. Set your project name as "Data" and choose where you want to save it, as shown in Figure 12-2. Now click Save.

Figure 12-1. Choosing a universal window-based application

Figure 12-2. Choosing where to save the project

Now in Xcode, find the file named *AppDelegate_Shared.h*. This is the shared delegate of our application, since our application is universal. Both the iPad and iPhone application delegates will use this delegate as their superclass. If you have a look at the contents of this file, you will find that three properties were added to the declaration of the application delegate for you. These properties are:

- `managedObjectContext` (of type `NSManagedObjectContext`)
- `managedObjectModel` (of type `NSManagedObjectModel`)
- `persistentStoreCoordinator` (of type `NSPersistentStoreCoordinator`)

I know this is new and probably confusing to you, but by comparing these new concepts to existing database concepts, it will become easier for you to digest:

Persistent store coordinator
  This is the bridge or the connection between the physical file that stores our data and our application. This bridge will be responsible for managing different object contexts.

Managed object model
  This is the same concept as a schema in a database. This could represent the tables in a database or the different types of managed objects we can create in our database.

Managed object context
  This is the bridge between the programmer and the managed object model. Using the managed object context, you can insert a new row into a new table, read rows from a certain table, and so on. (Actually, Core Data doesn't use the concept of a "table," but I'm using the term here because it's familiar and will help you understand how Core Data works.)

Managed object
  This is similar to a row in a table. We insert managed objects into the managed object context and save the context. This way, we create a new row in a table in our database.

In Recipe 12.1, you will learn how to create a Core Data model using Xcode. This is the first step toward creating a database *schema*.

12.1 Creating a Core Data Model with Xcode

Problem

You want to visually design the data model of your iOS application using Xcode.

Solution

Follow the instructions in this chapter's Introduction to create a Core Data project. Then find the file with the extension of *xcdatamodel* in your application bundle in Xcode and double-click it to open the visual data editor, as shown in Figure 12-3.

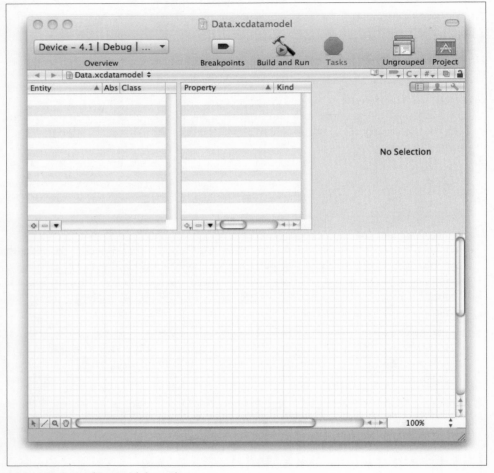

Figure 12-3. Xcode's visual data editor

Discussion

Xcode's visual data editor is a fantastic tool that allows programmers to design the data model of their applications with ease. There are two important definitions you need to learn before you can work with this tool:

Entity
> Is the same as a table in a database

Property
> Defines a column in the entity

Entities will later become objects (managed objects) when we generate the code based on our object model. This is explained in Recipe 12.2. For now, in this recipe, we will concentrate on creating the data model in this tool.

In the Entity pane, click the plus (+) button to create a new entity. The new entity will be named "Entity" by the editor. Change this name to "Person," as shown in Figure 12-4.

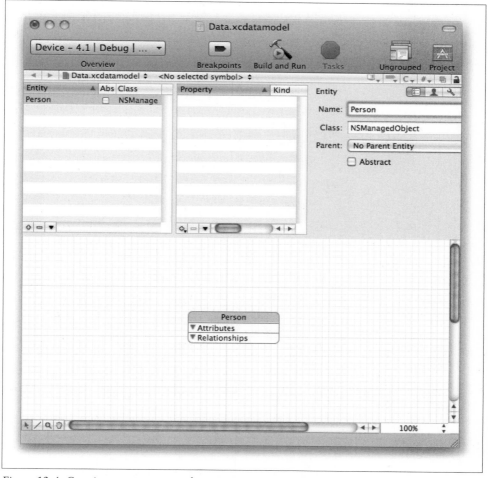

Figure 12-4. Creating a new entity and naming it "Person"; think of this entity as a new table in a database

Select the `Person` entity and create the following three properties for it (by selecting the plus [+] button in the Property pane and choosing Add Attribute from the menu):

- `firstName` (of type `String`, required)
- `lastName` (of type `String`, optional)
- `age` (of type `Int 32`, optional)

Now your data model in the editor must look similar to that depicted in Figure 12-5.

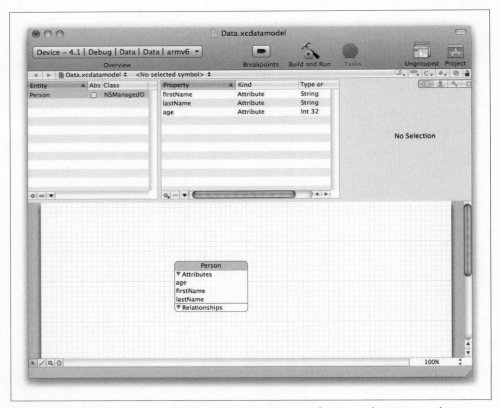

Figure 12-5. The Person managed object with three properties: firstName, lastName, and age

OK, we are done creating the model. Choose File→Save to make sure your changes are saved. To learn how to generate code based on the managed object you just created, refer to Recipe 12.2.

12.2 Creating and Using Core Data Model Classes

Problem

You followed the instructions in Recipe 12.1 and you want to know how to create code based on your object model.

Solution

Follow these steps:

1. In Xcode, find the file with the *xcdatamodel* extension that was created for your application when you created the application itself in Xcode. Click on the file, and you should see the editor on the righthand side of the Xcode window.
2. Select File→New File in Xcode.
3. Choose the iOS Cocoa Touch Class template on the lefthand side of the New File dialog and choose Managed Object Class on the righthand side, as shown in Figure 12-6. Once done, click Next.

Figure 12-6. Creating a managed object class based on the data model we created previously

4. Choose where you want to save the new managed object model class files (*.h* and *.m* files) and click Next.

5. Choose the Person entity from the list by checking the box next to its name, and click Finish (see Figure 12-7).

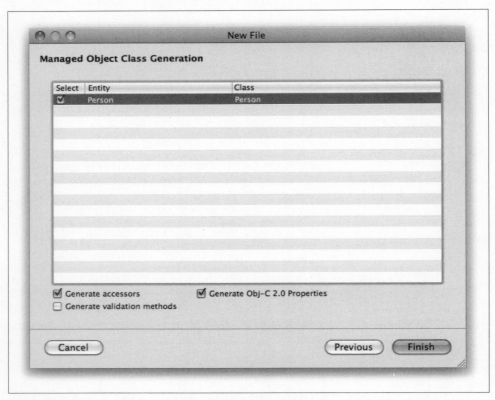

Figure 12-7. Choosing the Person entity for the managed object class generation process

Now you will see two new files in your project, called *Person.h* and *Person.m*. Open the contents of the *Person.h* file. It will look like the following:

```
#import <CoreData/CoreData.h>


@interface Person :  NSManagedObject
{
}

@property (nonatomic, retain) NSString * firstName;
@property (nonatomic, retain) NSNumber * age;
@property (nonatomic, retain) NSString * lastName;

@end
```

The *Person.m* file is implemented for us in this way:

```
#import "Person.h"


@implementation Person

@dynamic firstName;
@dynamic age;
@dynamic lastName;

@end
```

There you go! We turned our managed object into a real definition and implementation. In Recipe 12.3, you will learn how to instantiate and save a managed object of type Person into the managed object context of your application.

Discussion

When you create your data model using the editor in Xcode, you are creating the relationships, entities, properties, and so forth. However, to be able to use your model in your application, you must generate the code for your model. If you view the *.h* and *.m* files for your entities, you will realize that all the properties are being assigned dynamically. You can even see the @dynamic directive in the *.m* file of your entities to tell the compiler that you will fulfill the request of each property at runtime using dynamic method resolution.

None of the code that Core Data runs on your entities is visible to you, and there is no need for it to be visible to the programmer in the first place. All you have to know is that a Person entity has three properties named firstName, lastName, and age. You can assign values to these properties (if they are read/write properties) and you can save to and load them from the context, as we'll see in Recipe 12.3.

12.3 Creating and Saving Data Using Core Data

Problem

You want to create a new managed object and insert it into the managed object context.

Solution

Follow the instructions in Recipe 12.1 and Recipe 12.2. Now you can use the insert NewObjectForEntityForName:inManagedObjectContext: class method of NSEntityDe scription to create a new object of a type specified by the first parameter of this method. Once the new entity (the managed object) is created, you can modify it by changing its properties. After you are done, save your managed object context using the save: instance method of the managed object context.

Assuming that you have created a universal application in Xcode with the name "Data," follow these steps to insert a new managed object into the context:

1. Find the file named *AppDelegate_Shared.m*.

2. Import the *Person.h* file into the shared application delegate:

```
#import "AppDelegate_Shared.h"
#import "Person.h"

@implementation AppDelegate_Shared

@synthesize window;


... the rest of the code
```

Person is the entity we created in Recipe 12.1.

3. In the `application:didFinishLaunchingWithOptions:` method of your shared application delegate, write this code:

```
- (BOOL)            application:(UIApplication *)application
  didFinishLaunchingWithOptions:(NSDictionary *)launchOptions{

  /* First create an instance of Person in the context. At this point
   the new Person object has not been inserted into the database */
  Person *newPerson =
  [NSEntityDescription
   insertNewObjectForEntityForName:@"Person"
   inManagedObjectContext:self.managedObjectContext];

  /* Make sure we created an instance correctly */
  if (newPerson != nil){
    NSLog(@"Successfully created a new person.");

    /* Change the properties if we want to */
    newPerson.age = [NSNumber numberWithInteger:26];
    newPerson.firstName = @"Foo";
    newPerson.lastName = @"Bar";

    /* And eventually save any unsaved changes into the context */
    NSError *savingError = nil;
    if ([self.managedObjectContext save:&savingError] == YES){
      NSLog(@"Successfully saved the new person.");
    } else {
      NSLog(@"Failed to save the new person.");
    }

  } else {
    NSLog(@"Failed to create a new person.");
```

```
    }

    return(YES);

}
```

4. In the iPhone and iPad application delegates, make sure you call super in the application:didFinishLaunchingWithOptions: method so that this method gets called in the shared application delegate:

```
- (BOOL)              application:(UIApplication *)application
  didFinishLaunchingWithOptions:(NSDictionary *)launchOptions {

  // Override point for customization after application launch.

  [super                application:application
      didFinishLaunchingWithOptions:launchOptions];

  [window makeKeyAndVisible];

  return YES;
}
```

Discussion

Previous recipes showed how to create entities and generate code based on them using the editor in Xcode. The next thing we need to do is start using those entities and instantiate them. For this, we use NSEntityDescription and call its insertNewObjectFor EntityForName:inManagedObjectContext: class method. This will look up the given entity (specified by its name as NSString) in the given managed object context. If the entity is found, the method will return a new instance of that entity. This is similar to creating a new row (managed object) in a table (entity) in a database (managed object context).

Attempting to insert an unknown entity into a managed object context will raise an exception of type NSInternalInconsistencyException.

After inserting a new entity into the context, we must save the context. This will flush all the unsaved data of the context to the persistent store. We can do this using the save: instance method of our managed object context. If the BOOL return value of this method is YES, we can be sure that our context is saved. In Recipe 12.4, you will learn how to read the data back to memory.

12.4 Loading Data Using Core Data

Problem

You want to be able to read the contents of your entities (tables) using Core Data.

Solution

Use an instance of NSFetchRequest:

```
- (BOOL) createNewPersonWithFirstName:(NSString *)paramFirstName
                             lastName:(NSString *)paramLastName
                                  age:(NSUInteger)paramAge{

    BOOL result = NO;

    /* First create an instance of Person in the context. At this point
     the new Person object has not been inserted into the database */
    Person *newPerson =
    [NSEntityDescription
     insertNewObjectForEntityForName:@"Person"
     inManagedObjectContext:self.managedObjectContext];

    /* Make sure we created an instance correctly */
    if (newPerson != nil){
      NSLog(@"Successfully created a new person.");

      /* Change the properties if we want to */
      newPerson.age = [NSNumber numberWithInteger:paramAge];
      newPerson.firstName = paramFirstName;
      newPerson.lastName = paramLastName;

      /* And eventually save any unsaved changes into the context */
      NSError *savingError = nil;
      if ([self.managedObjectContext save:&savingError] == YES){
        NSLog(@"Successfully saved the new person.");
        result = YES;
      } else {
        NSLog(@"Failed to save the new person.");
      }

    } else {
      NSLog(@"Failed to create a new person.");
    }

    return(result);

}

- (BOOL)               application:(UIApplication *)application
    didFinishLaunchingWithOptions:(NSDictionary *)launchOptions{

    [self createNewPersonWithFirstName:@"Foo"
                              lastName:@"Foo Last Name"
```

```objc
                                age:25];

[self createNewPersonWithFirstName:@"Baz"
                          lastName:@"Baz Last Name"
                               age:29];


/* Create the fetch request first */
NSFetchRequest *fetchRequest =
[[NSFetchRequest alloc] init];

/* Here is the entity whose contents we want to read */
NSEntityDescription *entity =
[NSEntityDescription
 entityForName:@"Person"
 inManagedObjectContext:self.managedObjectContext];

/* Tell the request that we want to read the
 contents of the Person entity */
[fetchRequest setEntity:entity];

NSError *requestError = nil;

/* And execute the fetch request on the context */
NSArray *persons =
[self.managedObjectContext
 executeFetchRequest:fetchRequest
 error:&requestError];

/* Make sure we get the array */
if ([persons count] > 0){

  /* Go through the persons array one by one */
  NSUInteger counter = 1;
  for (Person *thisPerson in persons){

    NSLog(@"Person %lu First Name = %@",
          (unsigned long)counter,
          thisPerson.firstName);

    NSLog(@"Person %lu Last Name = %@",
          (unsigned long)counter,
          thisPerson.lastName);

    NSLog(@"Person %lu Age = %ld",
          (unsigned long)counter,
          (long)[thisPerson.age integerValue]);

    counter++;
  }

} else {
  NSLog(@"Failed to read the array of persons.");
}
```

```
/* Make sure to release the fetch request we allocated previously */
[fetchRequest release];

return(YES);

}
```

This code is being run on the shared delegate object of an application that was created with Core Data as its data storage mechanism, in Xcode. For more information about fetch requests, please refer to this recipe's Discussion.

Discussion

For those of you who are familiar with database terminology, a *fetch request* is similar to a SELECT statement. In the SELECT statement, you specify which rows, with which conditions, have to be returned from which table. With a fetch request, we do the same thing. We specify the entity (table) and the managed object context (the database layer). We can also specify sort descriptors for sorting the data we read. But first we'll focus on reading the data to make it simpler.

To be able to read the contents of the Person entity (we created this entity in Recipe 12.1 and turned it into code in Recipe 12.2), we first ask the NSEntityDescription class to search in our managed object context for an entity named Person. Once it is found, we will tell our fetch request what entity we want to read from. After this, all that's left to do is to execute the fetch request as we saw in this recipe's Solution:

```
NSError *requestError = nil;

/* And execute the fetch request on the context */
NSArray *persons =
[self.managedObjectContext executeFetchRequest:fetchRequest
                                          error:&requestError];
```

The return value of the executeFetchRequest:error: instance method of NSManagedObjectContext is either nil (in case of an error) or an array of Person managed objects. If no results are found for the given entity, the returned array will be empty.

See Also

Recipe 12.1; Recipe 12.2

12.5 Deleting Data Using Core Data

Problem

You want to delete a managed object from the managed object context.

Solution

Use the `deleteObject:` instance method of `NSManagedObjectContext`:

```
/* Get the person that has to be deleted from the array */
Person *thisPerson =
[self.arrayOfPersons objectAtIndex:indexPath.row];

AppDelegate_Shared *delegate = (AppDelegate_Shared *)
[[UIApplication sharedApplication] delegate];

/* Attempt to delete the person from the context */
[delegate.managedObjectContext deleteObject:thisPerson];

/* And we also have to save the context after deletion */
NSError *savingError = nil;
if ([delegate.managedObjectContext save:&savingError] == YES){

  /* Remove the person from the array and also make sure that
   we remove the corresponding cell from the table
   preferably with an animation */

  [self.arrayOfPersons removeObject:thisPerson];

  NSArray *cellsToDelete = [NSArray arrayWithObject:indexPath];
  [tableView deleteRowsAtIndexPaths:cellsToDelete
                  withRowAnimation:UITableViewRowAnimationFade];
```

Discussion

You can delete managed objects (records of a table in a database) using the `deleteObject:` instance method of `NSManagedObjectContext`. This is very simple, but let's put all that we have learned up to this point (adding, deleting, and fetching managed objects) into a real application.

The goal is to create an application with two view controllers:

- A root view controller with a table view that displays all the managed objects of type `Person` (designed in Recipe 12.1 and turned into code in Recipe 12.2) in our managed object context. This table view allows the user to delete rows (persons) using the editing features of the table view.

- A second view controller that will allow the user to create new `Person` managed objects by giving the first name, last name, and age of the person. Here we will just insert a new `Person` managed object into the context.

Let's begin with the delegate of our iPhone project. Here is the *.h* file of our iPhone application delegate (*AppDelegate_iPhone.h*):

```
#import <UIKit/UIKit.h>
#import "AppDelegate_Shared.h"

@interface AppDelegate_iPhone : AppDelegate_Shared {
@public
```

```
        UINavigationController  *navigationController;
}

@property (nonatomic, retain)
UINavigationController  *navigationController;

@end
```

We will implement the iPhone delegate (*AppDelegate_iPhone.m* file) like so:

```
#import "AppDelegate_iPhone.h"
#import "RootViewController_iPhone.h"

@implementation AppDelegate_iPhone

@synthesize navigationController;

- (BOOL)                    application:(UIApplication *)application
    didFinishLaunchingWithOptions:(NSDictionary *)launchOptions{

  /* Create a Navigation Controller and initialize it with
   the root View Controller */

  RootViewController_iPhone *controller =
  [[RootViewController_iPhone alloc]
   initWithNibName:@"RootViewController_iPhone"
   bundle:nil];

  UINavigationController *navController =
  [[UINavigationController alloc]
   initWithRootViewController:controller];

  self.navigationController = navController;

  [navController release];
  [controller release];

  [window addSubview:self.navigationController.view];

  [window makeKeyAndVisible];

  NSLog(@"%s", __FUNCTION__);

  return YES;
}

- (void)applicationWillResignActive:(UIApplication *)application {

  NSLog(@"%s", __FUNCTION__);

}

- (void)applicationDidEnterBackground:(UIApplication *)application {

    [super applicationDidEnterBackground:application];
```

```
    NSLog(@"%s", __FUNCTION__);

}

- (void)applicationWillEnterForeground:(UIApplication *)application {

    NSLog(@"%s", __FUNCTION__);

}

- (void)applicationDidBecomeActive:(UIApplication *)application {

    NSLog(@"%s", __FUNCTION__);

}

- (void)applicationWillTerminate:(UIApplication *)application {

    [super applicationWillTerminate:application];

    NSLog(@"%s", __FUNCTION__);

}

- (void)applicationDidReceiveMemoryWarning:(UIApplication *)application {

    [super applicationDidReceiveMemoryWarning:application];

    NSLog(@"%s", __FUNCTION__);

}

- (void)dealloc {

    [navigationController release];
    [super dealloc];

    NSLog(@"%s", __FUNCTION__);
}

@end
```

Now let's move on to the declaration of our root view controller, which will manage (delete and display) all the managed objects in the Person entity in our managed object context. Here is the .h file of our root view controller (*RootViewController_iPhone.h*):

```
#import <UIKit/UIKit.h>
#import <CoreData/CoreData.h>
#import "Person.h"

@interface RootViewController_iPhone : UIViewController
            <UITableViewDelegate, UITableViewDataSource> {
@public
  UITableView      *tableViewPersons;
  NSMutableArray   *arrayOfPersons;
```

```
      UIBarButtonItem *editButton;
      UIBarButtonItem *doneButton;
      UIBarButtonItem *addButton;
}

@property (nonatomic, retain)
IBOutlet UITableView  *tableViewPersons;

@property (nonatomic, retain)
NSMutableArray  *arrayOfPersons;

@property (nonatomic, retain)
UIBarButtonItem *editButton;

@property (nonatomic, retain)
UIBarButtonItem *doneButton;

@property (nonatomic, retain)
UIBarButtonItem *addButton;

- (BOOL) readAllPersons;
- (void) setDefaultNavigationBarButtons;
- (void) setEditingModeNavigationBarButtons;

@end
```

Before implementing this view controller, let's have a look at two of the properties that we have defined for this view controller and see what they are responsible for:

tableViewPersons

> This table view's responsibility is to display all the Person managed objects in the Person entity in our managed object context. In the database world, this can be translated to displaying all the rows (managed objects) of the Person entity (table) in our managed object context (database).

arrayOfPersons

> This is the array that will contain all the Person managed objects that we read from our managed object context.

> You must create a table view on the root view controller using Interface Builder and connect it to the tableViewPerson IBOutlet in the declaration (.h file) of your root view controller.

The root view controller that contains the tableViewPersons table view must become the delegate and the data source of the table view. Let's implement some of the important methods of the UITableViewDelegate protocol in the root view controller:

```
#pragma mark -
#pragma mark === Table View Delegate ===
#pragma mark -
```

```
- (void)              tableView:(UITableView*)tableView
  willBeginEditingRowAtIndexPath:(NSIndexPath *)indexPath{

  if ([tableView isEqual:self.tableViewPersons] == YES){
    if (self.tableViewPersons.editing == NO){
      [self setEditingModeNavigationBarButtons];
    }
  }

}

- (void)              tableView:(UITableView*)tableView
    didEndEditingRowAtIndexPath:(NSIndexPath *)indexPath{

  if ([tableView isEqual:self.tableViewPersons] == YES){
    if (self.tableViewPersons.editing == NO){
      [self setDefaultNavigationBarButtons];
    }
  }

}

- (UITableViewCellEditingStyle)tableView:(UITableView *)tableView
  editingStyleForRowAtIndexPath:(NSIndexPath *)indexPath{

  UITableViewCellEditingStyle result =
  UITableViewCellEditingStyleNone;

  if ([tableView isEqual:self.tableViewPersons] == YES){
    /* Allow the user to delete items from the table while we are
     in the editing mode */
    result = UITableViewCellEditingStyleDelete;
  }

  return(result);

}
```

This delegate method makes sure the editing style our table view allows is of type deletion. This way, the user will be able to delete the managed objects in the table view easily. Figure 12-8 depicts how our table view will look when it is in editing mode.

Now we will move on to the table view data source methods defined in the UITable ViewDataSource protocol, and we will implement the most important ones in the root view controller, like so:

```
#pragma mark -
#pragma mark === Table View Data Source ===
#pragma mark -

- (void)  tableView:(UITableView *)tableView
 commitEditingStyle:(UITableViewCellEditingStyle)editingStyle
  forRowAtIndexPath:(NSIndexPath *)indexPath{

  if ([tableView isEqual:self.tableViewPersons] == YES){
```

```
    /* Only allow deletion in the editing mode */
    if (editingStyle != UITableViewCellEditingStyleDelete){
      return;
    }

    /* Make sure our array contains the item that is about to
     be deleted from the table view */
    if ([self.arrayOfPersons count] <= indexPath.row){
      return;
    }

    /* Get the person that has to be deleted from the array */
    Person *thisPerson =
    [self.arrayOfPersons objectAtIndex:indexPath.row];

    AppDelegate_Shared *delegate = (AppDelegate_Shared *)
    [[UIApplication sharedApplication] delegate];

    /* Attempt to delete the person from the context */
    [delegate.managedObjectContext deleteObject:thisPerson];

    /* And we also have to save the context after deletion */
    NSError *savingError = nil;
    if ([delegate.managedObjectContext save:&savingError] == YES){

      /* Remove the person from the array and also make sure that
       we remove the corresponding cell from the table
       preferably with an animation */

      [self.arrayOfPersons removeObject:thisPerson];

      NSArray *cellsToDelete = [NSArray arrayWithObject:indexPath];

      [tableView
       deleteRowsAtIndexPaths:cellsToDelete
       withRowAnimation:UITableViewRowAnimationFade];

    } else {
      /* An error happened here */
    }

  }

}

- (NSInteger)tableView:(UITableView *)table
 numberOfRowsInSection:(NSInteger)section{

  NSInteger result = 0;

  if ([table isEqual:self.tableViewPersons] == YES){

    if (self.arrayOfPersons != nil){
      /* The number of cells we have is exactly equal
```

```
                    to the number of Person managed objects
                    that we have in our array */
              result = [self.arrayOfPersons count];
         }

    }

    return(result);

}

- (UITableViewCell *)tableView:(UITableView *)tableView
          cellForRowAtIndexPath:(NSIndexPath *)indexPath{

    UITableViewCell *result = nil;

    if ([tableView isEqual:self.tableViewPersons] == YES){

        static NSString *SimpleIdentifier = @"PersonCell";

        result =
        [tableView
         dequeueReusableCellWithIdentifier:SimpleIdentifier];

        /* A cell with subtitle */
        if (result == nil){
          result = [[[UITableViewCell alloc]
                     initWithStyle:UITableViewCellStyleSubtitle
                     reuseIdentifier:SimpleIdentifier] autorelease];
        }

        if (self.arrayOfPersons != nil){


            Person *thisPerson =
            [self.arrayOfPersons objectAtIndex:indexPath.row];

            /* The title will be "Last Name, First Name" */

            result.textLabel.text =
            [NSString stringWithFormat:@"%@, %@",
             thisPerson.lastName,
             thisPerson.firstName];

            /* The subtitle will be the age */

            result.detailTextLabel.text =
            [NSString stringWithFormat:@"Age %ld",
             (long)[thisPerson.age integerValue]];

            result.selectionStyle =
            UITableViewCellSelectionStyleNone;

        }
```

```
    }

    return(result);

}

- (NSInteger)numberOfSectionsInTableView:(UITableView *)tableView{

    NSInteger result = 0;

    if ([tableView isEqual:self.tableViewPersons] == YES){
        /* We only have one section */
        result = 1;
    }

    return(result);

}
```

Figure 12-8. Our table view in editing mode, displaying all managed objects of type Person in the managed object context

Perhaps a short explanation of each data source method would help you to understand what is really happening:

tableView:commitEditingStyle:forRowAtIndexPath:
> This gets called in the root view controller (as the data source of our table view) whenever the editing operation of the table view is committed. In our delegate, we set this style to UITableViewCellEditingStyleDelete so that users are only allowed to delete the cells, as shown in Figure 12-8. Here we have to do two things: delete that specific row from the cell, and remove the corresponding Person managed object from the managed object context. We can use the row property of the

forRowAtIndexPath parameter to determine which row (zero-based) inside the table view is being deleted by the user. We then get the corresponding `Person` object from `arrayOfPersons`. We delete the object from the managed object context and from the array. Then we will save the context, and if everything goes fine, we will delete the cell from the table view.

tableView:numberOfRowsInSection:

This returns the number of `Person` managed objects we have collected in the `arrayOfPersons` array. We will see how we collect these managed objects in this array shortly.

tableView:cellForRowAtIndexPath:

This retrieves a `Person` managed object out of the `arrayOfPersons` array (using the `row` property of the `cellForRowAtIndexPath` parameter) and constructs a table view cell with the `UITableViewCellStyleSubtitle` style. In the main label we will display the last name and first name of that person, and in the subtitle we will display the age of that person.

numberOfSectionsInTableView:

This tells the table view how many sections it must render. We will need no more than one section, as we are rendering all the instances of the `Person` entity in one section.

We have now fully implemented the data source and the delegate methods for our table view. Let's go ahead and implement the instance methods of our view controller:

```
#pragma mark -
#pragma mark === Instance Methods ===
#pragma mark -

- (id)initWithNibName:(NSString *)nibNameOrNil
              bundle:(NSBundle *)nibBundleOrNil {

  if ((self = [super initWithNibName:nibNameOrNil
                            bundle:nibBundleOrNil])) {

    self.title = NSLocalizedString(@"Persons", nil);
    /* Make sure we instantiate this array */
    NSMutableArray *newArray = [[NSMutableArray alloc] init];
    arrayOfPersons = [newArray retain];
    [newArray release];

  }

  return(self);
}

- (BOOL) readAllPersons{

  BOOL result = NO;

  if (self.arrayOfPersons == nil){
```

```objc
    return(NO);
  }

  AppDelegate_Shared *delegate = (AppDelegate_Shared *)
  [[UIApplication sharedApplication] delegate];

  NSManagedObjectContext *context = delegate.managedObjectContext;

  /* Find the Person entity in the context */
  NSEntityDescription *personEntity =
  [NSEntityDescription entityForName:@"Person"
              inManagedObjectContext:context];

  /* Now we want to read all the data in the Person entity */
  NSFetchRequest *request = [[NSFetchRequest alloc] init];

  [request setEntity:personEntity];

  NSError *fetchError = nil;

  /* Do the fetching here */
  NSArray *persons = [context executeFetchRequest:request
                                            error:&fetchError];

  /* Make sure we could read the data */
  if (persons != nil){

    /* Add all the persons from the array to our array */
    [self.arrayOfPersons removeAllObjects];
    [self.arrayOfPersons addObjectsFromArray:persons];
    [self.tableViewPersons reloadData];
    result = YES;

  }

  [request release];
  request = nil;

  return(result);

}

- (void) performAddNewPerson:(id)paramSender{

  /* Just display the Add New Person View Controller */

  AddNewPersonViewController_iPhone *controller =
  [[AddNewPersonViewController_iPhone alloc]
   initWithNibName:@"AddNewPersonViewController_iPhone"
   bundle:nil];

  [self.navigationController pushViewController:controller
                                      animated:YES];

  [controller release];
```

```
}

- (void) performDoneEditing:(id)paramSender{

  /* Take the table view out of the editing mode and re-create
     the default navigation bar buttons */

  [self.tableViewPersons setEditing:NO
                           animated:YES];

  [self setDefaultNavigationBarButtons];

}

- (void) performEditTable:(id)paramSender{

  /* Take the table to the editing mode */
  [self setEditingModeNavigationBarButtons];

  [self.tableViewPersons setEditing:YES
                           animated:YES];

}

- (void) viewDidAppear:(BOOL)animated{
  [super viewDidAppear:animated];

  /* Make sure we read the array of all the people in the
     managed object context and reload our table with
     the latest data. If in the Add New Person
     View Controller, we actually do add a person and
     come back to this screen, our table view will refresh
     itself automatically this way */

  [self readAllPersons];
}

- (void) setEditingModeNavigationBarButtons{

  /* In the editing mode, we just want a Done button on the
     top lefthand corner of the navigation bar to take the
     table view out of the editing mode */

  [self.navigationItem setLeftBarButtonItem:self.doneButton
                                   animated:YES];

  [self.navigationItem setRightBarButtonItem:nil
                                    animated:YES];

}

- (void) setDefaultNavigationBarButtons{

  /* Here we set an add button (+) on the top right and
```

an edit button on the top left that will take our table
view into editing mode where the user can delete items from
the table and our managed object context */

```
[self.navigationItem setRightBarButtonItem:self.addButton
                                  animated:YES];

[self.navigationItem setLeftBarButtonItem:self.editButton
                                 animated:YES];

}

- (void)viewDidLoad {
[super viewDidLoad];

  UIBarButtonItem *newAddButton =
  [[UIBarButtonItem alloc]
   initWithBarButtonSystemItem:UIBarButtonSystemItemAdd
   target:self
   action:@selector(performAddNewPerson:)];
  self.addButton = newAddButton;
  [newAddButton release];

  UIBarButtonItem *newEditButton =
  [[UIBarButtonItem alloc]
   initWithBarButtonSystemItem:UIBarButtonSystemItemEdit
   target:self
   action:@selector(performEditTable:)];
  self.editButton = newEditButton;
  [newEditButton release];

  UIBarButtonItem *newDoneButton =
  [[UIBarButtonItem alloc]
   initWithBarButtonSystemItem:UIBarButtonSystemItemDone
   target:self
   action:@selector(performDoneEditing:)];
  self.doneButton = newDoneButton;
  [newDoneButton release];

  [self setDefaultNavigationBarButtons];

  self.tableViewPersons.rowHeight = 60.0f;

}

- (void)viewDidUnload {
[super viewDidUnload];

  /* Free up some memory space */

  self.tableViewPersons = nil;
  self.arrayOfPersons = nil;
  self.editButton = nil;
  self.doneButton = nil;
  self.addButton = nil;
```

```
    self.navigationItem.rightBarButtonItem = nil;
    self.navigationItem.leftBarButtonItem = nil;

}

- (BOOL)shouldAutorotateToInterfaceOrientation:
    (UIInterfaceOrientation)interfaceOrientation {
    // Return YES for supported orientations
    return (YES);
}

- (void)didReceiveMemoryWarning {
    [super didReceiveMemoryWarning];
}

- (void)dealloc {

    [arrayOfPersons release];
    [tableViewPersons release];
    [editButton release];
    [doneButton release];
    [addButton release];

    [super dealloc];
}

@end
```

When we run our program for the first time, before implementing the AddNewPerson
ViewController_iPhone class that we use in the root view controller, we will see something similar to Figure 12-9. The add button (+) will be used to create new instances
of the Person entity in the managed object context, as we will see soon.

Pressing the Edit button will put the table view into editing mode, and the add button
(+) will disappear, as shown in Figure 12-8. The next thing we need to implement
is the AddNewPerson view controller (declared in the *AddNewPersonViewController_*
iPhone.h file and implemented in the *AddNewPersonViewController_iPhone.m* file). We
will begin with the declaration:

```
#import <UIKit/UIKit.h>
#import <CoreData/CoreData.h>
#import "Person.h"

@interface AddNewPersonViewController_iPhone :
            UIViewController <UITextFieldDelegate> {
@public
    UITextField     *textFieldFirstName;
    UITextField     *textFieldLastName;
    UITextField     *textFieldAge;
    UIBarButtonItem *doneButton;
    UIBarButtonItem *saveButton;
}
```

```
@property (nonatomic, retain)
IBOutlet UITextField    *textFieldFirstName;

@property (nonatomic, retain)
IBOutlet UITextField    *textFieldLastName;

@property (nonatomic, retain)
IBOutlet UITextField    *textFieldAge;

@property (nonatomic, retain)
UIBarButtonItem *doneButton;

@property (nonatomic, retain)
UIBarButtonItem *saveButton;

- (void) setDefaultNavigationBarButtons;
- (void) setEditingModeNavigationBarButtons;

@end
```

Figure 12-9. The root view controller with no Person managed objects to display

This file declares three Interface Builder outlets of type UITextField. We want this view controller to allow the user to enter the first name, last name, and age of a person and be able to insert that person into the managed object context through the GUI that we offer her (depicted in Figure 12-10).

Figure 12-10. The GUI of the AddNewPerson view controller with three text fields connected to the three IBOutlet instances of UITextField that we defined in the .h file of the AddNewPerson view controller

Please make sure all three text field instances (as shown in Figure 12-10) of the `AddNew Person` view controller are connected to their corresponding outlets in the *.h* file of this view controller and that they have their delegate set as the `AddNewPerson` view controller. We want to display the Done button on the righthand side of the navigation bar whenever the user starts, as shown in Figure 12-11.

Figure 12-11. The user editing the contents of the text fields that we placed on the AddNewPerson view controller; at this time, we are displaying the Done button on the navigation bar to facilitate a mechanism for hiding the keyboard when the user is done editing

Let's go ahead and implement the **AddNewPerson** view controller (*.m* file):

```objc
#import "AddNewPersonViewController_iPhone.h"
#import "AppDelegate_Shared.h"

@implementation AddNewPersonViewController_iPhone

@synthesize textFieldFirstName;
@synthesize textFieldLastName;
@synthesize textFieldAge;
@synthesize doneButton;
@synthesize saveButton;

- (void) performDone:(id)paramSender{
  [self.textFieldAge resignFirstResponder];
  [self.textFieldFirstName resignFirstResponder];
  [self.textFieldLastName resignFirstResponder];

  [self setDefaultNavigationBarButtons];
}

- (BOOL)textFieldShouldBeginEditing:(UITextField *)textField{

  BOOL result = YES;

  /* Make sure we create the Done button on the top
   righthand side of the navigation bar when we are
   editing the contents of our text fields */

  if (self.editing == NO){
    [self setEditingModeNavigationBarButtons];
  }

  return(result);

}

- (id)initWithNibName:(NSString *)nibNameOrNil
             bundle:(NSBundle *)nibBundleOrNil {

  if ((self = [super initWithNibName:nibNameOrNil
                              bundle:nibBundleOrNil])) {

    self.title = NSLocalizedString(@"Add New Person", nil);

  }

  return self;
}

- (void) displayAlertWithTitle:(NSString *)paramTitle
                       message:(NSString *)paramMessage{

  /* Simply create and display an alert view using
   the title and the message given to us */
  UIAlertView *alertView =
```

```objc
[[[UIAlertView alloc]
  initWithTitle:paramTitle
  message:paramMessage
  delegate:nil
  cancelButtonTitle:NSLocalizedString(@"OK", nil)
  otherButtonTitles:nil, nil] autorelease];

[alertView show];

}

- (void)  performAddNewPerson:(id)paramSender{

AppDelegate_Shared *delegate = (AppDelegate_Shared *)
[[UIApplication sharedApplication] delegate];

NSManagedObjectContext *context = delegate.managedObjectContext;

/* Get the values from the text fields */
NSString *firstName = self.textFieldFirstName.text;

NSString *lastName = self.textFieldLastName.text;

NSInteger ageAsInteger = [self.textFieldAge.text integerValue];
NSNumber *age = [NSNumber numberWithInteger:ageAsInteger];

/* Create a new instance of Person */
Person *newPerson = [NSEntityDescription
                     insertNewObjectForEntityForName:@"Person"
                     inManagedObjectContext:context];

if (newPerson != nil){

  /* Set the properties according to the values
   we retrieved from the text fields */

  newPerson.firstName = firstName;
  newPerson.lastName = lastName;
  newPerson.age = age;

  NSError *savingError = nil;

  /* Save the new person */
  if ([context save:&savingError] == YES){
    /* If successful, simply go back to the previous screen */
    [self.navigationController popViewControllerAnimated:YES];
  } else {
    /* If we failed to save, display a message */
    [self
     displayAlertWithTitle:NSLocalizedString(@"Saving", nil)
     message:NSLocalizedString(@"Failed to save the context", nil)];
  }

} else {
  /* We could not insert a new Person managed object */
```

```
      [self
        displayAlertWithTitle:NSLocalizedString(@"New Person", nil)
        message:NSLocalizedString(@"Failed To Insert A New Person", nil)];
    }

}

- (void) setDefaultNavigationBarButtons{

  /* By default, we have to have a Save button on the top
   righthand side of the navigation bar */

  [self.navigationItem setRightBarButtonItem:self.saveButton
                                    animated:YES];

  [self setEditing:NO];

}

- (void) setEditingModeNavigationBarButtons{

  /* When editing the text fields, we will replace the
   Save button with a Done button that will resign all
   text fields from being the first responder just
   so that the keyboard (if shown) will be hidden */

  [self.navigationItem setRightBarButtonItem:self.doneButton
                                    animated:YES];

  [self setEditing:YES];

}

- (void)viewDidLoad {
  [super viewDidLoad];

  UIBarButtonItem *newDoneButton =
  [[UIBarButtonItem alloc]
   initWithBarButtonSystemItem:UIBarButtonSystemItemDone
   target:self
   action:@selector(performDone:)];
  self.doneButton = newDoneButton;
  [newDoneButton release];

  UIBarButtonItem *newSaveButton =
  [[UIBarButtonItem alloc]
   initWithTitle:NSLocalizedString(@"Save", nil)
   style:UIBarButtonItemStylePlain
   target:self
   action:@selector(performAddNewPerson:)];
  self.saveButton = newSaveButton;
  [newSaveButton release];

  [self setDefaultNavigationBarButtons];
```

```
    }

- (void)viewDidUnload {
  [super viewDidUnload];

    /* Give up some memory used by our text fields
     in case of low memory */

    self.textFieldAge = nil;
    self.textFieldFirstName = nil;
    self.textFieldLastName = nil;
    self.doneButton = nil;
    self.saveButton = nil;

}

- (BOOL)shouldAutorotateToInterfaceOrientation:
  (UIInterfaceOrientation)interfaceOrientation {
  // Return YES for supported orientations
  return (YES);
}

- (void)didReceiveMemoryWarning {
  [super didReceiveMemoryWarning];
}

- (void)dealloc {
  [textFieldFirstName release];
  [textFieldLastName release];
  [textFieldAge release];
  [doneButton release];
  [saveButton release];
  [super dealloc];
}

@end
```

This view controller simply takes the values in the first name, last name, and age text
fields and creates a new managed object of type Person. After the managed object is
created, it will be inserted into the managed object context and the context will be
saved. If everything is successful (in the performAddNewPerson: instance method), the
view controller simply pops itself from the navigation controller to direct the user to
the root view controller where she can see all the entries in the context, as shown in
Figure 12-12.

12.6 Sorting Data Using Core Data

Problem

You want to sort the managed objects that you fetch from a managed object context.

Figure 12-12. Some dummy entries in the application managed object context, all of type Person

Solution

Create instances of `NSSortDescriptor` for each property (column, in the database world) of an entity that has to be sorted. Add the sort descriptors to an array and assign the array to an instance of `NSFetchRequest` using the `setSortDescriptors:` instance method. In this example code, `AppDelegate_Shared` is the class that represents the shared application delegate in a universal application. To understand how the `Person` entity is created, please refer to Recipe 12.1 and Recipe 12.2.

```
- (BOOL) readAllPersons{

  BOOL result = NO;

  if (self.arrayOfPersons == nil){
    return(NO);
  }

  AppDelegate_Shared *delegate = (AppDelegate_Shared *)
  [[UIApplication sharedApplication] delegate];

  NSManagedObjectContext *context =
  delegate.managedObjectContext;

  /* Find the Person entity in the context */
  NSEntityDescription *personEntity =
  [NSEntityDescription entityForName:@"Person"
              inManagedObjectContext:context];

  /* Now we want to read all the data in the Person entity */
  NSFetchRequest *request = [[NSFetchRequest alloc] init];

  [request setEntity:personEntity];
```

```
NSSortDescriptor *ageSortDescriptor =
[[NSSortDescriptor alloc] initWithKey:@"age"
                            ascending:YES];

NSSortDescriptor *firstNameSortDescriptor =
[[NSSortDescriptor alloc] initWithKey:@"firstName"
                            ascending:YES];

NSArray *sortDescriptors = [NSArray arrayWithObjects:
                            ageSortDescriptor,
                            firstNameSortDescriptor,
                            nil];

[firstNameSortDescriptor release];
[ageSortDescriptor release];

[request setSortDescriptors:sortDescriptors];

NSError *fetchError = nil;

/* Do the fetching here */
NSArray *persons = [context executeFetchRequest:request
                                          error:&fetchError];

/* Make sure we could read the data */
if (persons != nil){

  /* Add all the persons from the array to our array */
  [self.arrayOfPersons removeAllObjects];
  [self.arrayOfPersons addObjectsFromArray:persons];
  [self.tableViewPersons reloadData];
  result = YES;

}

[request release];
request = nil;

return(result);

}
```

 arrayOfPersons and tableViewPersons are properties of the object running this class. arrayOfPersons is of type NSMutableArray and tableView
Persons is of type UITableView. You can simply add this code in the small
program demonstrated in Recipe 12.5.

Discussion

An instance of NSFetchRequest can carry with itself an array of NSSortDescriptor instances. Each sort descriptor defines the property (column) on the current entity that

has to be sorted and whether the sorting has to be ascending or descending. For instance, the `Person` entity we created in Recipe 12.1 has `firstName`, `lastName`, and `age` properties. If we want to read all the persons in a managed object context and sort them from youngest to oldest, we would create an instance of `NSSortDescriptor` with the `age` key and set it to be `ascending`:

```
NSSortDescriptor *ageSortDescriptor =
[[NSSortDescriptor alloc] initWithKey:@"age"
                          ascending:YES];
```

You can assign more than one sort descriptor to one fetch request. If we replace the implementation of the `readAllPersons` method in Recipe 12.5 with the new implementation we saw in this recipe's Solution, instead of seeing the results shown in Figure 12-12, we will see the results shown in Figure 12-13.

Figure 12-13. Sorting managed objects by age and firstName properties (ascending)

See Also

Recipe 12.4

12.7 Boosting Data Access in Table Views

Problem

In an application that uses table views to present managed objects to the user, you want to be able to fetch and present the data in a more fluid and natural way than manipulating and repeatedly redisplaying arrays (as we did in Recipe 12.5).

Solution

Use fetched results controllers, which are instances of NSFetchedResultsController:

```
- (BOOL) readAllPersons{

    BOOL result = NO;

    if (self.personsFRC != nil){
      return(NO);
    }

    AppDelegate_Shared *delegate = (AppDelegate_Shared *)
    [[UIApplication sharedApplication] delegate];

    NSManagedObjectContext *context = delegate.managedObjectContext;

    /* Find the Person entity in the context */
    NSEntityDescription *personEntity =
    [NSEntityDescription entityForName:@"Person"
                inManagedObjectContext:context];

    NSSortDescriptor *ageSortDescriptor =
    [[NSSortDescriptor alloc] initWithKey:@"age"
                                ascending:YES];

    NSSortDescriptor *firstNameSortDescriptor =
    [[NSSortDescriptor alloc] initWithKey:@"firstName"
                                ascending:YES];

    NSArray *sortDescriptors = [NSArray arrayWithObjects:
                                ageSortDescriptor,
                                firstNameSortDescriptor,
                                nil];

    [firstNameSortDescriptor release];
    [ageSortDescriptor release];

    /* Now we want to read all the data in the Person entity */
    NSFetchRequest *request = [[NSFetchRequest alloc] init];

    /* set the entity of the request */
    [request setEntity:personEntity];

    /* And sort by age and then first name */
    [request setSortDescriptors:sortDescriptors];

    /* keep only 20 managed objects in the memory */
    [request setFetchBatchSize:20];

    NSFetchedResultsController *newFRC =
    [[NSFetchedResultsController alloc]
     initWithFetchRequest:request
     managedObjectContext:context
     sectionNameKeyPath:nil
```

```
                cacheName:@"PersonsCache"];

        /* Create the fetched results controller */
        self.personsFRC = newFRC;
        [newFRC release];

        /* whenever the fetched results controller (FRC) changes,
         meaning that a managed object inside it for instance gets deleted
         or modified, we will get notified and eventually we will reflect
         the changes in the GUI */
        self.personsFRC.delegate = self;

        NSError *fetchError = nil;
        /* Make sure to perform a fetch in order
         to read the managed objects */
        result = [self.personsFRC performFetch:&fetchError];

        [request release];

        return(result);

    }
```

Discussion

Fetched results controllers work in the same way as table views. Both have sections and rows. A fetched results controller can read managed objects from a managed object context and separate them into sections and rows. Each section is a group (if you specify it) and each row in a section is a managed object. You can then easily map this data to a table view and display it to the user. There are a few very important reasons why you might want to modify your application to use fetched results controllers:

- After a fetched results controller is created on a managed object context, any change (insertion, deletion, modification, etc.) will immediately be reflected on the fetched results controller as well. For instance, you could create your fetched results controller to read the managed objects of the Person entity. Then in some other place in your application, you insert a new Person managed object into the context (the same context the fetched results controller was created on). Immediately, the new managed object will become available in the fetched results controller. This is just magical!

- With a fetched results controller you can manage cache more efficiently. For instance, you can ask your fetched results controller to only keep N number of managed objects in memory per controller instance.

- Fetched results controllers are exactly like table views in the sense that they have sections and rows, as explained before. You can use a fetched results controller to present managed objects in the GUI of your application with table views with ease.

Here are some of the important properties and instance methods of fetched results controllers (all are objects of type NSFetchedResultsController):

sections *(property, of type* `NSArray`*)*

A fetched results controller can group data together using a key path. The designated initializer of the `NSFetchedResultsController` class accepts this grouping filter through the `sectionNameKeyPath` parameter. The `sections` array will then contain each grouped section. Each object in this array conforms to the `NSFetchedResultsSectionInfo` protocol.

objectAtIndexPath: *(instance method, returns a managed object)*

Objects fetched with a fetched results controller can be retrieved using their section and row index. Each section contains rows 0 to $N-1$, where N is the total number of items in that section. An index path object comprises a section and row index, and perfectly matches the information needed to retrieve objects from a fetched results controller. The `objectAtIndexPath:` instance method accepts index paths. Each index path is of type `NSIndexPath`. If you need to construct a table view cell using a managed object in a fetched results controller, simply pass the index path object in the `cellForRowAtIndexPath` parameter of the `tableView:cellForRowAtIndexPath:` delegate method of a table view. If you want to construct an index path yourself anywhere else in your application, use the `indexPathForRow:inSection:` class method of `NSIndexPath`.

fetchRequest *(property, of type* `NSFetchRequest`*)*

If at any point in your application, you believe you have to change the fetch request object for your fetched results controllers, you can do so using the `fetchRequest` property of an instance of `NSFetchedResultsController`. This is useful, for example, if you want to change the sort descriptors (refer to Recipe 12.6 for sort descriptors) of the fetch request object after you have allocated and initialized your fetched results controllers.

To use a fetched results controller in our next example, let's change the code we wrote in Recipe 12.5 and use a fetched results controller instead of an array. We'll start, as usual, with the *.h* file of our root view controller:

```
#import <UIKit/UIKit.h>
#import <CoreData/CoreData.h>
#import "Person.h"

@interface RootViewController_iPhone : UIViewController
            <UITableViewDelegate, UITableViewDataSource,
             NSFetchedResultsControllerDelegate> {
@public
  UITableView  *tableViewPersons;
  NSFetchedResultsController  *personsFRC;
  UIBarButtonItem *doneButton;
  UIBarButtonItem *addButton;
  UIBarButtonItem *editButton;
}

@property (nonatomic, retain)
IBOutlet UITableView  *tableViewPersons;
```

```
@property (nonatomic, retain)
NSFetchedResultsController  *personsFRC;

@property (nonatomic, retain)
UIBarButtonItem *doneButton;

@property (nonatomic, retain)
UIBarButtonItem *addButton;

@property (nonatomic, retain)
UIBarButtonItem *editButton;

- (BOOL) readAllPersons;
- (void) setDefaultNavigationBarButtons;
- (void) setEditingModeNavigationBarButtons;

@end
```

No changes are visible in the .h file, except that the personsFRC fetched results controller replaces the arrayOfPersons mutable array that we had before.

The implementation (.m file) of the root view controller is as follows:

```
#import "RootViewController_iPhone.h"
#import "AppDelegate_Shared.h"
#import "AddNewPersonViewController_iPhone.h"

@implementation RootViewController_iPhone

@synthesize tableViewPersons;
@synthesize personsFRC;
@synthesize doneButton;
@synthesize addButton;
@synthesize editButton;

#pragma mark -
#pragma mark === Fetched Results Controller Delegate ===
#pragma mark -

- (void)controller:(NSFetchedResultsController *)controller
  didChangeObject:(id)anObject
      atIndexPath:(NSIndexPath *)indexPath
    forChangeType:(NSFetchedResultsChangeType)type
     newIndexPath:(NSIndexPath *)newIndexPath{

  [self.tableViewPersons reloadData];

}

#pragma mark -
#pragma mark === Table View Delegate ===
#pragma mark -

- (void)              tableView:(UITableView*)tableView
 willBeginEditingRowAtIndexPath:(NSIndexPath *)indexPath{
```

```
    if ([tableView isEqual:self.tableViewPersons] == YES){
      if (self.tableViewPersons.editing == NO){
        [self setEditingModeNavigationBarButtons];
      }
    }

  }

- (void)            tableView:(UITableView*)tableView
    didEndEditingRowAtIndexPath:(NSIndexPath *)indexPath{

  if ([tableView isEqual:self.tableViewPersons] == YES){
    if (self.tableViewPersons.editing == NO){
      [self setDefaultNavigationBarButtons];
    }
  }

}

- (UITableViewCellEditingStyle)tableView:(UITableView *)tableView
  editingStyleForRowAtIndexPath:(NSIndexPath *)indexPath{

  UITableViewCellEditingStyle result =
  UITableViewCellEditingStyleNone;

  if ([tableView isEqual:self.tableViewPersons] == YES){
    /* Allow the user to delete items from the
     table while we are in the editing mode */
    result = UITableViewCellEditingStyleDelete;
  }

  return(result);

}

#pragma mark -
#pragma mark === Table View Data Source ===
#pragma mark -

- (void)  tableView:(UITableView *)tableView
 commitEditingStyle:(UITableViewCellEditingStyle)editingStyle
  forRowAtIndexPath:(NSIndexPath *)indexPath{

  if ([tableView isEqual:self.tableViewPersons] == YES){

    /* Only allow deletion in the editing mode */
    if (editingStyle != UITableViewCellEditingStyleDelete){
      return;
    }

    /* Make sure our array contains the item that is about to
     be deleted from the table view */

    /* Get the person that has to be deleted from the array */
    Person *thisPerson =
```

```
    [self.personsFRC objectAtIndexPath:indexPath];

    AppDelegate_Shared *delegate = (AppDelegate_Shared *)
    [[UIApplication sharedApplication] delegate];

    /* Attempt to delete the person from the context */
    [delegate.managedObjectContext deleteObject:thisPerson];

    /* And we also have to save the context after deletion */
    NSError *savingError = nil;
    if ([delegate.managedObjectContext
        save:&savingError] == YES){

    } else {
      /* An error happened here */
    }

  }

}


- (NSInteger)tableView:(UITableView *)table
 numberOfRowsInSection:(NSInteger)section{

  NSInteger result = 0;

  if ([table isEqual:self.tableViewPersons] == YES){

    if (self.personsFRC != nil){

      /* find the corresponding section in
       the Fetched Results Controller */

      id<NSFetchedResultsSectionInfo> thisSection =
      [self.personsFRC.sections objectAtIndex:section];

      /* And get the number of managed objects that are
       present in this section */
      result = [thisSection numberOfObjects];
    }

  }

  return(result);

}

- (UITableViewCell *)tableView:(UITableView *)tableView
        cellForRowAtIndexPath:(NSIndexPath *)indexPath{

  UITableViewCell *result = nil;

  if ([tableView isEqual:self.tableViewPersons] == YES){
```

```objc
    static NSString *SimpleIdentifier = @"PersonCell";

    result =
    [tableView
     dequeueReusableCellWithIdentifier:SimpleIdentifier];

    /* A cell with subtitle */
    if (result == nil){
      result = [[[UITableViewCell alloc]
                  initWithStyle:UITableViewCellStyleSubtitle
                  reuseIdentifier:SimpleIdentifier] autorelease];
    }

    if (self.personsFRC != nil){

      Person *thisPerson =
      [self.personsFRC objectAtIndexPath:indexPath];

      /* The title will be "Last Name, First Name" */

      result.textLabel.text =
      [NSString stringWithFormat:@"%@, %@",
       thisPerson.lastName,
       thisPerson.firstName];

      /* The subtitle will be the age */

      result.detailTextLabel.text =
      [NSString stringWithFormat:@"Age %ld",
       (long)[thisPerson.age integerValue]];

      result.selectionStyle =
      UITableViewCellSelectionStyleNone;

    }

  }

  return(result);

}

- (NSInteger)numberOfSectionsInTableView:(UITableView *)tableView{

  NSInteger result = 0;

  if ([tableView isEqual:self.tableViewPersons] == YES){
    if (self.personsFRC != nil){
      result = [self.personsFRC.sections count];
    }
  }

  return(result);

}
```

```
#pragma mark -
#pragma mark === Instance Methods ===
#pragma mark -

- (id)initWithNibName:(NSString *)nibNameOrNil
               bundle:(NSBundle *)nibBundleOrNil {

  if ((self = [super initWithNibName:nibNameOrNil
                              bundle:nibBundleOrNil])) {

    self.title = NSLocalizedString(@"Persons", nil);
    /* Make sure we instantiate this array */

  }

  return self;
}

- (BOOL) readAllPersons{

  BOOL result = NO;

  if (self.personsFRC != nil){
    return(NO);
  }

  AppDelegate_Shared *delegate = (AppDelegate_Shared *)
  [[UIApplication sharedApplication] delegate];

  NSManagedObjectContext *context =
  delegate.managedObjectContext;

  /* Find the Person entity in the context */
  NSEntityDescription *personEntity =
  [NSEntityDescription entityForName:@"Person"
               inManagedObjectContext:context];

  NSSortDescriptor *ageSortDescriptor =
  [[NSSortDescriptor alloc] initWithKey:@"age"
                              ascending:YES];

  NSSortDescriptor *firstNameSortDescriptor =
  [[NSSortDescriptor alloc] initWithKey:@"firstName"
                              ascending:YES];

  NSArray *sortDescriptors = [NSArray arrayWithObjects:
                              ageSortDescriptor,
                              firstNameSortDescriptor,
                              nil];

  [firstNameSortDescriptor release];
  [ageSortDescriptor release];

  /* Now we want to read all the data in the Person entity */
```

```
    NSFetchRequest *request = [[NSFetchRequest alloc] init];

    /* set the entity of the request */
    [request setEntity:personEntity];

    /* And sort by age and then first name */
    [request setSortDescriptors:sortDescriptors];

    /* keep only 20 managed objects in the memory */
    [request setFetchBatchSize:20];

    NSFetchedResultsController *newFRC =
    [[NSFetchedResultsController alloc]
     initWithFetchRequest:request
     managedObjectContext:context
     sectionNameKeyPath:nil
     cacheName:@"PersonsCache"];

    /* Create the fetched results controller */
    self.personsFRC = newFRC;
    [newFRC release];

    /* whenever the fetched results controller (FRC)
     changes, meaning that a managed object inside
     it for instance gets deleted or modified, we will
     get notified and eventually we will reflect
     the changes in the GUI */
    self.personsFRC.delegate = self;

    NSError *fetchError = nil;
    /* Make sure to perform a fetch in order
     to read the managed objects */
    result = [self.personsFRC performFetch:&fetchError];

    [request release];

    return(result);

}

- (void) performAddNewPerson:(id)paramSender{

    /* Just display the Add New Person View Controller */

    AddNewPersonViewController_iPhone *controller =
    [[AddNewPersonViewController_iPhone alloc]
     initWithNibName:@"AddNewPersonViewController_iPhone"
     bundle:nil];

    [self.navigationController pushViewController:controller
                                        animated:YES];

    [controller release];

}
```

```objc
- (void) performDoneEditing:(id)paramSender{

  /* Take the table view out of the editing mode and reset
   the default navigation bar buttons */

  [self.tableViewPersons setEditing:NO
                          animated:YES];

  [self setDefaultNavigationBarButtons];

}

- (void) performEditTable:(id)paramSender{

  /* Take the table to the editing mode */
  [self setEditingModeNavigationBarButtons];

  [self.tableViewPersons setEditing:YES
                          animated:YES];

}

- (void) viewDidAppear:(BOOL)animated{
  [super viewDidAppear:animated];
}

- (void) setEditingModeNavigationBarButtons{

  /* In the editing mode, we just want a Done button on the
   top lefthand corner of the navigation bar to take the
   table view out of the editing mode */

  [self.navigationItem setRightBarButtonItem:nil
                                    animated:YES];

  [self.navigationItem setLeftBarButtonItem:self.doneButton
                                    animated:YES];

}

- (void) setDefaultNavigationBarButtons{

  /* Here we set an add button (+) on the top right and
   an edit button on the top left that will take our table
   view into editing mode where the user can delete items from
   the table and our managed object context */

  [self.navigationItem setRightBarButtonItem:self.addButton
                                    animated:YES];

  [self.navigationItem setLeftBarButtonItem:self.editButton
                                    animated:YES];

}
```

```objc
- (void)viewDidLoad {
  [super viewDidLoad];

  UIBarButtonItem *newAddButton =
  [[UIBarButtonItem alloc]
   initWithBarButtonSystemItem:UIBarButtonSystemItemAdd
   target:self
   action:@selector(performAddNewPerson:)];
  self.addButton = newAddButton;
  [newAddButton release];

  UIBarButtonItem *newEditButton =
  [[UIBarButtonItem alloc]
   initWithBarButtonSystemItem:UIBarButtonSystemItemEdit
   target:self
   action:@selector(performEditTable:)];
  self.editButton = newEditButton;
  [newEditButton release];

  UIBarButtonItem *newDoneButton =
  [[UIBarButtonItem alloc]
   initWithBarButtonSystemItem:UIBarButtonSystemItemDone
   target:self
   action:@selector(performDoneEditing:)];
  self.doneButton = newDoneButton;
  [newDoneButton release];


  [self setDefaultNavigationBarButtons];
  [self readAllPersons];

  self.tableViewPersons.rowHeight = 60.0f;

}

- (void)viewDidUnload {
  [super viewDidUnload];

  /* Free up some memory space */


  self.tableViewPersons = nil;
  self.personsFRC = nil;
  self.addButton = nil;
  self.editButton = nil;
  self.doneButton = nil;

  self.navigationItem.rightBarButtonItem = nil;
  self.navigationItem.leftBarButtonItem = nil;

}

- (BOOL)shouldAutorotateToInterfaceOrientation:
  (UIInterfaceOrientation)interfaceOrientation {
```

```
    // Return YES for supported orientations
    return (YES);
}

- (void)didReceiveMemoryWarning {
    [super didReceiveMemoryWarning];
}

- (void)dealloc {

    [personsFRC release];
    [tableViewPersons release];
    [doneButton release];
    [addButton release];
    [editButton release];

    [super dealloc];
}

@end
```

In this example, we will keep the `AddNewPerson` view controller intact and will not modify it. Any new managed object of type `Person` that this view controller inserts into the managed object context will be reflected on our fetched results controller (`personsFRC`), and as soon as changes occur, we will receive delegate messages from the fetched results controller. The delegate object (in our example, the root view controller) of a fetched results controller must conform to the `NSFetchedResultsControllerDele gate` protocol. In our example, we used the `controller:didChangeObject:atIndex Path:forChangeType:newIndexPath:` method in this protocol to stay informed when objects in the managed object context that are read by our fetched results controller are changed (inserted, deleted, modified, etc.).

12.8 Implementing Relationships with Core Data

Problem

You want to be able to link your managed objects to each other: for instance, linking a `Person` to the `Home` she lives in.

Solution

Use inverse relationships in the model editor. Before continuing, please first create the `Person` entity by following the instructions in Recipe 12.1.

Discussion

Relationships in Core Data can be one-way, inverse one-to-many, or inverse many-to-many. Here is an example for each type of relationship:

One-way relationship

For instance, the relationship between a person and her nose. Each person can have only one nose and each nose can belong to only one person.

Inverse one-to-many relationship

For instance, the relationship between an employee and his manager. The employee can have only one direct manager, but his manager can have multiple employees working for her. Here the relationship of the employee with the manager is one-to-one, but from the manager's perspective, the relationship is one (manager) to many (employees); hence the word *inverse*.

Inverse many-to-many relationship

For instance, the relationship between a person and a car. One car can be used by more than one person, and one person can have more than one car.

In Core Data, you can create one-way relationships, but I highly recommend that you avoid doing so because, going back to the example in the preceding list, the person will know what nose she has but the nose will not know who it belongs to. In an object-oriented programming language such as Objective-C, it is always best to create inverse relationships so that children elements can refer to parent elements of that relationship.

Let's go ahead and create a data model that takes advantage of an inverse one-to-many relationship:

1. In Xcode, find the *xcdatamodel* file that was created for you when you started your Core Data project, as shown earlier in Figure 12-1 (refer to Recipe 12.1 to create such a project).

2. Open the data model file in the editor by double-clicking on it.

3. Remove any entities that were created for you previously by selecting them and pressing the Delete key on your keyboard.

4. Create a new entity and name it `Employee`. Create three properties for this entity, named `firstName` (of type `String`), `lastName` (of type `String`), and `age` (of type `Int 32`), as shown in Figure 12-14.

5. Create another entity named `Manager` with the same properties you created for the `Employee` entity (`firstName` of type `String`, `lastName` of type `String`, and `age` of type `Int 32`). See Figure 12-15.

6. Create a new relationship for the `Manager` entity by pressing the plus (+) button in the bottom of the Property box and choosing Add Relationship (see Figure 12-16).

7. Set the name of the new relationship to `FKManagerToEmployees` (see Figure 12-17).

8. Select the `Employee` entity and create a new relationship for it. Name the relationship `FKEmployeeToManager` (see Figure 12-18).

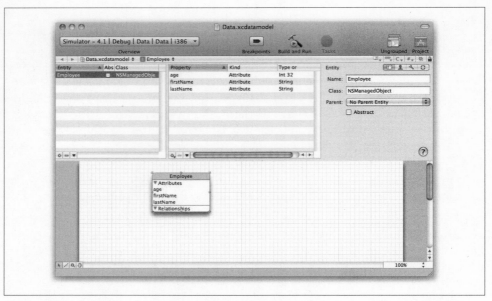

Figure 12-14. Creating a new entity named Employee with three properties, namely firstName, lastName, and age

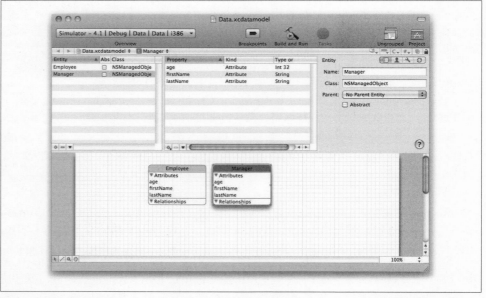

Figure 12-15. Creating an entity named Manager with the same properties as the Employee entity

Figure 12-16. Adding a new relationship to the Manager entity

Figure 12-17. Setting the name of the new Manager entity's relationship to FKManagerToEmployees

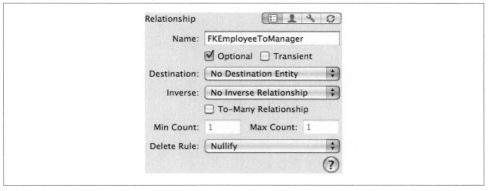

Figure 12-18. Setting the name of the new relationship for our Employee entity to FKEmployeeToManager

9. Choose the Manager entity, and then select the FKManagerToEmployees relationship for the Manager. In the Relationship properties box, choose Employee in the Destination drop-down menu (because we want to connect a Manager to an Employee entity through this relationship), set the Inverse box's value to FKEmployeeToManager (because the FKEmployeeToManager relationship of the Employee will link an

employee to her **Manager**), and tick the To-Many Relationship box (see Figure 12-19). The double arrowhead pointing to the Employee box at the bottom of the screen signifies that many employees can be associated with one manager. The single arrowhead pointing to the Manager box signifies that there can be only one manager for an employee.

10. Select both your **Employee** and **Manager** entities, select File→New File, and create the managed object classes for your model, as described in Recipe 12.2.

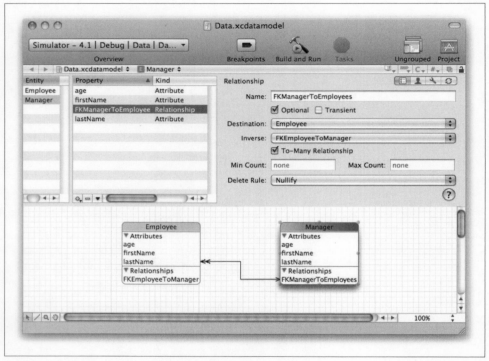

Figure 12-19. An inverse one-to-many relationship created between Manager and Employee entities

After creating the inverse one-to-many relationship, open the *.h* file of your **Employee** entity:

```
#import <CoreData/CoreData.h>

@class Manager;

@interface Employee :  NSManagedObject
{
}

@property (nonatomic, retain) NSString * firstName;
@property (nonatomic, retain) NSNumber * age;
```

```
@property (nonatomic, retain) NSString * lastName;
@property (nonatomic, retain) Manager * FKEmployeeToManager;

@end
```

You can see that a new property is added to this file. The property is named `FKEmployee``ToManager` and its type is `Manager`, meaning that from now on, if we have a reference to any object of type `Employee`, we can access its `FKEmployeeToManager` property to access that specific employee's `Manager` object (if any). Let's have a look at the *.h* file of the `Manager` entity:

```
#import <CoreData/CoreData.h>


@interface Manager : NSManagedObject
{
}

@property (nonatomic, retain) NSString * firstName;
@property (nonatomic, retain) NSNumber * age;
@property (nonatomic, retain) NSString * lastName;
@property (nonatomic, retain) NSSet* FKManagerToEmployees;

@end


@interface Manager (CoreDataGeneratedAccessors)
- (void)addFKManagerToEmployeesObject:(NSManagedObject *)value;
- (void)removeFKManagerToEmployeesObject:(NSManagedObject *)value;
- (void)addFKManagerToEmployees:(NSSet *)value;
- (void)removeFKManagerToEmployees:(NSSet *)value;

@end
```

The `FKManagerToEmployees` property is also created for the `Manager` entity. The data type of this object is `NSSet`. This simply means the `FKManagerToEmployees` property of any instance of the `Manager` entity can contain 1 to N number of `Employee` entities (a one-to-many relationship: one manager, many employees).

Another type of relationship that you might want to create is a many-to-many relationship. Going back to the `Manager` to `Employee` relationship, with a many-to-many relationship, any manager could have N number of employees and one employee could have N number of managers. To do this, follow the same instructions for creating a one-to-many relationship, but after step 9 and just before step 10, select the `Employee` entity and then the `FKEmployeeToManager` relationship. Change this name to `FKEmployee``ToManagers` and tick the To-Many Relationship box, as shown in Figure 12-20. Now the arrow has double arrowheads on both sides.

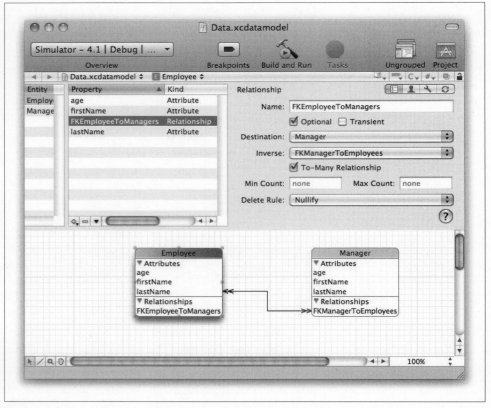

Figure 12-20. Creating a many-to-many relationship between the Employee and Manager entities

In your code, for a one-to-many relationship, you can simply create a new `Manager` managed object (read how you can insert objects to a managed object context in Recipe 12.3), save it to the managed object context, and then create a couple of `Employee` managed objects and save them to the context as well. Now, to associate the manager with an employee, set the value of the `FKEmployeeToManager` property of an instance of `Employee` to an instance of the `Manager` managed object. Core Data will then create the relationship for you.

CHAPTER 13

Event Kit

13.0 Introduction

The Event Kit and Event Kit UI frameworks, introduced in iOS SDK 4.0, allow iOS developers to access the Calendar database on an iOS device. You can insert, read, and modify events using the Event Kit framework. The Event Kit UI framework allows you to present built-in SDK GUI elements that allow the user to manipulate the Calendar database manually. In this chapter, we will focus on the Event Kit framework first, and then learn about the Event Kit UI framework.

With the Event Kit framework, a programmer can modify the user's Calendar database without him knowing. However, this is not a very good practice. In fact, Apple prohibits programmers from doing so and asks us to always notify users about any changes that the program might make to the Calendar database. Here is a quote from Apple:

> If your application modifies a user's Calendar database programmatically, it must get confirmation from the user before doing so. An application should never modify the Calendar database without specific instruction from the user.

iOS comes with a built-in Calendar application. The Calendar application can work with different types of calendars, such as local, CalDAV, and so forth. In this chapter, we will be working with different types of calendars as well. To make sure you are prepared to run the code in some of the recipes in this chapter, please create a Google account and associate it with Google Calendar. To get started, head over to *http://www .google.com/calendar/*.

Once there, create a Google account. After you are done, add Google Calendar to your iOS device by following these steps:

1. Go to the home screen of your iOS device.
2. Go to Settings.
3. Select Mail, Contacts, Calendars.
4. Select Add Account.
5. Select Other.

6. Select Add CalDAV Account.

7. On the CalDAV screen, enter the following (see Figure 13-1):

    a. **www.google.com** for the Server

    b. Your Google username for User Name

    c. Your Google password for Password

    d. A description of your choice for Description

8. Click Next.

Figure 13-1. Adding a Google account to an iOS device; we will use Google Calendar in the recipes in this chapter

Once you add the new Google account with Calendar access to your iOS device, you can see this calendar appear in the Calendars list in the Calendar application, as shown in Figure 13-2.

CalDAV is a protocol that allows access to the standard Calendar format used by Google Calendar and is supported by iOS. For more information about CalDAV, please refer to RFC 4791 at *http://tools.ietf.org/html/rfc4791*.

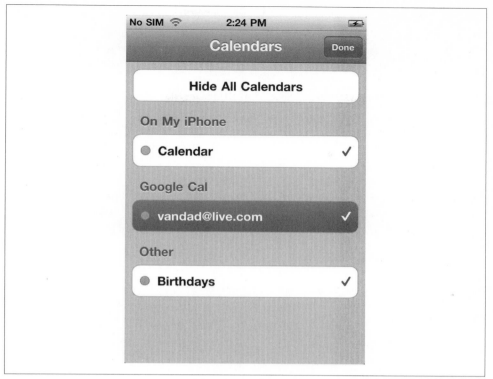

Figure 13-2. The new Google Calendar added to the list of calendars on an iPhone 4 device

Almost all the recipes in this chapter demonstrate the use of the Event Kit and Event Kit UI frameworks using a CalDAV calendar. Before reading the recipes, please take a few minutes and follow the instructions provided here to create a new CalDAV calendar and link it to your iOS device. Here are some of the benefits of using a CalDAV calendar:

- It's easy to set up.
- It can be shared among different platforms, so changes to a local instance of a CalDAV calendar object will be reflected—automatically by iOS—to the CalDAV server. This will give you a better understanding of how calendars work in iOS, and you can simply check your Google Calendar online and make sure the local changes are reflected there.
- You can add participants to an event using a CalDAV calendar. This is explained in Recipe 13.6.

To run the example code in this chapter, you must add the Event Kit framework, and in some cases the Event Kit UI framework, to your application by following these steps:

1. Find the *Frameworks* virtual folder in your application folder hierarchy in Xcode.

2. Right-click on the *Frameworks* virtual folder and choose Add→Existing Frameworks.

3. Choose the Event Kit and Event Kit UI frameworks and click Add. As mentioned before, these frameworks are only available for iOS SDK 4.0 and later.

 Neither the iPhone nor the iPad simulators on Mac OS X simulate the Calendar application on an iOS device. To test the recipes in this chapter, you must run and debug your program on a real iOS device. All examples in this chapter have been tested on the iPhone 3GS, iPhone 4, and iPad.

In most of the example code in this chapter, we will focus on manually reading and manipulating events in a calendar. If you want to use the built-in iOS capabilities to allow your users to quickly access their calendar events, please refer to Recipe 13.9 and Recipe 13.10.

13.1 Retrieving the List of Calendars

Problem

You want to retrieve the list of calendars available on the user's device before you attempt to insert new events into them.

Solution

Access the `calendars` array property of an instance of `EKEventStore`. Each calendar is of type `EKCalendar`:

```
EKEventStore *eventStore = [[EKEventStore alloc] init];

/* These are the calendar types an iOS Device can have. Please note
   that the "type" property of an object of type EKCalendar
   is of type EKCalendarType. The values in the "CalendarTypes"
   array reflect the exact same values in the EKCalendarType
   enumeration, but as NSString values */
NSArray *calendarTypes = [NSArray arrayWithObjects:
                          @"Local",
                          @"CalDAV",
                          @"Exchange",
                          @"Subscription",
                          @"Birthday",
                          nil];

/* Go through the calendars one by one */
NSUInteger counter = 1;
for (EKCalendar *thisCalendar in eventStore.calendars){

    /* The title of the calendar */
```

```
NSLog(@"Calendar %lu Title = %@",
      (unsigned long)counter, thisCalendar.title);

/* The type of the calendar */
NSLog(@"Calendar %lu Type = %@",
      (unsigned long)counter,
      [calendarTypes objectAtIndex:thisCalendar.type]);

/* The color that is associated with the calendar */
NSLog(@"Calendar %lu Color = %@",
      (unsigned long)counter,
      [UIColor colorWithCGColor:thisCalendar.CGColor]);

/* And whether the calendar can be modified or not */
if (thisCalendar.allowsContentModifications == YES){
  NSLog(@"Calendar %lu can be modified.",
        (unsigned long)counter);
} else {
  NSLog(@"Calendar %lu cannot be modified.",
        (unsigned long)counter);
}

counter++;
}

[eventStore release];
```

Running this code on an iPhone 4 with three calendars (Figure 13-2) will print results similar to this in the console window:

```
Calendar 1 Title = Calendar
Calendar 1 Type = Local
Calendar 1 Color = UIDeviceRGBColorSpace 0.054902 0.380392 0.72549 1
Calendar 1 can be modified.
Calendar 2 Title = Birthdays
Calendar 2 Type = Birthday
Calendar 2 Color = UIDeviceRGBColorSpace 0.72549 0.054902 0.156863 1
Calendar 2 cannot be modified.
Calendar 3 Title = vandad@live.com
Calendar 3 Type = CalDAV
Calendar 3 Color = UIDeviceRGBColorSpace 0.521569 0.192157 0.0156863 1
Calendar 3 can be modified.
```

Discussion

By allocating and initializing an object of type EKEventStore, you can access different types of calendars that are available on an iOS device. iOS supports common calendar formats such as CalDAV and Exchange. The calendars property of an instance of EKEventStore is of type NSArray and contains the array of calendars that are on an iOS device. Each object in this array is of type EKCalendar and each calendar has properties that allow us to determine whether, for instance, we can insert new events into that calendar.

As we will see in Recipe 13.2, a calendar object allows modifications only if its allowsContentModifications property has a YES value.

 You can use the colorWithCGColor: instance method of UIColor to retrieve an object of type UIColor from CGColorRef.

See Also

Recipe 13.2

13.2 Adding Events to Calendars

Problem

You would like to be able to create new events in users' calendars.

Solution

Find the calendar you want to insert your event into (please refer to Recipe 13.1). Create an object of type EKEvent using the eventWithEventStore: class method of EKEvent and save the event into the user's calendar using the saveEvent:span:error: instance method of EKEventStore:

```
- (BOOL)    createEventWithTitle:(NSString *)paramTitle
                       startDate:(NSDate *)paramStartDate
                         endDate:(NSDate *)paramEndDate
              inCalendarWithTitle:(NSString *)paramCalendarTitle
               inCalendarWithType:(EKCalendarType)paramCalendarType
                       withNotes:(NSString *)paramNotes{

    BOOL result = NO;

    EKEventStore *eventStore = [[EKEventStore alloc] init];

    /* Are there any calendars available to the event store? */
    if (eventStore.calendars == nil ||
        [eventStore.calendars count] == 0){
        [eventStore release];
        eventStore = nil;
        NSLog(@"No calendars are found.");
        return(NO);
    }

    EKCalendar *targetCalendar = nil;

    /* Try to find the calendar that the user asked for */
    for (EKCalendar *thisCalendar in eventStore.calendars){
        if ([thisCalendar.title
```

```
        isEqualToString:paramCalendarTitle] == YES &&
      thisCalendar.type == paramCalendarType){
    targetCalendar = thisCalendar;
    break;
  }
}

/* Make sure we found the calendar that we were asked to find */
if (targetCalendar == nil){
  NSLog(@"Could not find the requested calendar.");
  [eventStore release];
  return(NO);
}

/* If a calendar does not allow modification of its contents
 then we cannot insert an event into it */
if (targetCalendar.allowsContentModifications == NO){
  NSLog(@"The selected calendar does not allow modifications.");
  [eventStore release];
  return(NO);
}

/* Create an event */
EKEvent *event = [EKEvent eventWithEventStore:eventStore];
event.calendar = targetCalendar;

/* Set the properties of the event such as its title,
 start date/time, end date/time, etc. */
event.title = paramTitle;
event.notes = paramNotes;
event.startDate = paramStartDate;
event.endDate = paramEndDate;

/* Finally, save the event into the calendar */
NSError *saveError = nil;

result = [eventStore saveEvent:event
                          span:EKSpanThisEvent
                         error:&saveError];

if (result == NO){
  NSLog(@"An error occurred = %@", saveError);
}

[eventStore release];

return(result);

}
```

You can use the method we just implemented to insert new events into a user's calendar:

```
/* The event starts from today, right now */
NSDate *startDate = [NSDate date];

/* And the event ends this time tomorrow.
```

```
    24 hours, 60 minutes per hour and 60 seconds per minute
    hence 24 * 60 * 60 */
NSDate *endDate = [startDate
                    dateByAddingTimeInterval:24 * 60 * 60];

/* Create the new event */
[self createEventWithTitle:@"My event"
                 startDate:startDate
                   endDate:endDate
       inCalendarWithTitle:@"Calendar"
        inCalendarWithType:EKCalendarTypeLocal
                 withNotes:nil];
```

Discussion

To programmatically create a new event in a calendar on an iOS device, we must:

1. Allocate and initialize an instance of `EKEventStore`.

2. Find the calendar we want to save the event to (please refer to Recipe 13.1). We must make sure the target calendar supports modifications by checking that the calendar object's `allowsContentModifications` property is `YES`. If it is not, you must choose a different calendar or forego saving the event.

3. Once you find your target calendar, create an event of type `EKEvent` using the `eventWithEventStore:` class method of `EKEvent`.

4. Set the properties of the new event such as its `title`, `startDate`, and `endDate`.

5. Associate your event with the calendar that you found in step 2 using the `calendars` property of an instance of `EKEvent`.

6. Once you are done setting the properties of your event, add that event to the calendar using the `saveEvent:span:error:` instance method of `EKEventStore`. The return value of this method (a `BOOL` value) indicates whether the event was successfully inserted into the Calendar database. If the operation fails, the `NSError` object passed to the `error` parameter of this method will contain the error that has occurred in the system while inserting this event.

7. Release the event store object that you allocated in step 1.

If you attempt to insert an event without specifying a target calendar or if you insert an event into a calendar that cannot be modified, the `saveEvent:span:error:` instance method of `EKEventStore` will fail with an error similar to this:

```
Error Domain=EKErrorDomain Code=1 "The event has no calendar set."
UserInfo=0x1451f0 {NSLocalizedDescription=The event has no calendar set.}s
```

Running our code on an iPhone 4, we will see an event created in the Calendar database, as shown in Figure 13-3.

iOS syncs online calendars with the iOS calendar. These calendars could be Exchange, CalDAV, and other common formats. Creating an event on a CalDAV calendar on an

Figure 13-3. Programmatically adding an event to a calendar using the Event Kit framework

iOS device will create the same event on the server. The server changes are also reflected in the iOS Calendar database when the Calendar database is synced with the server.

See Also

Recipe 13.1

13.3 Accessing the Contents of Calendars

Problem

You want to retrieve events of type EKEvent from a calendar of type EKCalendar on an iOS device.

Solution

Follow these steps:

1. Instantiate an object of type EKEventStore.

2. Using the `calendars` property of the event store (instantiated in step 1), find the calendar you want to read from.

3. Determine the time and date where you want to start the search in the calendar and the time and date where the search must stop.

4. Pass the calendar object (found in step 2), along with the two dates you found in step 3, to the `predicateForEventsWithStartDate:endDate:calendars:` instance method of `EKEventStore`.

5. Pass the predicate created in step 4 to the `eventsMatchingPredicate:` instance method of `EKEventStore`. The result of this method is an array of `EKEvent` objects (if any) that fell between the given dates (step 3) in the specified calendar (step 2).

Code illustrating these steps follows:

```
- (EKCalendar *) getFirstAvailableCalDAVCalendar{

  EKCalendar *result = nil;

  EKEventStore *eventStore = [[EKEventStore alloc] init];

  for (EKCalendar *thisCalendar in eventStore.calendars){

    if (thisCalendar.type == EKCalendarTypeCalDAV){
      [eventStore release];
      return(thisCalendar);
    }

  }

  [eventStore release];

  return(result);

}

- (void)viewDidLoad {
  [super viewDidLoad];

  /* Find a calendar to base our search on */
  EKCalendar *targetCalendar =
  [self getFirstAvailableCalDAVCalendar];

  /* If we could not find a CalDAV calendar that we were looking for,
   then we will abort the operation */
  if (targetCalendar == nil){
    NSLog(@"No CalDAV calendars were found.");
    return;
  }

  /* We have to pass an array of calendars to the event store to search */
  NSArray *targetCalendars =
  [NSArray arrayWithObject:targetCalendar];
```

```
/* Instantiate the event store */
EKEventStore *eventStore = [[EKEventStore alloc] init];

/* We use the Calendar object to construct
 a starting and ending date */
NSCalendar *calendar = [NSCalendar currentCalendar];

/* Using the components in this object
 we can construct a starting date */
NSDateComponents *startDateComponents =
[[NSDateComponents alloc] init];

/* The start date will be 31st of July 2010 at 00:00:00 */
[startDateComponents setHour:0];
[startDateComponents setMinute:0];
[startDateComponents setSecond:0];
[startDateComponents setYear:2010];
[startDateComponents setMonth:7];
[startDateComponents setDay:31];

/* Construct the starting date using the components */
NSDate *startDate =
[calendar dateFromComponents:startDateComponents];

/* The end date will be 1st of August 2010 at 00:00:00.
 Exactly 24 hours after the starting date */
NSDate *endDate =
[startDate dateByAddingTimeInterval:24 * 60 * 60];

/* We do not need these components anymore */
[startDateComponents release];

/* Create the predicate that we can later pass to the
 event store in order to fetch the events */
NSPredicate *searchPredicate =
[eventStore predicateForEventsWithStartDate:startDate
                                    endDate:endDate
                                  calendars:targetCalendars];

/* Make sure we succeeded in creating the predicate */
if (searchPredicate == nil){
  NSLog(@"Could not create the search predicate.");
  [eventStore release];
  return;
}

/* Fetch all the events that fall between
 the starting and the ending dates */
NSArray *events =
[eventStore eventsMatchingPredicate:searchPredicate];

/* Go through all the events and print their information
 out to the console */
if (events != nil){
```

```
            NSUInteger counter = 1;
            for (EKEvent *event in events){

                NSLog(@"Event %lu Start Date = %@",
                        (unsigned long)counter,
                        event.startDate);

                NSLog(@"Event %lu End Date = %@",
                        (unsigned long)counter,
                        event.endDate);

                NSLog(@"Event %lu Title = %@",
                        (unsigned long)counter,
                        event.title);

                counter++;
            }

        } else {
            NSLog(@"The array of events for this start/end time is nil.");
        }

        [eventStore release];

    }
```

When we run this code on an iPhone 4 with three calendars set up (one of which is a CalDAV calendar), as shown in Figure 13-2, we will see the events that are available between July 31, 2010 at 00:00:00 and August 1, 2010 at 00:00:00:

```
Event 1 Start Date = 2010-07-31 09:00:00 +0100
Event 1 End Date = 2010-07-31 10:00:00 +0100
Event 1 Title = Go to the gym
Event 2 Start Date = 2010-07-31 11:00:00 +0100
Event 2 End Date = 2010-07-31 12:00:00 +0100
Event 2 Title = Do the laundry
Event 3 Start Date = 2010-07-31 13:00:00 +0100
Event 3 End Date = 2010-07-31 14:00:00 +0100
Event 3 Title = Learn about new iOS features.
```

The Calendar application in iOS displays the same events in the format shown in Figure 13-4. Please bear in mind that you will not see results similar to this unless you create the events in your Google Calendar just as I have created them, at the exact same date and time. However, if you do decide to create events in other calendars, such as the local calendar, on different dates, make sure you change the starting and ending dates of the event predicate and the calendar in which you are performing the search. For more information, please refer to this recipe's Discussion.

Discussion

As mentioned in this chapter's Introduction, an iOS device can be configured with different types of calendars using CalDAV, Exchange, and so on. Each calendar that is

Figure 13-4. The calendar events of a CalDAV account appearing in the Calendar application on iOS; these are the same events we can retrieve using the Event Kit framework

accessible by the Event Kit framework is encompassed within an `EKCalendar` object that can be accessed using the `calendars` array property of an instance of `EKEventStore`. You can fetch events inside a calendar in different ways, but the easiest way is to create and execute a specially formatted specification of dates and times, called a *predicate*, inside an event store.

A predicate of type `NSPredicate` that we can use in the Event Kit framework can be created using the `predicateForEventsWithStartDate:endDate:calendars:` instance method of an `EKEventStore`. The parameters to this method are:

`predicateForEventsWithStartDate`
    The starting date and time from when the events have to be fetched

`endDate`
    The ending date up until which the events will be fetched

`calendars`
    The array of calendars to search for events between the starting and ending dates

The dates passed to this method are of type NSDate. If you want to construct an NSDate object out of year, month, day, hour, and related values, you can use the NSDateComponents class, as demonstrated in our example code:

```
/* Using the components in this object
 we can construct a starting date */
NSDateComponents *startDateComponents =
[[NSDateComponents alloc] init];

/* The start date will be 31st of July 2010 at 00:00:00 */
[startDateComponents setHour:0];
[startDateComponents setMinute:0];
[startDateComponents setSecond:0];
[startDateComponents setYear:2010];
[startDateComponents setMonth:7];
[startDateComponents setDay:31];

/* Construct the starting date using the components */
NSDate *startDate =
[calendar dateFromComponents:startDateComponents];
```

See Also

Recipe 13.1

13.4 Removing Events from Calendars

Problem

You want to be able to delete a specific event or series of events from users' calendars.

Solution

Use the removeEvent:span:error: instance method of EKEventStore:

```
/* We have the index path of the row so now
 let's get the corresponding event from the
 array of events that we have */
EKEvent *event = [self.arrayOfEvents
                    objectAtIndex:indexPath.row];

NSError *removeError = nil;

/* Attempt to remove the event from the store */
if ([self.eventStore removeEvent:event
                        span:EKSpanThisEvent
                        error:&removeError] == YES){

  /* Successfully removed the event */

} else {
  NSLog(@"Failed to remove the event with error = %@",
```

```
                    removeError);
    }
```

Discussion

The `removeEvent:span:error:` instance method of `EKEventStore` can remove an instance of an event or all instances of a recurring event. For more information about recurring events, please refer to Recipe 13.5. In this recipe, we will only remove an instance of the event and not the other instances of the same event in the calendar.

The parameters that we can pass to this method are:

`removeEvent`
> This is the `EKEvent` instance to be removed from the calendar.

`span`
> This is the parameter that tells the event store whether we want to remove only this event or all the occurrences of this event in the calendar. To remove only the current event, specify the `EKSpanThisEvent` value for the `removeEvent` parameter. To remove all occurrences of the same event from the calendar, pass the `EKSpan FutureEvents` value for the parameter.

`error`
> This parameter can be given a reference to an `NSError` object that will be filled with the error (if any), when the return value of this method is `NO`.

Let's go ahead and create a simple application that mixes Recipe 13.3 and the current recipe. The application will list all events in modifiable calendars and will allow us to delete these events. We will populate an instance of `UITableView` with all the events that are present in any of the modifiable calendars, from last year until now. Then we will allow the user to delete these events from the table view, at which time we must delete the same events from the corresponding calendar.

To avoid using many preprocessor conditionals in the code, we will compile this code with just iOS SDK 4.0, since the Event Kit framework is not available on iOS SDKs earlier than 4.0. Let's have a look at the header file of our root view controller (.*h* file):

```
#import <UIKit/UIKit.h>
#import <EventKit/EventKit.h>

@interface RootViewController_iPhone : UIViewController
            <UITableViewDataSource, UITableViewDelegate> {
@public
  /* The table view that will contain the event cells */
  UITableView    *tableViewEvents;
  /* The array that will contain all the event objects */
  NSMutableArray *arrayOfEvents;
  /* And the event store that we will use to fetch
   and delete events */
  EKEventStore   *eventStore;

  UIBarButtonItem *editButton;
```

```
        UIBarButtonItem *doneButton;
}

@property (nonatomic, retain)
IBOutlet UITableView *tableViewEvents;

@property (nonatomic, retain)
NSMutableArray        *arrayOfEvents;

@property (nonatomic, retain)
EKEventStore          *eventStore;

@property (nonatomic, retain)
UIBarButtonItem       *editButton;

@property (nonatomic, retain)
UIBarButtonItem       *doneButton;

/* This method will display the Edit button on the top left
 corner of the navigation bar */
- (void) displayNormalNavigationItems;

/* This method will remove the edit button and will
 display a Done button on the top right corner of
 the navigation bar */
- (void) displayEditingNavigationItems;

@end
```

As you can see, we want to be able to retrieve all the events in the event store into an array named **arrayOfEvents**. After we retrieve the events, we also need to sort them by **startDate** so that the events at the bottom of the table view are the newest events (starting from oldest events to newest). Figure 13-5 depicts how our application will look once we are done implementing our root view controller.

 If you fetch events from an event store, add them to an array, and keep the array in memory but release your event store, the events in the array will point to empty values in memory and you will see empty values in all your **EKEvent** instances. This is a mistake that anybody could make since, as we know, when adding an object to an **NSArray** or **NSMutable Array** we are implicitly retaining that object. However, here the properties of each retained event get released once the event store that fetched them gets released. For this reason, please make sure the event store you use to fetch your events does not get released while you are still working with the events that are fetched by it.

If you retrieve the **startDate** property of an **EKEvent** instance, you will see the start date and time for that event. If the start date's hour, minute, and second values are all equal to zero, that event is an "All-Day" event.

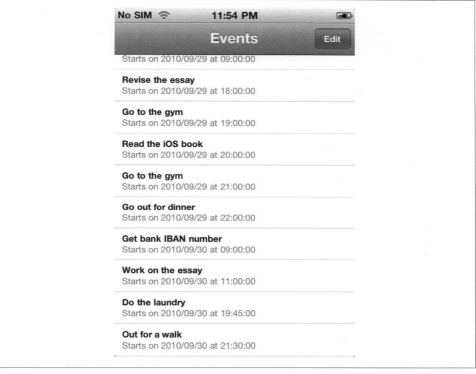

Figure 13-5. Displaying events from top to bottom, with the bottom items being the newest and the top items being the oldest

Here is how we will implement our view controller:

```
#import "RootViewController_iPhone.h"

@implementation RootViewController_iPhone

@synthesize tableViewEvents;
@synthesize arrayOfEvents;
@synthesize eventStore;
@synthesize editButton;
@synthesize doneButton;

- (void)              tableView:(UITableView*)tableView
    didEndEditingRowAtIndexPath:(NSIndexPath *)indexPath{

  if ([tableView isEqual:self.tableViewEvents] == YES){

    if (tableView.editing == NO){
      /* For swiping */
      [self displayNormalNavigationItems];
    }
```

```
    }

}

- (void)                     tableView:(UITableView*)tableView
 willBeginEditingRowAtIndexPath:(NSIndexPath *)indexPath{

  if ([tableView isEqual:self.tableViewEvents] == YES){

    if (tableView.editing == NO){
      /* For swiping */
      [self displayEditingNavigationItems];
    }

  }

}

- (void)  tableView:(UITableView *)tableView
 commitEditingStyle:(UITableViewCellEditingStyle)editingStyle
  forRowAtIndexPath:(NSIndexPath *)indexPath{

  /* Make sure this is a deletion operation on the events table
   view that we have */

  if ([tableView isEqual:self.tableViewEvents] == YES &&
      editingStyle == UITableViewCellEditingStyleDelete){

    /* We have the index path of the row so now let's
     get the corresponding event from the array of events
     that we have */
    EKEvent *event = [self.arrayOfEvents
                       objectAtIndex:indexPath.row];

    NSError *removeError = nil;

    /* Attempt to remove the event from the store */
    if ([self.eventStore removeEvent:event
                                span:EKSpanThisEvent
                               error:&removeError] == YES){

      [self.arrayOfEvents removeObject:event];

      NSArray *indexPathsToDelete =
      [NSArray arrayWithObject:indexPath];

      /* Delete the row for the current event with a nice animation */
      [tableView
       deleteRowsAtIndexPaths:indexPathsToDelete
       withRowAnimation:UITableViewRowAnimationFade];

    } else {
      NSLog(@"Failed to remove the event with error = %@",
            removeError);
    }
```

```
      }

}

- (UITableViewCellEditingStyle)tableView:(UITableView *)tableView
            editingStyleForRowAtIndexPath:(NSIndexPath *)indexPath{

  UITableViewCellEditingStyle result =
  UITableViewCellEditingStyleNone;

  if ([tableView isEqual:self.tableViewEvents] == YES){
    /* We want the user to be able only to delete events, not
     any other editing style */
    result = UITableViewCellEditingStyleDelete;
  }

  return(result);

}

- (NSInteger)tableView:(UITableView *)table
 numberOfRowsInSection:(NSInteger)section{

  NSInteger result = 0;

  if ([table isEqual:self.tableViewEvents] == YES){

    if (self.arrayOfEvents != nil){
      /* We only have one section and the number of
       rows in it is exactly equal to the number of
       events that are in our array */
      result = [self.arrayOfEvents count];
    }

  }

  return(result);

}

- (UITableViewCell *)tableView:(UITableView *)tableView
        cellForRowAtIndexPath:(NSIndexPath *)indexPath{

  UITableViewCell *result = nil;

  if ([tableView isEqual:self.tableViewEvents] == YES){

    static NSString *EventsCellIdentifier = @"Events";

    /* We have the index path so let's get the corresponding
     event from the array of events */
    EKEvent *event = [self.arrayOfEvents
                      objectAtIndex:indexPath.row];
```

```
/* Try to get a reusable table cell */
result =
[tableView dequeueReusableCellWithIdentifier:EventsCellIdentifier];

if (result == nil){
  result = [[[UITableViewCell alloc]
            initWithStyle:UITableViewCellStyleSubtitle
            reuseIdentifier:EventsCellIdentifier] autorelease];
}

/* The title text of the cell will be the title of the event */
result.textLabel.text = event.title;
result.textLabel.font = [UIFont boldSystemFontOfSize:12.0f];
result.detailTextLabel.font = [UIFont systemFontOfSize:12.0f];

/* Now let's format the date and the time of the event
 and display it as the subtitle of the cell */
NSCalendar *calendar = [NSCalendar currentCalendar];
NSDateComponents *components =
[calendar components:
 NSYearCalendarUnit |
 NSMonthCalendarUnit |
 NSDayCalendarUnit |
 NSHourCalendarUnit |
 NSMinuteCalendarUnit |
 NSSecondCalendarUnit
            fromDate:event.startDate];

if ([components hour] == 0 &&
    [components minute] == 0 &&
    [components second] == 0){

  result.detailTextLabel.text =
  [NSString stringWithFormat:
   @"Starts on %02ld/%02ld/%02ld All Day",
   (long)[components year],
   (long)[components month],
   (long)[components day]];

} else {
  result.detailTextLabel.text =
  [NSString stringWithFormat:
   @"Starts on %02ld/%02ld/%02ld at %02ld:%02ld:%02ld",
   (long)[components year],
   (long)[components month],
   (long)[components day],
   (long)[components hour],
   (long)[components minute],
   (long)[components second]];
}

}

return(result);
```

```
}

- (id)initWithNibName:(NSString *)nibNameOrNil
              bundle:(NSBundle *)nibBundleOrNil {

  self = [super initWithNibName:nibNameOrNil
                        bundle:nibBundleOrNil];

  if (self != nil){

  }

  return(self);
}

- (void) fetchAllEventsSinceLastYearUntilNow{

  NSAutoreleasePool *pool = [[NSAutoreleasePool alloc] init];

  self.arrayOfEvents = nil;

  /* Instantiate the event store */
  EKEventStore *newEventStore = [[EKEventStore alloc] init];
  self.eventStore = newEventStore;
  [newEventStore release];

  NSMutableArray *targetCalendars = [[NSMutableArray alloc] init];

  for (EKCalendar *thisCalendar in self.eventStore.calendars){
    if (thisCalendar.allowsContentModifications == YES){
      [targetCalendars addObject:thisCalendar];
    }
  }

  /* Start from a year ago (startDate) and
   end on today's date (endDate) */

  NSDate *endDate = [NSDate date];

  NSDate *startDate =
  [endDate dateByAddingTimeInterval:-(1 * 365 * 24 * 60 * 60)];

  /* Create the predicate that we can later pass to the
   event store in order to fetch the events */
  NSPredicate *searchPredicate =
  [self.eventStore predicateForEventsWithStartDate:startDate
                                         endDate:endDate
                                       calendars:targetCalendars];

  [targetCalendars release];

  /* Make sure we succeeded in creating the predicate */
  if (searchPredicate == nil){
    NSLog(@"Could not create the search predicate.");
    return;
```

```
    }

    /* Fetch all the events that fall between
     the starting and the ending dates */
    NSArray *allEvents =
    [self.eventStore eventsMatchingPredicate:searchPredicate];

    if (allEvents != nil){
      /* Sort our events by their start date */
      allEvents =
      [allEvents sortedArrayUsingSelector:
       @selector(compareStartDateWithEvent:)];

      self.arrayOfEvents = [NSMutableArray arrayWithArray:allEvents];
    }

    /* We are on a separate thread now. Make sure that we call
     the reloadData method of the table view on the main thread */
    if (self.arrayOfEvents != nil){

      [self.tableViewEvents
       performSelectorOnMainThread:@selector(reloadData)
       withObject:nil
       waitUntilDone:NO];

    }

    [pool release];

}

- (void)viewDidLoad {
  [super viewDidLoad];

  UIBarButtonItem *newDoneButton =
  [[UIBarButtonItem alloc]
   initWithBarButtonSystemItem:UIBarButtonSystemItemDone
   target:self
   action:@selector(finishEditing:)];
  self.doneButton = newDoneButton;
  [newDoneButton release];

  UIBarButtonItem *newEditButton =
  [[UIBarButtonItem alloc]
   initWithTitle:NSLocalizedString(@"Edit", nil)
   style:UIBarButtonItemStylePlain
   target:self
   action:@selector(startEditing:)];
  self.editButton = newEditButton;
  [newEditButton release];


  [self displayNormalNavigationItems];

  self.title = NSLocalizedString(@"Events", nil);
```

```objc
  SEL selectorToCall =
  @selector(fetchAllEventsSinceLastYearUntilNow);

  [self
   performSelectorInBackground:selectorToCall
   withObject:nil];

}

- (void)viewDidUnload {
  [super viewDidUnload];


  /* Release some memory in case of a low memory warning */
  self.tableViewEvents = nil;
  self.eventStore = nil;
  self.arrayOfEvents = nil;
  self.doneButton = nil;
  self.editButton = nil;

  self.navigationItem.rightBarButtonItem = nil;
  self.navigationItem.leftBarButtonItem = nil;

}

- (void) startEditing:(id)paramSender{

  if (self.tableViewEvents.editing == NO){

    /* Take the table view into editing mode */
    [self.tableViewEvents setEditing:YES
                            animated:YES];

    [self displayEditingNavigationItems];

  }



}

- (void) finishEditing:(id)paramSender{

  if (self.tableViewEvents.editing == YES){
    /* Take the table view out of the editing mode */
    [self.tableViewEvents setEditing:NO
                            animated:YES];

    [self displayNormalNavigationItems];
  }

}

- (void) displayNormalNavigationItems{
```

```
    /* And nothing for the right navigation bar button */
    [self.navigationItem setRightBarButtonItem:self.editButton
                                      animated:YES];

}

- (void) displayEditingNavigationItems{

    /* Nothing for the left button */

    [self.navigationItem setRightBarButtonItem:self.doneButton
                                      animated:YES];

}

- (BOOL)shouldAutorotateToInterfaceOrientation:
  (UIInterfaceOrientation)interfaceOrientation {
    /* Support all interface orientations */
    return (YES);
}

- (void)didReceiveMemoryWarning {
    [super didReceiveMemoryWarning];
}

- (void)dealloc {

    [tableViewEvents release];
    [arrayOfEvents release];
    [eventStore release];
    [editButton release];
    [doneButton release];

    [super dealloc];
}

@end
```

As you can see, we have implemented a method called fetchAllEventsSinceLastYear
UntilNow and are invoking it on a background thread as soon as the view of our view
controller loads. This method retrieves all the events in all modifiable calendars on an
iOS device from one year ago until today. Since this operation could take a long time,
we are using a background thread to load all the events into our arrayofEvents array.
If you take a closer look at the implementation of this method, we are also performing
a sort operation to make sure the oldest events appear at the top of the array and the
newest ones at the bottom. This is also a time-consuming operation, and the background thread will keep our main UI thread responsive while we do all these calculations. Once we are done, we have to reload our table view using reloadData, but we
need to make sure we invoke any UIKit methods on the main thread using the perform
SelectorOnMainThread:withObject:waitUntilDone: instance method of NSObject.

Figure 13-6 and Figure 13-7 depict how the editing mode in our events table view is implemented. We are doing all this using the `tableView:editingStyleForRowAtIndex Path:` and `tableView:commitEditingStyle:forRowAtIndexPath:` delegate and data source methods of the table view.

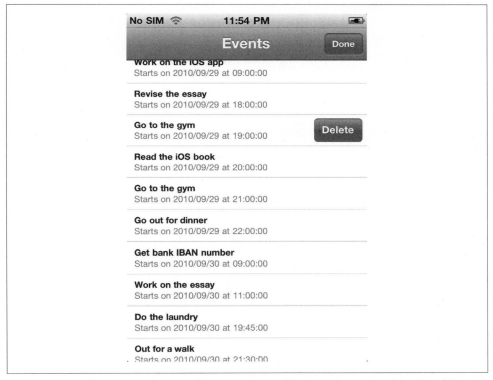

Figure 13-6. Implementing editing mode in our events table view; the user is allowed to delete any event just by swiping a finger on the cell that represents the event, from left to right or right to left

See Also

Recipe 13.1; Recipe 13.3

13.5 Adding Recurring Events to Calendars

Problem

You want to add a recurring event to a calendar.

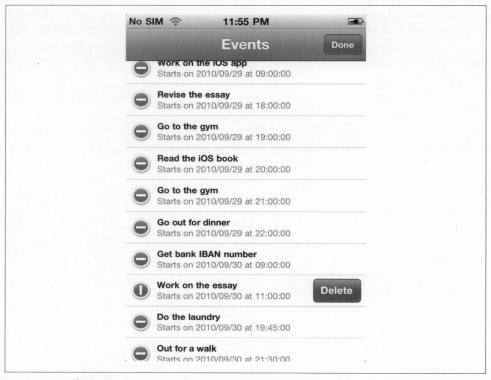

Figure 13-7. The Delete button, which the user can tap to delete any row; the user can also delete a row with a swipe gesture

Solution

The steps to create a recurring event follow. In this example, we are creating an event that occurs on the same day, every month, for an entire year:

1. Create an instance of `EKEventStore`.
2. Find a modifiable calendar inside the `calendars` array of the event store (for more information, refer to Recipe 13.1).
3. Create an object of type `EKEvent` (for more information, refer to Recipe 13.2).
4. Set the appropriate values for the event, such as its `startDate` and `endDate` (for more information, refer to Recipe 13.2).
5. Instantiate an object of type `NSDate` that contains the exact date when the recurrence of this event ends. In this example, this date is one year from today's date.
6. Use the `recurrenceEndWithEndDate:` class method of `EKRecurrenceEnd` and pass the `NSDate` you created in step 5 to create an object of type `EKRecurrenceEnd`.

7. Allocate and then instantiate an object of type `EKRecurrenceRule` using the `initRecurrenceWithFrequency:interval:end:` method of `EKRecurrenceRule`. Pass the recurrence end date that you created in step 6 to the `end` parameter of this method. For more information about this method, please refer to this recipe's Discussion.

8. Assign the recurring event that you created in step 7 to the `recurringRule` property of the `EKEvent` object that was created in step 3.

9. Invoke the `saveEvent:span:error:` instance method with the event (created in step 3) as the `saveEvent` parameter and the value `EKSpanFutureEvents` for the `span` parameter. This will create our recurring event for us.

Code illustrating these steps follows:

```objc
- (void)viewDidLoad {
  [super viewDidLoad];

  /* Step 1: And now the event store */
  EKEventStore *eventStore = [[EKEventStore alloc] init];

  /* Step 2: Find the first local calendar that is modifiable */
  EKCalendar *targetCalendar = nil;

  for (EKCalendar *thisCalendar in eventStore.calendars){
    if (thisCalendar.type == EKCalendarTypeLocal &&
        thisCalendar.allowsContentModifications == YES){
      targetCalendar = thisCalendar;
    }
  }

  /* The target calendar wasn't found? */
  if (targetCalendar == nil){
    NSLog(@"The target calendar is nil.");
    [eventStore release];
    return;
  }

  /* Step 3: Create an event */
  EKEvent *event = [EKEvent eventWithEventStore:eventStore];

  /* Step 4: Create an event that happens today and happens
   every month for a year from now */

  NSDate *eventStartDate = [NSDate date];

  /* Step 5: The event's end date is one hour
   from the moment it is created */
  NSDate *eventEndDate =
  [eventStartDate dateByAddingTimeInterval:1 * 60 * 60];

  /* Assign the required properties, especially
   the target calendar */
  event.calendar = targetCalendar;
  event.title = @"My Event";
```

```
event.startDate = eventStartDate;
event.endDate = eventEndDate;

/* The end date of the recurring rule
 is one year from now */
NSDate *oneYearFromNow =
[[NSDate date]
 dateByAddingTimeInterval:365 * 24 * 60 * 60];

/* Step 6: Create an Event Kit date from this date */
EKRecurrenceEnd *recurringEnd =
[EKRecurrenceEnd recurrenceEndWithEndDate:oneYearFromNow];

/* Step 7: And the recurring rule. This event happens every
 month (EKRecurrenceFrequencyMonthly), once a month (interval:1)
 and the recurring rule ends a year from now (end:RecurringEnd) */

EKRecurrenceRule *recurringRule =
[[EKRecurrenceRule alloc]
 initRecurrenceWithFrequency:EKRecurrenceFrequencyMonthly
 interval:1
 end:recurringEnd];

/* Step 8: Set the recurring rule for the event */
event.recurrenceRule = recurringRule;

[recurringRule release];

NSError *saveError = nil;

/* Step 9: Save the event */
if ([eventStore saveEvent:event
                      span:EKSpanFutureEvents
                     error:&saveError] == YES){
  NSLog(@"Successfully created the recurring event.");
} else {
  NSLog(@"Failed to create the recurring event %@", saveError);
}

[eventStore release];

}
```

Discussion

A recurring event is an event that happens more than once. We can create a recurring event just like a normal event. Please refer to Recipe 13.2 for more information about inserting normal events into the Calendar database. The only difference between a recurring event and a normal event is that you apply a recurring rule to a recurring event. A recurring rule tells the Event Kit framework how the event has to occur in the future.

We create a recurring rule by instantiating an object of type `EKRecurrenceRule` using the `initRecurrenceWithFrequency:interval:end:` initialization method. The parameters to this method are:

`initRecurrenceWithFrequency`

Specifies whether you want the event to be repeated daily (`EKRecurrenceFre quencyDaily`), weekly (`EKRecurrenceFrequencyWeekly`), monthly (`EKRecurrenceFre quencyMonthly`), or yearly (`EKRecurrenceFrequencyYearly`).

`interval`

A value greater than zero that specifies the interval between each occurrence's start and end period. For instance, if you want to create an event that happens every week, specify the `EKRecurrenceFrequencyWeekly` value with an `interval` of `1`. If you want this event to happen every other week, specify `EKRecurrenceFrequency Weekly` with an `interval` of `2`.

`end`

A date of type `EKRecurrenceEnd` that specifies the date when the recurring event ends in the specified calendar. This parameter is not the same as the event's end date (the `endDate` property of `EKEvent`). The end date of an event specifies when that specific event ends in the calendar, whereas the `end` parameter of the `initRecurrenceWithFrequency:interval:end:` method specifies the final occurrence of the event in the database.

Figure 13-8 depicts how our recurring event appears in the Calendar application.

Figure 13-8. The recurring event that was created as My Event, in the Calendar application

By editing this event (see Figure 13-9) in the Calendar application on an iOS device, you can see that the event is truly a recurring event that happens every month, on the same day the event was created, for a whole year, considering that this event was created at 10:40 a.m. on Saturday, July 31, 2010.

Figure 13-9. The Repeat and End Repeat fields in the Edit pane of the Calendar application on the recurring event that we created; these fields indicate how frequently our event occurs in the calendar

See Also

Recipe 13.2

13.6 Retrieving the Attendees of an Event

Problem

You want to retrieve the list of attendees for a specific event.

Solution

Use the `attendees` property of an instance of `EKEvent`. This property is of type `NSArray` and includes objects of type `EKParticipant`.

Example code follows that will retrieve all the events that happen on July 31 (the seventh month) of 2010, and will print out useful event information, including the attendees of that event, to the console window:

```objective-c
- (EKCalendar *)  getFirstAvailableCalDAVCalendar{

    EKCalendar *result = nil;

    EKEventStore *eventStore = [[EKEventStore alloc] init];

    for (EKCalendar *thisCalendar in eventStore.calendars){
      if (thisCalendar.type == EKCalendarTypeCalDAV){
        [eventStore release];
        return(thisCalendar);
      }
    }

    [eventStore release];
    return(result);

}

- (void)viewDidLoad {
  [super viewDidLoad];

  /* Find a calendar to base our search on */
  EKCalendar *targetCalendar =
  [self getFirstAvailableCalDAVCalendar];

  /* If we could not find a CalDAV calendar that
   we were looking for, then we will abort the operation */
  if (targetCalendar == nil){
    NSLog(@"No CalDAV calendars were found.");
    return;
  }

  /* We have to pass an array of calendars
   to the event store to search */
  NSArray *targetCalendars =
  [NSArray arrayWithObject:targetCalendar];

  /* Instantiate the event store */
  EKEventStore *eventStore = [[EKEventStore alloc] init];

  /* We use the Calendar object to
   construct a starting and ending date */
  NSCalendar *calendar = [NSCalendar currentCalendar];

  /* Using the components in this object
   we can construct a starting date */
  NSDateComponents *startDateComponents =
  [[NSDateComponents alloc] init];

  /* The start date will be 31st of July 2010 at 00:00:00 */
  [startDateComponents setHour:0];
  [startDateComponents setMinute:0];
  [startDateComponents setSecond:0];
  [startDateComponents setYear:2010];
  [startDateComponents setMonth:7];
```

```
[startDateComponents setDay:31];

/* Construct the starting date using the components */
NSDate *startDate =
[calendar dateFromComponents:startDateComponents];

/* The end date will be 1st of August 2010 at 00:00:00.
 Exactly 24 hours after the starting date */
NSDate *endDate =
[startDate dateByAddingTimeInterval:24 * 60 * 60];

/* We do not need these components anymore */
[startDateComponents release];

/* Create the predicate that we can later pass to
 the event store in order to fetch the events */
NSPredicate *searchPredicate =
[eventStore predicateForEventsWithStartDate:startDate
                                    endDate:endDate
                                  calendars:targetCalendars];

/* Make sure we succeeded in creating the predicate */
if (searchPredicate == nil){
  NSLog(@"Could not create the search predicate.");
  [eventStore release];
  return;
}

/* Fetch all the events that fall between the
 starting and the ending dates */
NSArray *events =
[eventStore eventsMatchingPredicate:searchPredicate];

/* Array of NSString equivalents of the values
 in the EKParticipantRole enumeration */
NSArray *attendeeRole = [NSArray arrayWithObjects:
                        @"Unknown",
                        @"Required",
                        @"Optional",
                        @"Chair",
                        @"Non Participant",
                        nil];

/* Array of NSString equivalents of the values
 in the EKParticipantStatus enumeration */
NSArray *attendeeStatus = [NSArray arrayWithObjects:
                        @"Unknown",
                        @"Pending",
                        @"Accepted",
                        @"Declined",
                        @"Tentative",
                        @"Delegated",
                        @"Completed",
                        @"In Process",
                        nil];
```

```objc
/* Array of NSString equivalents of the values
 in the EKParticipantType enumeration */
NSArray *attendeeType = [NSArray arrayWithObjects:
                         @"Unknown",
                         @"Person",
                         @"Room",
                         @"Resource",
                         @"Group",
                         nil];

/* Go through all the events and print their information
 out to the console */
if (events != nil){

  NSUInteger eventCounter = 0;
  for (EKEvent *thisEvent in events){

    eventCounter++;

    NSLog(@"Event %lu Start Date = %@",
          (unsigned long)eventCounter,
          thisEvent.startDate);

    NSLog(@"Event %lu End Date = %@",
          (unsigned long)eventCounter,
          thisEvent.endDate);

    NSLog(@"Event %lu Title = %@",
          (unsigned long)eventCounter,
          thisEvent.title);

    if (thisEvent.attendees == nil ||
        [thisEvent.attendees count] == 0){
      NSLog(@"Event %lu has no attendees",
            (unsigned long)eventCounter);
      continue;
    }

    NSUInteger attendeeCounter = 1;
    for (EKParticipant *participant in thisEvent.attendees){

      NSLog(@"Event %lu Attendee %lu Name = %@",
            (unsigned long)eventCounter,
            (unsigned long)attendeeCounter,
            participant.name);

      NSLog(@"Event %lu Attendee %lu Role = %@",
            (unsigned long)eventCounter,
            (unsigned long)attendeeCounter,
            [attendeeRole objectAtIndex:
             participant.participantRole]);

      NSLog(@"Event %lu Attendee %lu Status = %@",
            (unsigned long)eventCounter,
            (unsigned long)attendeeCounter,
```

```
                [attendeeStatus objectAtIndex:
                 participant.participantStatus]);

        NSLog(@"Event %lu Attendee %lu Type = %@",
              (unsigned long)eventCounter,
              (unsigned long)attendeeCounter,
              [attendeeType objectAtIndex:
               participant.participantType]);

        NSLog(@"Event %lu Attendee %lu URL = %@",
              (unsigned long)eventCounter,
              (unsigned long)attendeeCounter,
              participant.URL);

        attendeeCounter++;

      }

    } /* for (EKEvent *Event in Events){ */

  } else {
    NSLog(@"The array of events is nil.");
  }

  [eventStore release];

}
```

When we run this code on an iPhone 4 with a couple of events set up on its first CalDAV
calendar (refer to this chapter's Introduction for more information about CalDAV calendars and how you can set one up on your iOS device), we get results similar to these
in the console window:

```
Event 1 Start Date = 2010-07-31 10:00:00 +0100
Event 1 End Date = 2010-07-31 11:00:00 +0100
Event 1 Title = Write the essay...
Event 1 has no attendees
Event 2 Start Date = 2010-07-31 13:00:00 +0100
Event 2 End Date = 2010-07-31 14:00:00 +0100
Event 2 Title = Go to the gym.
Event 2 has no attendees
Event 3 Start Date = 2010-07-31 15:30:00 +0100
Event 3 End Date = 2010-07-31 16:30:00 +0100
Event 3 Title = Get Bank IBAN number.
Event 3 has no attendees
Event 4 Start Date = 2010-07-31 17:00:00 +0100
Event 4 End Date = 2010-07-31 18:00:00 +0100
Event 4 Title = Skype Call
Event 4 Attendee 1 Name = Vandad NP
Event 4 Attendee 1 Role = Required
Event 4 Attendee 1 Status = Accepted
Event 4 Attendee 1 Type = Person
Event 4 Attendee 1 URL = mailto:vandad@someserver.com
Event 4 Attendee 2 Name = brianjepson@someserver.com
Event 4 Attendee 2 Role = Required
```

```
Event 4 Attendee 2 Status = Pending
Event 4 Attendee 2 Type = Person
Event 4 Attendee 2 URL = mailto:brianjepson@someserver.com
Event 4 Attendee 3 Name = andyoram@someserver.com
Event 4 Attendee 3 Role = Required
Event 4 Attendee 3 Status = Pending
Event 4 Attendee 3 Type = Person
Event 4 Attendee 3 URL = mailto:andyoram@someserver.com
Event 5 Start Date = 2010-07-31 18:30:00 +0100
Event 5 End Date = 2010-07-31 19:30:00 +0100
Event 5 Title = Do the laundry.
Event 5 has no attendees
```

Discussion

Different types of calendars, such as CalDAV, can include participants in an event. iOS allows users to add participants to a calendar on the server, although not to the calendar on the iOS device. You can do this using Google Calendar, for instance.

Once the user adds participants to an event, you can use the `attendees` property of an instance of `EKEvent` to access the participant objects of type `EKParticipant`. Each participant has properties such as:

name
> This is the name of the participant. If you just specified the email address of a person to add him to an event, this field will be that email address.

URL
> This is usually the "mailto" URL for the attendee.

participantRole
> This is the role the attendee plays in the event. Different values that can be applied to this property are listed in the `EKParticipantRole` enumeration.

participantStatus
> This tells us whether this participant has accepted or declined the event request. This property could have other values, all specified in the `EKParticipantStatus` enumeration.

participantType
> This is of type `EKParticipantType`, which is an enumeration and, as its name implies, specifies the type of participant, such as group (`EKParticipantTypeGroup`) or individual person (`EKParticipantTypePerson`).

It is worth noting that events with attendees on calendars of types such as CalDAV are not editable on iOS, as shown in Figure 13-10.

See Also

Recipe 13.2; Recipe 13.3

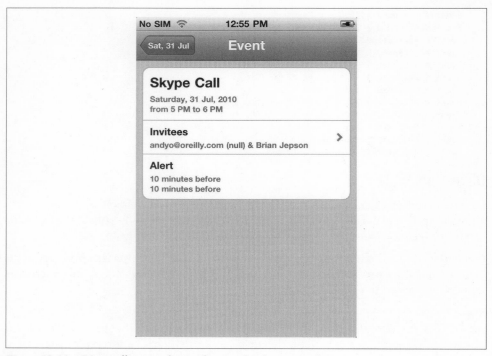

Figure 13-10. iOS not allowing editing of events that have attendees

13.7 Adding Alarms to Calendars

Problem

You want to add alarms to the events in a calendar.

Solution

Use the `alarmWithRelativeOffset:` class method of `EKAlarm` to create an instance of `EKAlarm`. Add the alarm to an event using the `addAlarm:` instance method of `EKEvent`, like so:

```
- (EKCalendar *)  getFirstModifiableLocalCalendar{

    EKCalendar *result = nil;

    EKEventStore *eventStore = [[EKEventStore alloc] init];

    for (EKCalendar *thisCalendar in eventStore.calendars){
      if (thisCalendar.type == EKCalendarTypeLocal &&
          thisCalendar.allowsContentModifications == YES){
        [eventStore release];
        return(thisCalendar);
      }
```

```
        }

    [eventStore release];
    return(result);

}

- (void)viewDidLoad {
    [super viewDidLoad];

    EKCalendar *targetCalendar =
    [self getFirstModifiableLocalCalendar];

    if (targetCalendar == nil){
        NSLog(@"Could not find the target calendar.");
        return;
    }

    EKEventStore *eventStore = [[EKEventStore alloc] init];

    /* The event starts 60 seconds from now */
    NSDate *startDate =
    [NSDate dateWithTimeIntervalSinceNow:60.0f];

    /* And end the event 20 seconds after its start date */
    NSDate *endDate =
    [startDate dateByAddingTimeInterval:20.0f];

    EKEvent *eventWithAlarm =
    [EKEvent eventWithEventStore:eventStore];

    eventWithAlarm.calendar = targetCalendar;
    eventWithAlarm.startDate = startDate;
    eventWithAlarm.endDate = endDate;

    /* The alarm goes off 2 seconds before the event happens */
    EKAlarm *alarm = [EKAlarm alarmWithRelativeOffset:-2.0f];

    eventWithAlarm.title = @"Event with Alarm";
    [eventWithAlarm addAlarm:alarm];

    NSError *saveError = nil;

    if ([eventStore saveEvent:eventWithAlarm
                          span:EKSpanThisEvent
                         error:&saveError] == YES){
        NSLog(@"Saved an event that fires 60 seconds from now.");
    } else {
        NSLog(@"Failed to save the event. Error = %@", saveError);
    }

    [eventStore release];

}
```

Discussion

An event of type `EKEvent` can have multiple alarms. Simply create the alarm using either the `alarmWithAbsoluteDate:` or `alarmWithRelativeOffset:` class method of `EKAlarm`. The former method requires an absolute date and time (you can use the `CFAbsoluteTimeGet Current` function to get the current absolute time), whereas the latter method requires a number of seconds relative to the start date of the event when the alarm must be fired. For instance, if the event is scheduled for August 1, 2010 at 6:00 a.m., and we go ahead and create an alarm with the relative offset of -60 (which is counted in units of seconds), our alarm will be fired at 5:59 a.m. the same day. Only zero and negative numbers are allowed for this offset. Positive numbers will be changed to zero automatically by iOS. Once an alarm is fired, iOS will display the alarm to the user, as shown in Figure 13-11.

Figure 13-11. iOS displaying an alert on the screen when an alarm is fired

You can use the `removeAlarm:` instance method of `EKEvent` to remove an alarm associated with that event instance.

See Also

Recipe 13.1

13.8 Handling Event Changed Notifications

Problem

You want to get notified in your application when the user changes the contents of the Calendar database.

Solution

Register for the EKEventStoreChangedNotification notification:

```objc
- (void) handleEventStoreChanged:(NSNotification *)paramNotification{

    NSLog(@"The calendar database has been changed.");

    NSLog(@"Object = %@", [paramNotification object]);

    if (self.tableViewEvents.editing == YES){
        [self finishEditing:nil];
    }

    NSString *alertTitle =
    NSLocalizedString(@"Calendar Database Has Been Changed", nil);

    NSString *alertMessage =
    NSLocalizedString(@"Please Press OK To Refresh", nil);

    UIAlertView *alertView =
    [[[UIAlertView alloc]
        initWithTitle:alertTitle
        message:alertMessage
        delegate:self
        cancelButtonTitle:NSLocalizedString(@"OK", nil)
        otherButtonTitles:nil, nil] autorelease];

    alertView.tag = REFRESH_ARRAY_ALERTVIEW_ID;

    [alertView show];

}

- (void)viewDidLoad {
    [super viewDidLoad];

    UIBarButtonItem *newEditButton =
    [[UIBarButtonItem alloc]
        initWithTitle:NSLocalizedString(@"Edit", nil)
        style:UIBarButtonItemStylePlain
        target:self
        action:@selector(startEditing:)];
    self.editButton = newEditButton;
    [newEditButton release];

    UIBarButtonItem *newDoneButton =
```

```
      [[UIBarButtonItem alloc]
       initWithBarButtonSystemItem:UIBarButtonSystemItemDone
       target:self
       action:@selector(finishEditing:)];
      self.doneButton = newDoneButton;
      [newDoneButton release];

      [[NSNotificationCenter defaultCenter]
       addObserver:self
       selector:@selector(handleEventStoreChanged:)
       name:EKEventStoreChangedNotification
       object:nil];

      [self displayNormalNavigationItems];

      self.title = NSLocalizedString(@"Events", nil);

      SEL selectorToCall =
      @selector(fetchAllEventsSinceLastYearUntilNow);

      [self
       performSelectorInBackground:selectorToCall
       withObject:nil];

    }

  - (void)viewDidUnload {
      [super viewDidUnload];

      [[NSNotificationCenter defaultCenter]
       removeObserver:self
       name:EKEventStoreChangedNotification
       object:nil];

      self.tableViewEvents = nil;
      self.arrayOfEvents = nil;
      self.eventStore = nil;
      self.editButton = nil;
      self.doneButton = nil;

      self.navigationItem.rightBarButtonItem = nil;
      self.navigationItem.leftBarButtonItem = nil;

    }
```

This recipe uses the same code that we implemented in Recipe 13.4. For more information about the details of this code, and especially of rendering the events into a table view, please refer to Recipe 13.4.

Discussion

Multitasking is possible on iOS 4 and later. Imagine you have fetched a series of events from EKEventStore into an array and you allow your user to work with them (edit them, add to them, and remove from them). The user could simply switch from your application to the Calendar application and delete the same event she is trying to delete in your application. On iOS 4 and later, such a sequence of activities will generate an EKEventStoreChangedNotification notification that you can choose to receive.

The EKEventStoreChangedNotification notification will be sent to your application (only if you subscribe to this notification) even if your application is in the foreground. Because of this, you must make sure you treat this notification differently depending on whether your application is in the background or the foreground. Here are a couple of things to consider:

- If you receive the EKEventStoreChangedNotification notification while your application is in the foreground, it is best to implement a mechanism to find out whether the changes to the event store originated inside your own application or came from someone else outside the application. If they came from outside the application, you must make sure you are retaining the latest version of the events in the store, and not the old events. If for any reason you copied one of the events in the event store and kept the copy somewhere, you must call the refresh instance method of that event of type EKEvent. If the return value of this method is YES, you can keep the object in memory. If the return value is NO, you must dispose of the object, because someone outside your application has deleted the event.

- If you receive the EKEventStoreChangedNotification notification while your application is in the background, according to documentation from Apple, your application should not attempt to do any GUI-related processing and should, in fact, use as little processing power as possible. You must therefore refrain from adding new screens to, or modifying in any way, the GUI of your application.

- If you receive the EKEventStoreChangedNotification notification while your application is in the background, you must make note of it inside the application (perhaps store this in a property of type BOOL) and react to this change when the application is brought to the foreground again. Normally, if you receive any notification about a change to an event while you are in the background, you should retrieve all events stored in the application when you return to the foreground.

The objects attached to this notification are of type EKEventStore. Here is an example of the data printed to the console window using our EKEventStoreChangedNotification handler method when the user deleted an event using the Calendar application and came back to our application:

```
The calendar database has been changed.
Object = <EKEventStore: 0x138d00>
The calendar database has been changed.
Object = <EKEventStore: 0x12d010>
The calendar database has been changed.
Object = <EKEventStore: 0x12d010>
The calendar database has been changed.
Object = <EKEventStore: 0x138d00>
The calendar database has been changed.
Object = <EKEventStore: 0x138d00>
```

Coalescing is not enabled on the `EKEventStoreChangedNotification` event store notification. In other words, you can receive multiple notifications of the same type if a single event changes in the Calendar database. It is up to you to determine how and when you need to refetch your retained events.

13.9 Presenting Event View Controllers

Problem

You want to use the built-in iOS SDK view controllers to display the properties of an event in the Calendar database.

Solution

Create an instance of `EKEventViewController` and push it into a navigation controller:

```
/* Construct the event view controller */
EKEventViewController *controller =
[[EKEventViewController alloc] init];

/* Make sure to set the event that has to be viewed */
controller.event = event;

/* Do not allow the user to edit this event */
controller.allowsEditing = NO;
controller.allowsCalendarPreview = YES;
```

The Event object in the code is of type `EKEvent`. For more information about this object and how we retrieved it, please refer to this recipe's Discussion.

Discussion

Users of iOS devices are already familiar with the interface they see on the Calendar application. When they select an event, they can see that event's properties and they might be allowed to modify the event. To present a view to a user using built-in iOS SDK event view controllers, we can instantiate an object of type `EKEventViewController` and assign an event of type `EKEvent` to its event property. Once that's done, we can push the event view controller into our navigation controller and let iOS take care of the rest.

Let's go ahead and write an application that demonstrates how this feature works. We want to be able to fetch all events in an event store (of type `EKEventStore`) from this date last year up to now and grab the last event we find (the newest event). We will then use `EKEventViewController` to present that event to the user. Here is the *.h* file of our view controller:

```
#import <UIKit/UIKit.h>
```

```objc
#import <EventKit/EventKit.h>
#import <EventKitUI/EventKitUI.h>

@interface RootViewController_iPhone : UIViewController {
@public
  EKEventStore   *eventStore;
}

@property (nonatomic, retain) EKEventStore  *eventStore;

@end
```

Here is the implementation of the same view controller (.*m* file):

```objc
#import "RootViewController_iPhone.h"

@implementation RootViewController_iPhone

@synthesize eventStore;

- (id)initWithNibName:(NSString *)nibNameOrNil
               bundle:(NSBundle *)nibBundleOrNil {

  self = [super initWithNibName:nibNameOrNil
                         bundle:nibBundleOrNil];

  if (self != nil){

  }

  return(self);

}

- (void) pushController:(UIViewController *)paramController{

  [self.navigationController
   pushViewController:paramController
   animated:YES];

}

- (void) displayLastEvent{

  NSAutoreleasePool *pool = [[NSAutoreleasePool alloc] init];

  NSDate *startDate =
  [NSDate
   dateWithTimeIntervalSinceNow:-(365 * 24 * 60 * 60)];

  NSDate *endDate = [NSDate date];

  NSPredicate *predicate =
  [self.eventStore
   predicateForEventsWithStartDate:startDate
   endDate:endDate
```

```
   calendars:self.eventStore.calendars];

NSArray *events =
[self.eventStore eventsMatchingPredicate:predicate];

if (events == nil ||
    [events count] == 0){
  NSLog(@"No events were found.");
} else {

  /* Get the newest events at the end of the array */
  SEL sortingSelector = @selector(compareStartDateWithEvent:);
  events =
  [events
   sortedArrayUsingSelector:sortingSelector];

  /* And pick the last event (The newest) */
  EKEvent *event = [events lastObject];

  /* Construct the event view controller */
  EKEventViewController *controller =
  [[EKEventViewController alloc] init];

  /* Make sure to set the event that has to be viewed */
  controller.event = event;

  /* Do not allow the user to edit this event */
  controller.allowsEditing = NO;
  controller.allowsCalendarPreview = YES;

  /* And make sure pushing the view controller happens on the
   main UI thread */
  [self performSelectorOnMainThread:@selector(pushController:)
                         withObject:controller
                      waitUntilDone:NO];

  [controller release];

}

[pool release];

}

- (void)viewDidLoad {
  [super viewDidLoad];

  EKEventStore *newEventStore =
  [[EKEventStore alloc] init];

  self.eventStore = newEventStore;
  [newEventStore release];

  [self
   performSelectorInBackground:@selector(displayLastEvent)
```

```
        withObject:nil];

}

- (void)viewDidUnload {
  [super viewDidUnload];

  self.eventStore = nil;

}

- (BOOL)shouldAutorotateToInterfaceOrientation:
  (UIInterfaceOrientation)interfaceOrientation {
  // Return YES for supported orientations
  return (YES);
}

- (void)didReceiveMemoryWarning {
  // Releases the view if it doesn't have a superview.
  [super didReceiveMemoryWarning];
}

- (void)dealloc {
  [eventStore release];
  [super dealloc];
}

@end
```

Running this example requires that you have at least one event since today's date a year ago in one of the calendars available on your iOS device. Once we run this application on an iPhone 4 as an example, we can see the built-in event view controller displaying the contents of the event that we have found (see Figure 13-12).

Different properties of an instance of `EKEventViewController` that we can use to change the behavior of this object are as follows:

`allowsEditing`
> If this property's value is set to `YES`, the Edit button will appear on the navigation bar of the event view controller, allowing the user to edit the event. This happens only on modifiable calendars and only for events that have been created by the user on this device. For instance, if you create an event on the Web using Google Calendar and the event appears in your iOS device, you are not allowed to edit that event.

`allowsCalendarPreview`
> If this property's value is set to `YES` and the event the user is viewing is an invitation, the user will be given the option to view this current event in a calendar with other events that have been scheduled on the same date.

event

This property must be set before presenting the event view controller. This will be the event that the event view controller will display to the user.

Figure 13-12. The built-in event view controller displaying the contents of an event

Please refer to Recipe 2.10 for more information about pushing view controllers into the screen. When you push the event view controller, the Back button will appear with the title "Back" by default, so you do not have to change it manually. However, if you decide to change the Back button, you can do so by assigning a new object of type UIBarButtonItem to the backBarButtonItem property of your navigation item. In our example code, we can modify the pushController: method to give our root view controller a custom Back button before pushing the event view controller, like so:

```
- (void) pushController:(UIViewController *)paramController{

    UIBarButtonItem *backButton =
    [[UIBarButtonItem alloc]
     initWithTitle:NSLocalizedString(@"Go Back", nil)
     style:UIBarButtonItemStylePlain
     target:nil
     action:nil];
```

```
    self.navigationItem.backBarButtonItem = backButton;

    [backButton release];

    [self.navigationController
     pushViewController:paramController
     animated:YES];

}
```

The results of this modification are depicted in Figure 13-13 (please note that in this example, editing is enabled for the event view controller).

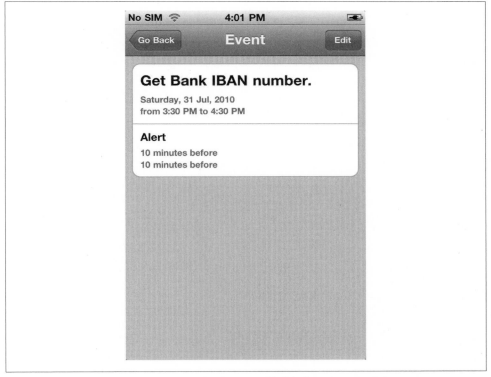

Figure 13-13. An edit view controller with editing enabled and a custom Back button

See Also

Recipe 13.10

13.10 Presenting Event Edit View Controllers

Problem

You want to allow your users to edit (insert, delete, and modify) events in the Calendar database from inside your application, using built-in SDK view controllers.

Solution

Instantiate an object of type EKEventEditViewController and present it on a navigation controller using the presentModalViewController:animated: instance method of UINavigationController. Here is the *.h* file of our view controller:

```
#import <UIKit/UIKit.h>

#import <EventKit/EventKit.h>
#import <EventKitUI/EventKitUI.h>

@interface RootViewController_iPhone : UIViewController
                                    <EKEventEditViewDelegate> {
@public
  EKEventStore   *eventStore;
}

@property (nonatomic, retain) EKEventStore  *eventStore;

@end
```

The implementation of the same view controller is as follows (*.m* file):

```
#import "RootViewController_iPhone.h"

@implementation RootViewController_iPhone

@synthesize eventStore;

- (id)initWithNibName:(NSString *)nibNameOrNil
              bundle:(NSBundle *)nibBundleOrNil {

  self = [super initWithNibName:nibNameOrNil
                        bundle:nibBundleOrNil];

  if (self != nil){

  }

  return(self);

}

- (void)eventEditViewController:(EKEventEditViewController *)controller
        didCompleteWithAction:(EKEventEditViewAction)action{

  switch (action){
```

```objc
      case EKEventEditViewActionCanceled:{
        /* Handle this case here */
        break;
      }
      case EKEventEditViewActionSaved:{
        /* Handle this case here */
        break;
      }
      case EKEventEditViewActionDeleted:{
        /* Handle this case here */
        break;
      }
    }

    [controller dismissModalViewControllerAnimated:YES];

}

- (void) presentController:(UIViewController *)paramController{

    [self.navigationController
     presentModalViewController:paramController
     animated:YES];

}

- (void) displayLastEvent{

    NSAutoreleasePool *pool = [[NSAutoreleasePool alloc] init];

    NSDate *startDate =
    [NSDate
     dateWithTimeIntervalSinceNow:-(365 * 24 * 60 * 60)];

    NSDate *endDate = [NSDate date];

    NSPredicate *predicate =
    [self.eventStore
     predicateForEventsWithStartDate:startDate
     endDate:endDate
     calendars:self.eventStore.calendars];

    NSArray *events =
    [self.eventStore eventsMatchingPredicate:predicate];

    if (events == nil ||
        [events count] == 0){
      NSLog(@"No events were found.");
    } else {

      /* Get the newest events at the end of the array */
      SEL sortingSelector = @selector(compareStartDateWithEvent:);
      events =
      [events
       sortedArrayUsingSelector:sortingSelector];
```

```objc
    /* And pick the last event (The newest) */
    EKEvent *event = [events lastObject];

    EKEventEditViewController *controller =
    [[EKEventEditViewController alloc] init];

    /* Make sure to set the event that has to be viewed.
     If we set this property to nil, the view controller
     allows us to add an event to the system */
    controller.event = event;

    controller.eventStore = self.eventStore;
    controller.editViewDelegate = self;

    /* And make sure pushing the view controller happens on the
     main UI thread */
    [self
    performSelectorOnMainThread:@selector(presentController:)
    withObject:controller
    waitUntilDone:NO];

    [controller release];

  }

  [pool release];

}

- (void)viewDidLoad {
  [super viewDidLoad];

  EKEventStore *newEventStore =
  [[EKEventStore alloc] init];

  self.eventStore = newEventStore;
  [newEventStore release];

  SEL selectorToCall = @selector(displayLastEvent);
  [self performSelectorInBackground:selectorToCall
                         withObject:nil];

}

- (void)viewDidUnload {
  [super viewDidUnload];

  self.eventStore = nil;
}

- (BOOL)shouldAutorotateToInterfaceOrientation:
  (UIInterfaceOrientation)interfaceOrientation {
  // Return YES for supported orientations
  return (YES);
```

```
}

- (void)didReceiveMemoryWarning {
  // Releases the view if it doesn't have a superview.
  [super didReceiveMemoryWarning];
}

- (void)dealloc {
  [eventStore release];
  [super dealloc];
}

@end
```

Discussion

An instance of the EKEventEditViewController class allows us to present an event edit view controller to the user. This view controller, depending on how we set it up, can allow the user to either edit an existing event or create a new event. If you want this view controller to edit an event, set the **event** property of this instance to an event object. If you want the user to be able to insert a new event into the system, set the **event** property of this instance to nil.

The editViewDelegate property of an instance of EKEventEditViewController is the object that will receive delegate messages from this view controller telling the programmer about the action the user has taken. One of the most important delegate messages your delegate object must handle (a required delegate selector) is the eventEditViewController:didCompleteWithAction: method. This delegate method will be called whenever the user dismisses the event edit view controller in one of the possible ways indicated by the didCompleteWithAction parameter. This parameter can have values such as the following:

EKEventEditViewActionCanceled
    The user pressed the Cancel button on the view controller.

EKEventEditViewActionSaved
    The user saved (added/modified) an event in the Calendar database.

EKEventEditViewActionDeleted
    The user deleted an event from the Calendar database.

Please make sure to dismiss the event edit view controller after receiving this delegate message.

See Also

Recipe 13.9

CHAPTER 14

Graphics

14.0 Introduction

The iOS SDK allows programmers to work with Quartz Core, which is encapsulated in the Core Graphics framework. In this chapter, we will use the Core Graphics framework extensively, so please make sure you have this framework added to your Xcode project by following these steps:

1. Find the Frameworks item in your Xcode project hierarchy and right-click on it.
2. Choose Add→Existing Frameworks.
3. Hold down the Command key and select the CoreGraphics.framework and QuartzCore.framework frameworks.
4. Click the Add button.

In this chapter, we will be using functions and methods available in the Core Graphics and Quartz Core frameworks. Quartz 2D is the engine in the Core Graphics framework that allows us to draw sophisticated shapes, paths, images, and so on.

Make sure you import both frameworks into your source files whenever needed, like so:

```
#import <QuartzCore/QuartzCore.h>
#import <CoreGraphics/CoreGraphics.h>
```

Almost all the recipes in this chapter make use of the `drawRect:` method of an instance of `UIView`. This method gets called whenever a view has to be drawn. The only parameter to this method is of type `CGRect`, which tells you the rectangular area where you are supposed to be doing your painting. Please refrain from painting anything that is off this rectangle, as this will decrease the performance of your application.

To be able to use the code samples in these recipes, make sure you subclass a `UIView` object by following these steps:

1. In Xcode, choose File→New File.
2. Choose Cocoa Touch Classes as the template.

3. Choose the Objective-C class as the object type and choose `UIView` in the "Subclass of" drop-down menu. Click Next.

4. Give a name to your file and make sure the "Also Create '*foo*.h' file" option is selected (where *foo* is the name you've chosen for your file).

5. Click the Finish button.

6. Open the XIB file for your view controller and select the view object for your view controller. Change the class name of this view object to *foo*, where *foo* is what you chose as the name of your new object in step 4.

Many programmers who have worked with graphics are used to measuring pixels when drawing shapes and paths on a context or a canvas. In iOS, we will be working with logical points instead. One way to explain this is if you draw a rectangle from point (0, 0) to (320, 480) on an iPhone 3GS (whose screen resolution is 320 × 480), the whole screen will be covered by the rectangle. If you draw the exact same rectangle on an iPhone 4 (whose screen resolution is 640 × 960), you will see the exact same effect: the whole screen will be covered by your rectangle, even though you drew a rectangle 320 points in width and 480 points in height. This is because iOS works with logical points, and every coordinate value you use in Quartz Core when drawing shapes, paths, and images on a context is specified in logical points. The "Points Versus Pixels" section of "iOS Application Programming Guide" on Apple's website explains this concisely. Here is an important extract from this guide:

> In iPhone OS there is a distinction between the coordinates you specify in your drawing code and the pixels of the underlying device. When using native drawing technologies such as Quartz, UIKit, and Core Animation, you specify coordinate values using a logical coordinate space, which measures distances in points. This logical coordinate system is decoupled from the device coordinate space used by the system frameworks to manage the pixels on the screen. The system automatically maps points in the logical coordinate space to pixels in the device coordinate space, but this mapping is not always one-to-one. This behavior leads to an important fact that you should always remember: **One point does not necessarily correspond to one pixel on the screen**.[*]

14.1 Drawing Basic Shapes on a Graphics Context

Problem

You want to be able to draw shapes such as an ellipse or a rectangle on a graphics context inside an instance of `UIView`.

---

[*] *http://developer.apple.com/library/ios/#documentation/WindowsViews/Conceptual/ViewPG_iPhoneOS/ WindowsandViews/WindowsandViews.html*

Solution

Use Core Graphics functions such as `CGContextFillRect` and `CGContextFillEllipseIn Rect`. We are using the `drawRect:` instance method of `UIView` to do the paintings on the view. The only parameter to this method is of type `CGRect` and it specifies that the rectangle has to be repainted. You can use this rectangle to determine whether the object you are attempting to paint is on or off the screen. Painting objects that are not visible on the screen consumes CPU power and energy and will result in fewer frames per second (fps).

```
- (void)drawRect:(CGRect)rect {

    /* Get the current context */
    CGContextRef context = UIGraphicsGetCurrentContext();

    /* The points of the rectangle that needs to be drawn */

    CGRect drawingRect = CGRectMake(0.0,     /* X */
                                    0.0f,    /* Y */
                                    100.0f,  /* Width */
                                    200.0f); /* Height */

    /* Get the red color */
    const CGFloat *rectColorComponents =
    CGColorGetComponents([UIColor redColor].CGColor);

    /* Draw with red fill color */
    CGContextSetFillColor(context, rectColorComponents);

    /* Now draw the rectangle */
    CGContextFillRect(context, drawingRect);

    /* The rectangular space in which the ellipse has to be drawn */
    CGRect ellipseRect = CGRectMake(160.0f,  /* X */
                                    200.0f,  /* Y */
                                    150.0f,  /* Width */
                                    200.0f); /* Height */

    /* The blue color's components */
    const CGFloat *ellipseColorComponents =
    CGColorGetComponents([UIColor blueColor].CGColor);

    /* Set the blue color as the current fill color */
    CGContextSetFillColor(context, ellipseColorComponents);

    /* And finally draw the ellipse */
    CGContextFillEllipseInRect(context, ellipseRect);

}
```

Figure 14-1 depicts this code running on an iPhone 4 device with the Retina Display.

Figure 14-1. Drawing a rectangle and an ellipse on a view

Discussion

After we have the handle to a graphics context, we can use different Core Graphics methods to draw basic shapes. Two of the most important methods that we can use are:

`CGContextFillRect`
> Draws a rectangle. The rectangle's color will reflect the fill color you previously specified.

`CGContextFillEllipseInRect`
> Draws an ellipse.

As you can see from the code, both methods draw a shape using the current fill color on the destination graphics context. The destination graphics context specifies a variety of parameters about the shapes you draw, such as the color and the border. The graphics context's fill color can be changed using the `CGContextSetFillColor` method, which takes two parameters:

Context

> This is the destination graphics context where the fill color must be set. This takes a value of type `CGContextRef`. We can use the `UIGraphicsGetCurrentContext` function to retrieve the current graphics context.

FillColor

> This is an array of values that specify the intensity of each color component that makes up the final color value, plus the alpha. All values are in the range +0.0 to +1.0. If you are more familiar with RGBA values of 0 to 255, with 255 being the most intense, you can simply specify your RGBA values in the range +0.0 to +255.0 and then divide the final result by +255.0.

The fill color (second parameter) of the `CGContextSetFillColor` method can easily be constructed using `UIColor` and the `CGColorGetComponents` function. To construct the fill color for the `CGContextSetFillColor` method or any other method that requires an array of values representing the colors, using `UIColor`, simply follow these steps:

1. Construct your `UIColor` from one of the many class methods of `UIColor`, such as `redColor` or `blueColor`, or use the `colorWithRed:green:blue:alpha:` class method to specify RGBA values.

2. Pass the `CGColor` property of the instance of `UIColor` that you constructed in step 1 to the `CGColorGetComponents` function. The return value will be a `const CGFloat *`.

See Also

"Getting Started with Graphics and Animation," *http://developer.apple.com/library/ios/ #referencelibrary/GettingStarted/GS_Graphics_iPhone/*

14.2 Drawing Paths on a Graphics Context

Problem

You want to draw on a graphics context by connecting lines, points, rectangles, or other shapes to create your final shape.

Solution

Use graphics context paths:

1. Retrieve the graphics context using `UIGraphicsGetCurrentContext` or a similar function.

2. Invoke the `CGContextBeginPath` method to create an empty path on a graphics context.

3. Create your path using different functions that we will discuss in this recipe's Discussion.

4. At the end, invoke the `CGContextClosePath` method to close the path that you started in step 2.

Discussion

A *path* is a combination of different shapes on a graphics context. A line is a path, a rectangle is a path, and the combination of the two can also be a path. Think of a path as a line that a person follows to get to a destination. You start a path by calling the `CGContextBeginPath` method, and once you have reached the destination where you want to conclude the path, you need to call `CGContextClosePath`. Let's have a look at an example:

```
- (void)drawRect:(CGRect)rect {

  CGPoint trianglePoints[3] = {
    CGPointMake(100.0f, 100.0f),
    CGPointMake(200.0f, 200.0f),
    CGPointMake(0.0f, 200.0),
  };

  /* Get the current context */
  CGContextRef context = UIGraphicsGetCurrentContext();

  CGContextBeginPath(context);

  /* Move the initial point on the path to (100, 100) */
  CGContextMoveToPoint(context,
                       trianglePoints[0].x,
                       trianglePoints[0].y);

  /* Move from (100, 100) to (200, 200) and draw a line */
  CGContextAddLineToPoint(context,
                          trianglePoints[1].x,
                          trianglePoints[1].y);

  /* From point (200, 200), draw a line to (0, 200) */
  CGContextAddLineToPoint(context,
                          trianglePoints[2].x,
                          trianglePoints[2].y);

  /* Close the path */
  CGContextClosePath(context);

  /* Set the fill color to red */
  CGContextSetFillColorWithColor(context,
                                 [UIColor redColor].CGColor);

  /* And fill the path with red */
  CGContextFillPath(context);

}
```

Before we delve into details about the code, we must first see how the output of this code looks when we run it on an iOS device (see Figure 14-2).

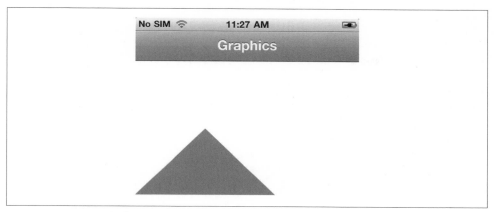

Figure 14-2. A triangle drawn on the view using paths

To get a better idea of how paths work, let's have a look at another example:

```
- (void) addFilledRectangleToPathOnContext:(CGContextRef)paramContext{

    /* The points of the rectangle that needs to be drawn */

    CGRect drawingRect = CGRectMake(0.0,      /* X */
                                    0.0f,     /* Y */
                                    100.0f,   /* Width */
                                    200.0f);  /* Height */

    UIColor *halfTransparentGreenColor =
    [UIColor colorWithRed:0.0f
                    green:1.0f
                     blue:0.0f
                    alpha:0.5f];

    const CGFloat *rectangleFillColor =
    CGColorGetComponents(halfTransparentGreenColor.CGColor);

    CGContextSetFillColor(paramContext, rectangleFillColor);

    CGContextAddRect(paramContext, drawingRect);

    CGContextFillPath(paramContext);

}

- (void) addStrokedEllipseToPathOnContext:(CGContextRef)paramContext{

    CGRect ellipseRect = CGContextGetClipBoundingBox(paramContext);

    CGContextAddEllipseInRect(paramContext, ellipseRect);
```

```
    CGContextSetStrokeColorWithColor(paramContext,
                                     [UIColor brownColor].CGColor);

    CGContextStrokePath(paramContext);

}

- (void)drawRect:(CGRect)rect {

    /* Get the current context */
    CGContextRef context = UIGraphicsGetCurrentContext();

    CGContextBeginPath(context);

    /* Add a filled rectangle to the current path */
    [self addFilledRectangleToPathOnContext:context];

    /* Add a stroked ellipse to the current path */
    [self addStrokedEllipseToPathOnContext:context];

    CGContextClosePath(context);

}
```

The output is shown in Figure 14-3.

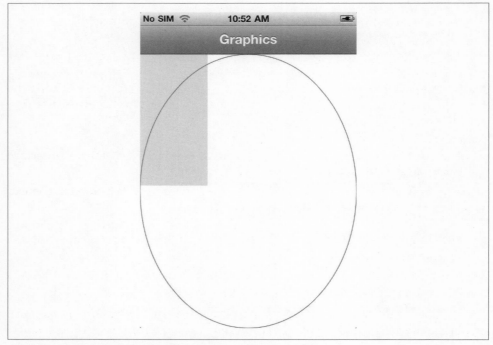

Figure 14-3. A rectangle and an ellipse drawn on a view using paths

There are a lot of Core Graphics methods and functions that you can use to work with paths. Discussing all these functions is impossible in this book, but I would like to list some of the most useful Core Graphics methods that allow you to work with paths:

CGContextBeginPath
> This creates an empty path on a specified graphics context.

CGContextMoveToPoint
> This moves the current virtual drawing point to the point you specify by the X and Y values. The next line segment drawn in the path will start from the point set by this method.

CGContextAddLineToPoint
> This takes the current point as the starting point and the X and Y values that you specify as the ending point, and draws a line segment between the two points. The specified X and Y values will become the new current point on the path.

CGContextSetFillColorWithColor
> This sets the current fill color on a given graphics context to a CGColorRef that you specify. You can use the CGColor property of an instance of UIColor to create a color of your choice.

CGContextSetStrokeColorWithColor
> This sets the stroke color on a given graphics context (the brown ellipse in Figure 14-3 is a stroked shape). The parameters to this method are similar to the parameters provided to the CGContextSetFillColorWithColor method.

CGPathCreateMutable
> This creates a new mutable path object of type CGMutablePathRef. You must then use CGPath methods to draw on the path. Once the path is constructed, use the CGContextAddPath method to add the path to the target graphics context. You must then use the CGPathRelease method to release the path you created with the CGMutablePathRef method.

CGContextAddRect
> This adds a rectangle to the current path on the graphics context that you specify.

CGContextAddEllipseInRect
> This adds an ellipse, inside a rectangle, to the current path on a graphics context.

To draw a path, you can either:

- Use the graphics context's path methods and functions, such as CGContextAddLineToPoint.
- Create a separate path object using CGPathCreateMutable, and add the CGPath methods such as CGPathAddLineToPoint to draw your path.

Once you are done drawing the path, you can add that path to the current path on a graphics context using the CGContextAddPath method. If we were to rewrite the example code whose output was depicted in Figure 14-2 to create a separate path and draw onto it, we could write the code in this way:

```
- (void)drawRect:(CGRect)rect {

    CGPoint trianglePoints[3] = {
      CGPointMake(100.0f, 100.0f),
      CGPointMake(200.0f, 200.0f),
      CGPointMake(0.0f, 200.0),
    };

    /* Get the current context */
    CGContextRef context = UIGraphicsGetCurrentContext();

    CGMutablePathRef path = CGPathCreateMutable();

    /* Move the initial point on the path to (100, 100) */
    CGPathMoveToPoint(path,
                      nil,
                      trianglePoints[0].x,
                      trianglePoints[0].y);

    /* Move from (100, 100) to (200, 200) and draw a line */
    CGPathAddLineToPoint(path,
                         nil,
                         trianglePoints[1].x,
                         trianglePoints[1].y);

    /* From point (200, 200), draw a line to (0, 200) */
    CGPathAddLineToPoint(path,
                         nil,
                         trianglePoints[2].x,
                         trianglePoints[2].y);

    /* Set the fill color to red */
    CGContextSetFillColorWithColor(context,
                                   [UIColor redColor].CGColor);


    /* Create a new path on the context */
    CGContextBeginPath(context);
    /* Add our path to the current path on the context */
    CGContextAddPath(context, path);
    /* Close the path */
    CGContextClosePath(context);
    /* And fill the path with red */
    CGContextFillPath(context);

    /* And make sure that we release the mutable path
     object that we created earlier */
    CGPathRelease(path);

}
```

The output of this code is exactly the same as what we saw in Figure 14-2.

14.3 Drawing Images on a Graphics Context

Problem

You have loaded an image into an instance of UIImage and you want to draw that image on a graphics context.

Solution

Use one of these methods:

- The drawInRect: instance method of UIImage, which draws the image on the current graphics context
- The drawInRect:blendMode:alpha: instance method of UIImage, which draws the image on the current graphics context and additionally specifies the blending mode and alpha transparency of the image

For sample code that demonstrates use of both methods, please refer to this recipe's Discussion.

Discussion

To demonstrate both methods mentioned here, we will draw an image named *Avatar.png* in our application bundle with both methods. The result contains two drawings on the current graphics context, as shown in Figure 14-4. The top-left drawing comes from the drawInRect: instance method of UIImage and the bottom-left drawing comes from the drawInRect:blendMode:alpha: instance method of UIImage.

The code for Figure 14-4 is as follows:

```
- (void) simpleUIImageDrawInRect:(CGRect)paramRect{

    UIImage *myImage = [UIImage imageNamed:@"Avatar.png"];
    [myImage drawInRect:paramRect];

}

- (void) complexUIImageDrawInRect:(CGRect)paramRect{

    UIImage *myImage = [UIImage imageNamed:@"Avatar.png"];
    [myImage drawInRect:paramRect
             blendMode:kCGBlendModeDarken
                 alpha:1.0f];

}

- (void) drawBackgroundImageInRect:(CGRect)paramRect{

    UIImage *backgroundImage =
    [UIImage imageNamed:@"Checkerboard.png"];
```

```
    [backgroundImage drawInRect:paramRect];

}

- (void)drawRect:(CGRect)rect {

    [self drawBackgroundImageInRect:rect];

    /* The rectangle in which the first image will be drawn */
    CGRect destinationRect = CGRectMake(0.0f,      /* X */
                                        0.0f,      /* Y */
                                        128.0f,    /* W */
                                        128.0f);   /* H */

    [self simpleUIImageDrawInRect:destinationRect];

    /* For the second image, shift the destination rectangle down */
    destinationRect.origin.y += destinationRect.size.height;

    [self complexUIImageDrawInRect:destinationRect];

}
```

Figure 14-4. Two methods used to draw the same image

The `drawInRect:blendMode:alpha:` instance method of `UIImage` accepts many different types of blending modes that you can play with. All these values are declared inside the `CGBlendMode` enumeration.

See Also

"Quartz 2D Programming Guide," *http://developer.apple.com/library/ios/#documenta tion/GraphicsImaging/Conceptual/drawingwithquartz2d/Introduction/Introduction .html*

14.4 Capturing the Screen Contents into an Image

Problem

You want to capture the graphical representation of the contents of the screen into an instance of `UIImage`.

Solution

Follow these steps:

1. Make sure you have added the Quartz Core framework to your project.
2. Import the Quartz Core header files.
3. Create a new bitmap context using the `UIGraphicsBeginImageContext` method.
4. Retrieve the newly created context using `UIGraphicsGetCurrentContext`. The return value will be of type `CGContextRef`.
5. Find the `UIView` that you want to take a screenshot of. This could be the `view` property of an instance of `UIViewController`. Use the `renderInContext:` instance method of the `layer` property of the instance of `UIView` to render the contents of the layer into the context you created in step 3.
6. Retrieve the image from the current context (created in step 3) using the `UIGraphicsGetImageFromCurrentImageContext` function. The return value of this function is of type `UIImage`.
7. End the image context created in step 3 using the `UIGraphicsEndImageContext` method.

Here is sample code in the *.m* file of a view controller that takes a screenshot of the view object and stores it in an instance of `UIImage`:

```
- (IBAction) performTakeScreenShot:(id)paramSender{

    /* Create a new bitmap context. This context becomes the
     current graphics context */
    UIGraphicsBeginImageContext(self.view.bounds.size);

    /* Retrieve the handle to the new context that we created */
```

```
        CGContextRef context = UIGraphicsGetCurrentContext();

        /* Render the contents of the view of this View Controller
         into the context we created */
        [self.view.layer renderInContext:context];

        /* Construct an image out of the current context's contents. */
        UIImage *image = UIGraphicsGetImageFromCurrentImageContext();
        UIGraphicsEndImageContext();

        /* Now push a new View Controller into the stack where
         we will display the image to the user */
        SecondViewController *controller =
        [[SecondViewController alloc]
         initWithNibName:@"SecondViewController"
         bundle:nil];

        [self.navigationController pushViewController:controller
                                            animated:YES];

        /* Set the image inside the image view of the second
         View Controller so that the user can see the screenshot */
        controller.imageViewShot.image = image;

        [controller release];

    }
```

Please bear in mind that the size of the resultant image is equal to the size of the layer that was drawn into the context. This size is measured in logical points and not actual pixels, as explained in this chapter's Introduction. If you print out (to the console screen) the size of the image of type UIImage in this example code on an iPhone 4 with the Retina Display (640 × 960) on a view controller without the status bar and the navigation bar, the resultant size will be 320 × 480. This is simply because the UIImage object is using logical points instead of actual screen pixels for its size property. For more information about logical points versus pixels, please refer to this chapter's Introduction.

Discussion

Using the renderInContext: instance method of CALayer, you can draw the contents of a layer inside a context object of type CGContextRef. If you want to draw the contents of a layer into a new context, you can use the UIGraphicsBeginImageContext method to create a new context. All you have to do is to pass the size of the new context to the UIGraphicsBeginImageContext method. After calling this method, your new context will become the current graphics context until you call the UIGraphicsEndImageContext method.

After creating a new bitmap context, you can retrieve its handle using the UIGraphics
GetCurrentContext function. After this, rendering the contents of a CALayer instance to
this context is as easy as calling the renderInContext: instance method of CALayer and
passing the handle of the context we just created.

After the layer is done drawing itself to the new context, you can call the UIGraphics
GetImageFromCurrentImageContext function to retrieve the UIImage representation
of the current graphics context. Once you are done with the image context, call the
UIGraphicsEndImageContext method to pop the context out of the stack.

To demonstrate our code in this recipe's Solution, let's create two view controllers with
their corresponding XIB files. We'll call the first one our root view controller and the
second one our second view controller. We will then declare (.h file) the root view
controller like so:

```
#import <UIKit/UIKit.h>
#import <QuartzCore/QuartzCore.h>

@interface RootViewController_iPhone : UIViewController {

}

- (IBAction) performTakeScreenShot:(id)paramSender;

@end
```

The implementation (.m file) of the same view controller is also very simple:

```
#import "RootViewController_iPhone.h"
#import "SecondViewController.h"

@implementation RootViewController_iPhone

- (id)initWithNibName:(NSString *)nibNameOrNil
              bundle:(NSBundle *)nibBundleOrNil {

  self = [super initWithNibName:nibNameOrNil
                        bundle:nibBundleOrNil];

  if (self != nil){

  }

  return(self);

}

- (IBAction) performTakeScreenShot:(id)paramSender{

  /* Create a new bitmap context. This context becomes the
   current graphics context */
  UIGraphicsBeginImageContext(self.view.bounds.size);

  /* Retrieve the handle to the new context that we created */
```

```
    CGContextRef context = UIGraphicsGetCurrentContext();

    /* Render the contents of the view of this View Controller
     into the context we created */
    [self.view.layer renderInContext:context];

    /* Construct an image out of the current context's contents. */
    UIImage *image = UIGraphicsGetImageFromCurrentImageContext();
    UIGraphicsEndImageContext();

    /* Now push a new View Controller into the stack where
     we will display the image to the user */
    SecondViewController *controller =
    [[SecondViewController alloc]
     initWithNibName:@"SecondViewController"
     bundle:nil];

    [self.navigationController pushViewController:controller
                                        animated:YES];

    /* Set the image inside the image view of the second
     View Controller so that the user can see the screenshot */
    controller.imageViewShot.image = image;

    [controller release];

}

- (BOOL)shouldAutorotateToInterfaceOrientation:
  (UIInterfaceOrientation)interfaceOrientation {
 return (YES);
}

- (void)didReceiveMemoryWarning {
  [super didReceiveMemoryWarning];
}

- (void)viewDidUnload {
  [super viewDidUnload];
}

- (void)dealloc {
  [super dealloc];
}

@end
```

As you can see from the implementation, we must put a button on the XIB file of our view controller and connect its TouchUpInside method to the performTakeScreenShot: IBAction we defined in the .h file of our view controller. You can design this view controller as simply as the one shown in Figure 14-5.

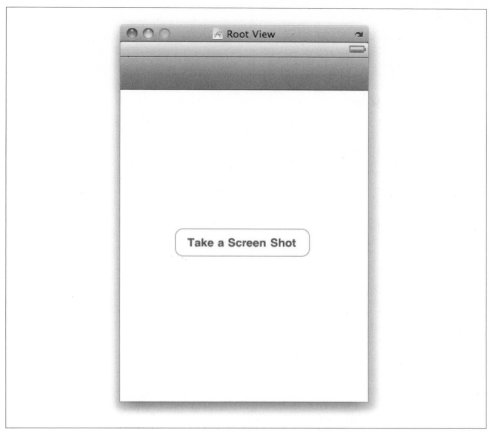

Figure 14-5. A very simple view controller with a button on it that is connected to the performTakeScreenShot: IBAction defined in the .h file of the view controller; this method is responsible for taking a screenshot and displaying it in another view controller

Our second view controller has only one important feature, which is its `imageView Shot` property of type `UIImageView`:

```
#import <UIKit/UIKit.h>

@interface SecondViewController : UIViewController {
@public
  UIImageView     *imageViewShot;
}

@property (nonatomic, retain) IBOutlet UIImageView  *imageViewShot;

@end
```

This image view's image will be changed by the root view controller when the user presses the Take a Screen Shot button shown in Figure 14-5. Pressing the button will capture the contents of the current view controller view's layer and display the captured image in the second view controller, as shown in Figure 14-6.

Figure 14-6. The contents of the root view controller's view captured into an instance of UIImage and displayed on a second view controller; you can see that the button was tapped at the time the contents of the view's layer were saved

14.5 Drawing Text with Core Graphics

Problem

You want to use Core Graphics to draw text at a specific place on your context. This is useful because Core Graphics gives you more options when drawing text, compared to Cocoa Touch `NSString` methods.

Solution

Use the `CGContextShowTextAtPoint` method to draw a C String at a specific place (specified by points) on a context:

```
- (void)drawRect:(CGRect)rect {

    /* Get the current context */
    CGContextRef context = UIGraphicsGetCurrentContext();

    /* Save the state of the context */
    CGContextSaveGState(context);

    /* This is the font we want to use */
    UIFont *systemFont = [UIFont systemFontOfSize:16.0f];

    /* Get the name of the font */
    const char *fontNameAsCString =
    [systemFont.fontName UTF8String];

    /* Using the font name, select the font into the context */
    CGContextSelectFont(context,
                        fontNameAsCString,
                        16.0f,
                        kCGEncodingMacRoman);

    /* Here is what we want to draw */
    const char *textToDraw = "Hello, World!";

    /* Now translate the context's coordinate and then
     rotate the context around the x-axis (-1 for y)
     to access the Cartesian coordinate system */
    CGContextTranslateCTM(context,
                          0.0f,
                          self.bounds.size.height);

    CGContextScaleCTM(context,
                      1.0,
                      -1.0f);

    /* Now draw at X = 200 and Y = 100 */
    CGContextShowTextAtPoint(context,
                             200.0f,
                             100.0,
                             textToDraw,
                             strlen(textToDraw));

    /* Restore the state of the context */
    CGContextRestoreGState(context);

}
```

The results are depicted in Figure 14-7.

Figure 14-7. Text drawn on the current graphics context using the Cartesian coordinate system

Discussion

To be able to draw text using Core Graphics, we must first select our font on the context that will display the text. We can select the font using the `CGContextSelectFont` method, which accepts four parameters:

`DestinationContext`
A context object of type `CGContextRef` on which the font will be selected.

`FontName`
The name of the font as a C String.

`FontSize`
The font size as a floating-point value.

`FontEncoding`
The encoding for the font. You can choose either `kCGEncodingMacRoman` or `kCGEncodingFontSpecific` for this parameter.

After selecting the font, we can start drawing it on a context using the `CGContextShow TextAtPoint` method. Please bear in mind that the coordinate system maintained by

Core Graphics might not be the same coordinate system being used by the views and Cocoa Touch in general. Cocoa Touch uses the window coordinate system, where the top-left corner of the screen is (0, 0). Using the `CGContextTranslateCTM` and `CGContext ScaleCTM` methods, as demonstrated in the code, we can modify the default coordinate system used by Core Graphics and change it to the Cartesian coordinate system. The bottom-lefthand corner of the screen will then become the point (0, 0). You can do the appropriate mapping of the screen coordinate system to the Cartesian coordinate system with simple calculations. For more information about these coordinate systems, please refer to the "View Geometry and Coordinate Systems" section of the "View Programming Guide for iOS" at:

http://developer.apple.com/library/ios/documentation/WindowsViews/Conceptual/ ViewPG_iPhoneOS/WindowsandViews/WindowsandViews.html#//apple_ref/doc/ uid/TP40009503-CH2-SW5

CHAPTER 15

Core Motion

15.0 Introduction

iOS devices such as the iPhone and iPad are usually equipped with accelerometer hardware. Starting with the iPhone 4, some iOS devices might also include a gyroscope. Before attempting to use either the accelerometer or the gyroscope in your iOS applications, you must check the availability of these devices on the iPhone or iPad. Recipe 15.1 and Recipe 15.2 include techniques you can use to detect the availability of the accelerometer and gyroscope. With a gyroscope, iOS devices such as the iPhone 4 are able to detect motion in six axes.

Let's set up a simple example you can try out to see the value of the gyroscope. The accelerometer cannot detect the rotation of the device around its vertical axis if you are holding the device perfectly still in your hands, sitting in a computer chair, and rotating your chair in a clockwise or counterclockwise fashion. From the standpoint of the floor or the Earth, the device is rotating around the vertical axis, but it's not rotating around its own y-axis, which is the vertical center of the device. So, the accelerometer does not detect any motion.

However, the gyroscope included with iOS devices starting with the iPhone 4 allows us to detect such movements. This allows more fluid and flawless movement detection routines. This is typically useful in games, where the developers need to know not only whether the device is moving on the x-, y-, and z-axes—information they can get from the accelerometer—but also whether it is changing in relation to the Earth along these directions, which requires a gyroscope.

Prior to the release of iOS SDK 4, accelerometer data could be retrieved using the `sharedAccelerometer` class method of `UIAccelerometer`. After the shared instance of the accelerometer object is retrieved, you can set an update rate and a delegate for it. Delegate messages indicating motion on the device will then be delivered to the delegate object.

With iOS SDK 4, programmers can use the Core Motion framework to access both the accelerometer and the gyroscope data (if available). All recipes in this chapter make use of the Core Motion framework. Please follow these steps to add this framework to your project:

1. Find Frameworks in your Xcode project hierarchy and right-click on it.
2. Choose Add→Existing Frameworks.
3. Choose CoreMotion.framework from the list and click Add (see Figure 15-1).

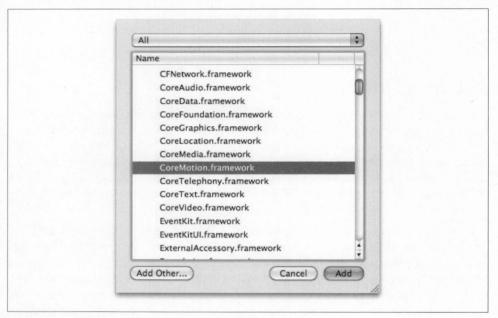

Figure 15-1. Adding the Core Motion framework to our project

iPhone Simulator does not simulate the accelerometer or the gyroscope hardware. However, you can generate a *shake* with iPhone Simulator using Hardware→Shake Gesture (see Figure 15-2).

15.1 Detecting the Availability of an Accelerometer

Problem

In your program, you want to detect whether the accelerometer hardware is available.

Figure 15-2. The Shake Gesture option in iPhone Simulator's Hardware menu

Solution

Use the `isAccelerometerAvailable` method of `CMMotionManager` to detect the accelerometer hardware. The `isAccelerometerActive` method can also be used to detect
whether the accelerometer hardware is currently sending updates to the program.

```
- (id)initWithNibName:(NSString *)nibNameOrNil
              bundle:(NSBundle *)nibBundleOrNil {

  self = [super initWithNibName:nibNameOrNil
                         bundle:nibBundleOrNil];

  if (self != nil){

    CMMotionManager *newMotionManager =
    [[CMMotionManager alloc] init];

    motionManager = [newMotionManager retain];
    [newMotionManager release];

    NSString *accelerometerAvailability = @"is not";

    if ([motionManager isAccelerometerAvailable] == YES){
      accelerometerAvailability = @"is";
    }

    NSLog(@"Accelerometer %@ available",
          accelerometerAvailability);

    NSString *accelerometerIsOrIsNotActive = @"is not";

    if ([motionManager isAccelerometerActive] == YES){
      accelerometerIsOrIsNotActive = @"is";
    }
```

```
    NSLog(@"Accelerometer %@ active",
        accelerometerIsOrIsNotActive);

}

return(self);

}
```

 In our example code, motionManager is the property (of type CMMotion
Manager) of an instance of UIViewController.

Accelerometer hardware might be available on the iOS device that is running your
program. This, however, does not mean the accelerometer hardware is sending updates
to your program. If the accelerometer or gyroscope is sending updates to your program,
we say it is *active*, which requires you to define a delegate object, as we will soon see.

If you run this code on iPhone Simulator, you will get values similar to these in the
console window:

```
Accelerometer is not available
Accelerometer is not active
```

Running the same code on an iPhone 4 device, you will get values similar to these:

```
Accelerometer is available
Accelerometer is not active
```

Discussion

An iOS device could have a built-in accelerometer. As we don't know yet which iOS
devices might have accelerometer hardware built into them and which ones won't, it
is best to test the availability of the accelerometer before using it.

You can detect the availability of this hardware by instantiating an object of type
CMMotionManager and accessing its isAccelerometerAvailable method. This method is
of type BOOL and returns YES if the accelerometer hardware is available and NO if not.

In addition, you can detect whether the accelerometer hardware is currently sending
updates to your application (whether it is active) by issuing the isAccelerometerAc
tive method of CMMotionManager. You will learn about retrieving accelerometer data in
Recipe 15.3.

See Also

Recipe 15.3

15.2 Detecting the Availability of a Gyroscope

Problem

You want to find out whether the current iOS device that is running your program has gyroscope hardware available.

Solution

Use the `isGyroAvailable` method of an instance of `CMMotionManager` to detect the gyroscope hardware. The `isGyroActive` method is also available if you want to detect whether the gyroscope hardware is currently sending updates to your program (in other words, whether it is active).

```
- (id)initWithNibName:(NSString *)nibNameOrNil
              bundle:(NSBundle *)nibBundleOrNil {

    self = [super initWithNibName:nibNameOrNil
                      bundle:nibBundleOrNil];

    if (self != nil){

        CMMotionManager *newMotionManager =
        [[CMMotionManager alloc] init];

        self.motionManager = newMotionManager;
        [newMotionManager release];

        NSString *gyroscropeAvailability = @"is not";

        if ([motionManager isGyroAvailable] == YES){
          gyroscropeAvailability = @"is";
        }

        NSLog(@"Gyroscope %@ available",
              gyroscropeAvailability);

        NSString *gyroIsOrIsNotActive = @"is not";

        if ([motionManager isGyroActive] == YES){
          gyroIsOrIsNotActive = @"is";
        }

        NSLog(@"Gyroscope %@ active",
              gyroIsOrIsNotActive);


    }
    return self;
}
```

In our example code, `motionManager` is the property (of type `CMMotion Manager`) of an instance of `UIViewController`, just as in Recipe 15.1.

iPhone Simulator does not have gyroscope simulation in place. If you run this code on the simulator, you will receive results similar to these in the console window:

```
Gyroscope is not available
Gyroscope is not active
```

If you run the same code on an iPhone 3GS, you will get the same results as you get from iPhone Simulator. However, if you run this code on an iOS device with a gyroscope, such as the iPhone 4, the results could be different:

```
Gyroscope is available
Gyroscope is not active
```

Discussion

If you plan to release an application that makes use of the gyroscope, you must make sure other iOS devices without this hardware can run your application. For instance, if you compile your application with the target deployment OS of iOS 4.0, an iPhone 3GS with iOS 4.0 could run your application. If you are using the gyroscope as part of a game, for instance, you must make sure other iOS devices that are capable of running your application can play the game, although they might not have a gyroscope installed.

To achieve this, you must first instantiate an object of type `CMMotionManager`. After this, you must access the `isGyroAvailable` method (of type `BOOL`) and see whether the gyroscope is available on the device running your code. You can also use the `isGyroActive` method of the `CMMotionManager` instance to find out whether the gyroscope is currently sending your application any updates. For more information about this, please refer to Recipe 15.5.

See Also

Recipe 15.5

15.3 Retrieving Accelerometer Data

Problem

You want to ask iOS to send accelerometer data to your application.

Solution

Use the `startAccelerometerUpdatesToQueue:withHandler:` instance method of `CMMotion Manager`. Here is the *.h* file of a view controller that utilizes `CMMotionManager` to get accelerometer updates:

```
#import <UIKit/UIKit.h>
#import <CoreMotion/CoreMotion.h>

@interface RootViewController : UIViewController {
@public
  CMMotionManager *motionManager;
}

@property (nonatomic, retain) CMMotionManager *motionManager;

@end
```

The implementation of the same view controller (*.m* file) is as follows:

```
#import "RootViewController.h"

@implementation RootViewController

@synthesize motionManager;

- (id)initWithNibName:(NSString *)nibNameOrNil
               bundle:(NSBundle *)nibBundleOrNil {

  self = [super initWithNibName:nibNameOrNil
                         bundle:nibBundleOrNil];
  if (self != nil){

    /* Allocate and initialize the motion manager here */
    CMMotionManager *newMotionManager =
    [[CMMotionManager alloc] init];

    motionManager = [newMotionManager retain];
    [newMotionManager release];

  }

  return(self);

}

void (^accelerometerHandler)(CMAccelerometerData *, NSError *) =
    ^(CMAccelerometerData *accelerometerData, NSError *error){

  NSLog(@"X = %.04f, Y = %.04f, Z = %.04f",
        accelerometerData.acceleration.x,
        accelerometerData.acceleration.y,
        accelerometerData.acceleration.z);

};
```

```
- (void)viewDidLoad {
  [super viewDidLoad];

  /* Is the accelerometer available? */
  if ([self.motionManager isAccelerometerAvailable] == YES){

    /* Start the accelerometer if it is not active already */
    if ([self.motionManager isAccelerometerActive] == NO){
      /* Update us twice a second */
      [self.motionManager
        setAccelerometerUpdateInterval:1.0f / 2.0f];
      /* And on a handler block object */
      [self.motionManager
        startAccelerometerUpdatesToQueue:[NSOperationQueue mainQueue]
        withHandler:accelerometerHandler];
    }
  }

}


- (void)viewDidUnload {
  [super viewDidUnload];

  if ([self.motionManager isAccelerometerAvailable] == YES &&
      [self.motionManager isAccelerometerActive] == YES){
    [self.motionManager stopAccelerometerUpdates];
  }

}

- (BOOL)shouldAutorotateToInterfaceOrientation:
  (UIInterfaceOrientation)interfaceOrientation {
  return (YES);
}

- (void)didReceiveMemoryWarning {
  [super didReceiveMemoryWarning];
}

- (void)dealloc {

  [motionManager release];

  [super dealloc];
}

@end
```

Discussion

The accelerometer reports three-dimensional data (three axes) that iOS reports to your program as *x*, *y*, and *z* values. These values are encapsulated in a `CMAcceleration` structure:

```
typedef struct {
    double x;
    double y;
    double z;
} CMAcceleration __OSX_AVAILABLE_STARTING(__MAC_NA,__IPHONE_4_0);
```

If you hold your iOS device in front of your face with the screen facing you in portrait mode:

- The x-axis runs from left to right at the horizontal center of the device with values ranging from −1 to +1 from left to right.
- The y-axis runs from bottom to top at the vertical center of the device with values ranging from −1 to +1 from bottom to top.
- The z-axis runs from the back of the device, through the device, toward you, with values ranging from −1 to +1 from back through to the front.

The best way to understand the values reported from the accelerometer hardware is by taking a look at a few examples. Here is one: let's assume you have your iOS device facing you with the bottom of the device pointing to the ground and the top pointing up. If you hold it perfectly still without tilting it in any specific direction, the values you have for the x-, y-, and z-axes at this moment will be (x: 0.0, y: -1.0, z: 0.0). Now try the following while the screen is facing you and the bottom of the device is pointing to the ground:

1. Turn the device 90 degrees clockwise. The values you have at this moment are (x: +1.0, y: 0.0, z: 0.0).
2. Turn the device a further 90 degrees clockwise. Now the top of the device must be pointing to the ground. The values you have at this moment are (x: 0.0, y: +1.0, z: 0.0).
3. Turn the device a further 90 degrees clockwise. Now the top of the device must be pointing to the left side. The values you have at this moment are (x: -1.0, y: 0.0, z: 0.0).
4. Finally, if you rotate the device a further 90 degrees clockwise, where the top of the device once again points to the sky and the bottom of the device points to the ground, the values will be as they were originally (x: 0.0, y: -1.0, z: 0.0).

So, from these values, we can conclude that rotating the device around the z-axis changes the x and y values reported by the accelerometer, but not the z value.

Let's conduct another experiment. Hold the device again while it's facing you with its bottom pointing to the ground and its top pointing to the sky. The values that a program will get from the accelerometer, as you already know, are (x: 0.0, y: -1.0, z: 0.0). Now try these movements:

1. Tilt the device backward 90 degrees around the x-axis so that its top will be pointing backward. In other words, hold it as though it is sitting faceup on a table. The values you get at this moment will be (x: 0.0, y: 0.0, z: -1.0).

2. Now tilt the device backward 90 degrees again so that its back is facing you, its top is facing the ground, and its bottom is facing the sky. The values you get at this moment will be (x: 0.0, y: 1.0, z: 0.0).

3. Tilt the device backward 90 degrees so that it's facing the ground with its back facing the sky and its top pointing toward you. The reported values at this moment will be (x: 0.0, y: 0.0, z: 1.0).

4. And finally, if you tilt the device one more time in the same direction, so the device is facing you and its top is facing the sky, the values you get will be the same values you started with.

Therefore, we can observe that rotating the device around the x-axis changes the values of the y- and z-axes, but not x. I encourage you to try the third type of rotation—around the y-axis (pointing from top to bottom)—and observe the changes in the values reported for the x- and the z-axes.

To be able to receive accelerometer updates, you have two options:

1. The `startAccelerometerUpdatesToQueue:withHandler:` instance method of `CMMotionManager`

   This method will deliver accelerometer updates on an operation queue (of type `NSOperationQueue`) and will require a basic knowledge of blocks that are used extensively in Grand Central Dispatch (GCD). For more information about blocks, please refer to "Blocks Programming Topics" on the Apple Developer website (*http://developer.apple.com/library/ios/#documentation/Cocoa/Conceptual/Blocks/Articles/00_Introduction.html*).

2. The `startAccelerometerUpdates` instance method of `CMMotionManager`

   Once you call this method, the accelerometer (if available) will start updating accelerometer data in the motion manager object. You need to set up your own thread to continuously read the value of the `accelerometerData` property (of type `CMAccelerometerData`) of `CMMotionManager`.

In this recipe, we are using the first method (with blocks). I highly recommend that you read Apple's Developer website documentation about blocks. The block we provide to the `startAccelerometerUpdatesToQueue:withHandler:` instance method of `CMMotionManager` must be of type `CMAccelerometerHandler`:

```
typedef void (^CMAccelerometerHandler)(CMAccelerometerData *accelerometerData,
NSError *error)
```

In other words, we must accept two parameters on the block. The first one must be of type CMAccelerometerData and the second must be of type NSError, as implemented in our example code:

```
void (^accelerometerHandler)(CMAccelerometerData *, NSError *) =
    ^(CMAccelerometerData *accelerometerData, NSError *error){

    NSLog(@"X = %.04f, Y = %.04f, Z = %.04f",
        accelerometerData.acceleration.x,
        accelerometerData.acceleration.y,
        accelerometerData.acceleration.z);

};
```

If you use the startAccelerometerUpdatesToQueue:withHandler: instance method of CMMotionManager and provide a handle to a block you have defined somewhere else in your implementation, you cannot refer to self or objects that you have defined in the current context of your code. To circumvent this limitation, you can use an inline block, like so:

```
- (void)viewDidLoad {
    [super viewDidLoad];



    /* Is the accelerometer available? */
    if ([self.motionManager isAccelerometerAvailable] == YES){

        /* Start the accelerometer if it is not active already */
        if ([self.motionManager isAccelerometerActive] == NO){
            /* Update us forty times a second */
            [self.motionManager setAccelerometerUpdateInterval:1.0f / 40.0f];
            /* And on a handler block object */
            [self.motionManager
             startAccelerometerUpdatesToQueue:[NSOperationQueue mainQueue]
             withHandler:^(CMAccelerometerData *accelerometerData,
                           NSError *error){

                /* Move the button with the accelerometer data */
                CGRect buttonRect = self.myButton.frame;
                buttonRect.origin.x += accelerometerData.acceleration.x;
                buttonRect.origin.y += -accelerometerData.acceleration.y;
                self.myButton.frame = buttonRect;

            }];

        }
    }

}

- (void)viewDidUnload {
    [super viewDidUnload];
```

```
    if ([self.motionManager isAccelerometerAvailable] == YES &&
        [self.motionManager isAccelerometerActive] == YES){
      [self.motionManager stopAccelerometerUpdates];
    }


  }

  - (BOOL)shouldAutorotateToInterfaceOrientation:
    (UIInterfaceOrientation)interfaceOrientation {
    return (NO);
  }
```

In this example, we have placed an IBAction object of type UIButton in the current view controller's XIB file. After running the application on an iOS device, we will see the button moving as we tilt the device in different directions.

See Also

Recipe 15.1

15.4 Detecting a Shake on an iOS Device

Problem

You want to know when the user shakes an iOS device.

Solution

Follow these steps to be able to detect shake events:

1. Subclass UIWindow and set the new class as the main window of your application. In the following code, the new class of type UIWindow is called MainWindow:

```
#import <UIKit/UIKit.h>

@class MainWindow;

@interface AppDelegate_iPhone : NSObject
          <UIApplicationDelegate> {

  MainWindow *window;
  UINavigationController  *navigationController;
}

@property (nonatomic, retain)
IBOutlet MainWindow *window;

@property (nonatomic, retain)
UINavigationController  *navigationController;

@end
```

2. In your main window XIB file, make sure you change the class name of the window object to the class name of your new window (in this example, `MainWindow`), as shown in Figure 15-3.

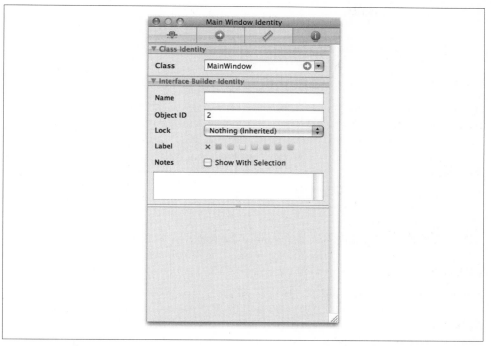

Figure 15-3. Changing the class name of the main window (to MainWindow) in the window XIB file that is created for us when we create a new window-based application in Xcode

3. Implement the `motionEnded:withEvent:` instance method in your `MainWindow` object and deliver the motions of type `UIEventSubtypeMotionShake` with a notification:

```
#import "MainWindow.h"

@implementation MainWindow

- (void) motionEnded:(UIEventSubtype)motion
          withEvent:(UIEvent *)event{

  if (motion == UIEventSubtypeMotionShake){
    [[NSNotificationCenter defaultCenter]
     postNotificationName:NOTIFICATION_SHAKE
     object:event];
  }

}

@end
```

The `NOTIFICATION_SHAKE` constant value is defined in our project's *.pch* file that is accessible to all files inside the project. This constant is defined in this way:

```
#ifdef __OBJC__
    #import <Foundation/Foundation.h>
    #import <UIKit/UIKit.h>
#endif

#define NOTIFICATION_SHAKE  @"NOTIFICATION_SHAKE"
```

Discussion

When an object of type `UIResponder` is the first responder, it can detect the motion events by implementing the following methods:

`motionBegan:withEvent:`
> This method is called in the responder object whenever a motion event has begun. For instance, if the user starts shaking the device continuously from time 00:00 to 00:10 (10 seconds), the `motionBegan:withEvent:` method will get called at 00:00 only.

`motionEnded:withEvent:`
> This method gets called when the motion has ended. For instance, if the user starts shaking the device continuously from time 00:00 to 00:10 (10 seconds), the `motionEnded:withEvent:` method will get called at 00:10 only.

`motionCancelled:withEvent:`
> This method gets called whenever a motion event is cancelled. Motions could be cancelled for a variety of reasons: for instance, if the responder is no longer the first responder, or if the system determines that an interruption has occurred and terminates the delivery of motion events to the application.

 To learn how you can generate shake events in iPhone Simulator, please refer to this chapter's Introduction.

Now that your window object is delivering notification messages about shake events, you can create a whole application out of it. Let's quickly create a view controller with a table view. The view controller will allocate and initialize an array of type `NSMutableArray`. Whenever a shake occurs, the current date and time will be retrieved and an instance of `NSDate` will be placed inside this array. The table view will just display the times and dates of each shake, as shown in Figure 15-4. Dates are formatted as HH:MM:SS DD/MM/YYYY.

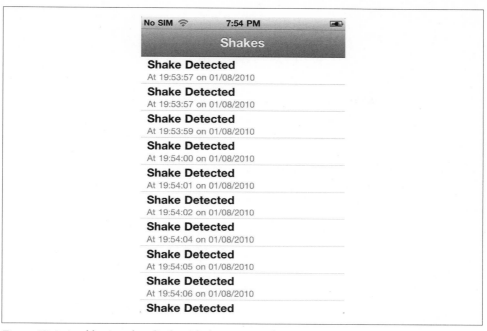

Figure 15-4. A table view that displays shake events as they occur

Here is the declaration of our view controller (.*h* file):

```
#import <UIKit/UIKit.h>
#import <CoreMotion/CoreMotion.h>

@interface RootViewController : UIViewController
          <UITableViewDelegate, UITableViewDataSource> {
@public
  NSMutableArray   *arrayOfShakeDateAndTimes;
  UITableView      *tableViewShakes;
}

@property (nonatomic, retain)
IBOutlet UITableView     *tableViewShakes;

@property (nonatomic, copy)
NSMutableArray  *arrayOfShakeDateAndTimes;

@end
```

The implementation of the same view controller (.*m* file) is as follows:

```
#import "RootViewController.h"

@implementation RootViewController

@synthesize tableViewShakes;
@synthesize arrayOfShakeDateAndTimes;
```

```
- (NSInteger)tableView:(UITableView *)table
 numberOfRowsInSection:(NSInteger)section{

  NSInteger result = 0;

  if ([table isEqual:self.tableViewShakes] == YES){
    result = [self.arrayOfShakeDateAndTimes count];
  }

  return(result);

}

- (UITableViewCell *)tableView:(UITableView *)tableView
       cellForRowAtIndexPath:(NSIndexPath *)indexPath{

  UITableViewCell *result = nil;

  if ([tableView isEqual:self.tableViewShakes] == YES){

    static NSString *CellIdentifier = @"SimpleCells";

    result = [tableView
              dequeueReusableCellWithIdentifier:CellIdentifier];

    if (result == nil){
      result = [[[UITableViewCell alloc]
                 initWithStyle:UITableViewCellStyleSubtitle
                 reuseIdentifier:CellIdentifier] autorelease];
    }

    /* Get the corresopnding date object for the current row */
    NSDate *dateOfShake =
    [self.arrayOfShakeDateAndTimes objectAtIndex:indexPath.row];

    /* Start formatting the date */
    NSCalendar *calendar = [NSCalendar currentCalendar];

    /* We want Year, Month, Day, Hour, Minute and Second
     out of this NSDate */
    NSDateComponents *dateComponents =
    [calendar components:
     NSYearCalendarUnit |
     NSMonthCalendarUnit |
     NSDayCalendarUnit |
     NSHourCalendarUnit |
     NSMinuteCalendarUnit |
     NSSecondCalendarUnit
                 fromDate:dateOfShake];

    /* Place the components into an NSString */
    NSString *detailsText =
    [NSString stringWithFormat:
     @"At %02ld:%02ld:%02ld on %02ld/%02ld/%04ld",
```

```
                (long)[dateComponents hour],
                (long)[dateComponents minute],
                (long)[dateComponents second],
                (long)[dateComponents day],
                (long)[dateComponents month],
                (long)[dateComponents year]];

        /* The title is the same */
        result.textLabel.text = @"Shake Detected";

        /* The detailed text is the date and time for every shake */
        result.detailTextLabel.text = detailsText;

    }

    return(result);

}

- (id)initWithNibName:(NSString *)nibNameOrNil
                bundle:(NSBundle *)nibBundleOrNil{

    self = [super initWithNibName:nibNameOrNil
                           bundle:nibBundleOrNil];

    if (self != nil){

        NSMutableArray *newArray =
        [[NSMutableArray alloc] init];

        arrayOfShakeDateAndTimes = [newArray mutableCopy];
        [newArray release];

        self.title = @"Shakes";

    }

    return(self);

}

- (void) detectShakes:(NSNotification *)paramNotification{

    if (self.arrayOfShakeDateAndTimes != nil){

        [self.arrayOfShakeDateAndTimes addObject:[NSDate date]];
        [self.tableViewShakes reloadData];

    } else {
        /* The user shook the device but our array is nil */
    }

}

- (void)viewDidLoad {
```

```
    [super viewDidLoad];

    /* Listen for shake notifications that are
     getting sent by the window */
    [[NSNotificationCenter defaultCenter]
     addObserver:self
     selector:@selector(detectShakes:)
     name:NOTIFICATION_SHAKE
     object:nil];

}

- (void)viewDidUnload {
    [super viewDidUnload];

    self.tableViewShakes = nil;
    self.arrayOfShakeDateAndTimes = nil;

    /* Stop listening to shake notifications */
    [[NSNotificationCenter defaultCenter]
     removeObserver:self
     name:NOTIFICATION_SHAKE
     object:nil];

}

- (BOOL)shouldAutorotateToInterfaceOrientation:
    (UIInterfaceOrientation)interfaceOrientation {
    return (YES);
}

- (void)didReceiveMemoryWarning {
    [super didReceiveMemoryWarning];
}

- (void)dealloc {

    [arrayOfShakeDateAndTimes release];
    [super dealloc];
}

@end
```

15.5 Retrieving Gyroscope Data

Problem

You want to be able to retrieve information about the device's motion from the gyroscope hardware on an iOS device.

Solution

Follow these steps:

1. Find out whether the gyroscope hardware is available on the iOS device. Please refer to Recipe 15.2 for directions on how to do this.

2. If the gyroscope hardware is available, make sure it is not sending you updates already. Please refer to Recipe 15.2 for directions.

3. Use the `setGyroUpdateInterval:` instance method of `CMMotionManager` to set the number of updates you want to receive per second. For instance, for 20 updates per second (one second), set this value to 1.0/20.0.

4. Invoke the `startGyroUpdatesToQueue:withHandler:` instance method of `CMMotionManager`. The queue object could simply be the main operation queue (as we will see later) and the handler block must follow the `CMGyroHandler` format.

The following code implements these steps:

```
- (void)viewDidLoad {
  [super viewDidLoad];

  if ([self.motionManager isGyroAvailable]){
    /* Start the gyroscope if it is not active already */
    if ([self.motionManager isGyroActive] == NO){
      /* Update us forty times a second */
      [self.motionManager setGyroUpdateInterval:1.0f / 40.0f];
      /* And on a handler block object */

      /* Receive the gyroscope data on this block */
      [self.motionManager
       startGyroUpdatesToQueue:[NSOperationQueue mainQueue]
       withHandler:^(CMGyroData *gyroData, NSError *error){

         NSLog(@"Gyro Rotation x = %.04f", gyroData.rotationRate.x);
         NSLog(@"Gyro Rotation y = %.04f", gyroData.rotationRate.y);
         NSLog(@"Gyro Rotation z = %.04f", gyroData.rotationRate.z);

       }];

    }
  } else {
    NSLog(@"Gyroscope is not available.");
  }

}


- (void)viewDidUnload {
  [super viewDidUnload];

  if ([self.motionManager isGyroAvailable] &&
      [self.motionManager isGyroActive]){
    [self.motionManager stopGyroUpdates];
  }

}
```

The `motionManager` object is the property of type `CMMotionManager` that we have defined for the current view controller running this code. For more information, please refer to this recipe's Discussion.

Discussion

With `CMMotionManager`, application programmers can attempt to retrieve gyroscope updates from iOS. You must first make sure the gyroscope hardware is available on the iOS device on which your application is running (please refer to Recipe 15.2). After doing so, you can call the `setGyroUpdateInterval:` instance method of `CMMotionManager` to set the number of updates you would like to receive per second on updates from the gyroscope hardware. For instance, if you want to be updated N times per second, set this value to $1.0/N$.

After you set the update interval, you can call the `startGyroUpdatesToQueue:withHandler:` instance method of `CMMotionManager` to set up a handler block for the updates. For more information about blocks, please refer to "Blocks Programming Topics" on the Apple Developer website (*http://developer.apple.com/library/ios/#documentation/ Cocoa/Conceptual/Blocks/Articles/00_Introduction.html*). Your block object must be of type `CMGyroHandler`, which accepts two parameters:

gyroData
> The data that comes from the gyroscope hardware, encompassed in an object of type `CMGyroData`. You can use the `rotationRate` property of `CMGyroData` (a structure) to get access to the x, y, and z values of the data, which represent all three Euler angles known as roll, pitch, and yaw. You can learn more about these by reading about flight dynamics.

error
> An error of type `NSError` that might occur when the gyroscope is sending us updates.

If you do not wish to use block objects, you must call the `startGyroUpdates` instance method of `CMMotionManager` instead of the `startGyroUpdatesToQueue:withHandler:` instance method, and set up your own thread to read the gyroscope hardware updates that are posted to the `gyroData` property of the instance of `CMMotionManager` you are using.

 After building a test application using iOS SDK 4.1 for the base SDK and target deployment SDK and running the program on an iPhone 4 with iOS 4.1 installed, I realized the `isGyroAvailable` instance method of `CMMotionManager` returns the integer value `64` instead of `YES` (which is defined as 1 in Objective-C). As in our example, please make sure you do not compare the return value of this method with `YES`.

Here is the *.h* file of our view controller that declares the instance of `CMMotionManager`:

```objectivec
#import <UIKit/UIKit.h>
#import <CoreMotion/CoreMotion.h>

@interface RootViewController : UIViewController {
@public
  CMMotionManager *motionManager;
}

@property (nonatomic, retain) CMMotionManager *motionManager;

@end
```

The *.m* file of the same view controller is as follows:

```objectivec
#import "RootViewController.h"

@implementation RootViewController

@synthesize motionManager;

- (id)initWithNibName:(NSString *)nibNameOrNil
               bundle:(NSBundle *)nibBundleOrNil{

  self = [super initWithNibName:nibNameOrNil
                         bundle:nibBundleOrNil];

  if (self != nil){

    /* Allocate and initialize the motion manager here */
    CMMotionManager *newMotionManager =
    [[CMMotionManager alloc] init];

    motionManager = [newMotionManager retain];
    [newMotionManager release];

  }

  return(self);

}

- (void)viewDidLoad {
  [super viewDidLoad];

  if ([self.motionManager isGyroAvailable]){

    /* Start the gyroscope if it is not active already */
    if ([self.motionManager isGyroActive] == NO){
      /* Update us forty times a second */
      [self.motionManager setGyroUpdateInterval:1.0f / 40.0f];
      /* And on a handler block object */

      /* Receive the gyroscope data on this block */
      [self.motionManager
```

```
        startGyroUpdatesToQueue:[NSOperationQueue mainQueue]
        withHandler:^(CMGyroData *gyroData, NSError *error){

            NSLog(@"Gyro Rotation x = %.04f", gyroData.rotationRate.x);
            NSLog(@"Gyro Rotation y = %.04f", gyroData.rotationRate.y);
            NSLog(@"Gyro Rotation z = %.04f", gyroData.rotationRate.z);

        }];

    }
  } else {
    NSLog(@"Gyroscope is not available.");
  }
}


- (void)viewDidUnload {
  [super viewDidUnload];

  if ([self.motionManager isGyroAvailable] &&
      [self.motionManager isGyroActive]){
    [self.motionManager stopGyroUpdates];
  }

}

- (BOOL)shouldAutorotateToInterfaceOrientation:
(UIInterfaceOrientation)interfaceOrientation {
  // Return YES for supported orientations
  return (YES);
}

- (void)didReceiveMemoryWarning {
  [super didReceiveMemoryWarning];
}

- (void)dealloc {
  [motionManager release];
  [super dealloc];
}

@end
```

See Also

Recipe 15.2

Index

We'd like to hear your suggestions for improving our indexes. Send email to *index@oreilly.com*.

Two-Stage Creation, 7

About the Author

Vandad Nahavandipoor has been developing software using Cocoa, Assembly, Delphi, .NET, and Cocoa Touch for many years. Vandad started developing for the iPhone OS just as the SDK was released to the public. From that moment on he became dedicated to developing applications for the iPhone, and now also for the iPad. One of Vandad's most viable assets in developing iOS applications is his real-life experience working with some of the world's biggest brands, such as Visa and U.S. Bank, to deliver mobile applications to their customers.

Colophon

The animal on the cover of *iOS 4 Programming Cookbook* is an Egyptian mongoose (*Herpestes ichneumon*), also known as an ichneumon. In ancient and medieval writings, the ichneumon is described as the enemy of the dragon, though it is more famous for battling snakes. Historic notables such as Pliny the Elder and Leonardo da Vinci recorded how the ichneumon would coat itself in several layers of mud, let it dry into a form of armor, and then attack a snake, eventually going for the reptile's throat. Later, Rudyard Kipling's short story *Rikki-Tikki-Tavi* describes the exploits of the eponymous Indian mongoose that saves his human family from the scheming cobras in their garden.

There are more than 30 species of mongoose, and all are skilled snake-killers, due in part to their resistance to venom. Because mongooses have chemical receptors shaped like those of snakes, it is difficult for neurotoxins to attach and paralyze them. Their agility, thick fur, and highly developed carnassial teeth (ideal for tearing) are also of use.

The Egyptian mongoose is the largest of all the African species, ranging from 19-23 inches long (with their black-tipped tail adding 13-21 inches more) and weighing 4-7 pounds. The animal's fur is coarse, generally colored gray with brown flecks. Despite its name, the Egyptian mongoose can also be found throughout most of sub-Saharan Africa, and has even been introduced to Madagascar, Spain, Portugal, and Italy. It lives in forest, savanna, and scrub habitats, never far from water.

The diet of the Egyptian mongoose is primarily carnivorous: it eats rodents, fish, birds, amphibians, reptiles (including venomous snakes), and insects, though eggs and fruit are also common. To crack eggs open, the mongoose will hold one between its legs and throw it at a rock or other hard surface. Mongooses are very versatile in action as well as diet; they can run backwards, roll over, swim, and stand on their hind feet.

The cover image is from Wood's *Animate Creation*. The cover font is Adobe ITC Garamond. The text font is Linotype Birka; the heading font is Adobe Myriad Condensed; and the code font is LucasFont's TheSansMonoCondensed.

Get even more for your money.

Join the O'Reilly Community, and register the O'Reilly books you own. It's free, and you'll get:

- $4.99 ebook upgrade offer
- 40% upgrade offer on O'Reilly print books
- Membership discounts on books and events
- Free lifetime updates to ebooks and videos
- Multiple ebook formats, DRM FREE
- Participation in the O'Reilly community
- Newsletters
- Account management
- 100% Satisfaction Guarantee

Signing up is easy:

1. Go to: oreilly.com/go/register
2. Create an O'Reilly login.
3. Provide your address.
4. Register your books.

Note: English-language books only

To order books online:
oreilly.com/store

For questions about products or an order:
orders@oreilly.com

To sign up to get topic-specific email announcements and/or news about upcoming books, conferences, special offers, and new technologies:
elists@oreilly.com

For technical questions about book content:
booktech@oreilly.com

To submit new book proposals to our editors:
proposals@oreilly.com

O'Reilly books are available in multiple DRM-free ebook formats. For more information:
oreilly.com/ebooks

O'REILLY®

Spreading the knowledge of innovators   oreilly.com

Buy this book and get access to the online edition for 45 days—for free!

Solutions and Examples for iPhone, iPad, and iPod touch Apps

iOS 4 Programming Cookbook

O'REILLY®   *Vandad Nahavandipoor*

iOS 4 Programming Cookbook
By Vandad Nahavandipoor
January 2011, $49.99
ISBN 9781449388225

Spreading the knowledge of innovators   safari.oreilly.com